T0250001

ENGINEERING TEXTILES

Research Methodologies,
Concepts, and Modern Applications

ENGINEERING TEXTILES

Research Methodologies, Concepts, and Modern Applications

Edited by

**Alexandr A. Berlin, DSc, Roman Joswik, PhD, and
Nikolai I. Vatin, DSc**

Reviewers and Advisory Members

Gennady E. Zaikov, DSc, and A. K. Haghi, PhD

APPLE
ACADEMIC
PRESS

Apple Academic Press Inc. | Apple Academic Press Inc.
3333 Mistwell Crescent | 9 Spinnaker Way
Oakville, ON L6L 0A2 | Waretown, NJ 08758
Canada | USA

©2016 by Apple Academic Press, Inc.

First issued in paperback 2021

Exclusive worldwide distribution by CRC Press, a member of Taylor & Francis Group
No claim to original U.S. Government works

ISBN 13: 978-1-77463-211-6 (pbk)
ISBN 13: 978-1-77188-078-7 (hbk)

All rights reserved. No part of this work may be reprinted or reproduced or utilized in any form or by any electric, mechanical or other means, now known or hereafter invented, including photocopying and recording, or in any information storage or retrieval system, without permission in writing from the publisher or its distributor, except in the case of brief excerpts or quotations for use in reviews or critical articles.

This book contains information obtained from authentic and highly regarded sources. Reprinted material is quoted with permission and sources are indicated. Copyright for individual articles remains with the authors as indicated. A wide variety of references are listed. Reasonable efforts have been made to publish reliable data and information, but the authors, editors, and the publisher cannot assume responsibility for the validity of all materials or the consequences of their use. The authors, editors, and the publisher have attempted to trace the copyright holders of all material reproduced in this publication and apologize to copyright holders if permission to publish in this form has not been obtained. If any copyright material has not been acknowledged, please write and let us know so we may rectify in any future reprint.

Trademark Notice: Registered trademark of products or corporate names are used only for explanation and identification without intent to infringe.

Library and Archives Canada Cataloguing in Publication

Engineering textiles : research methodologies, concepts, and modern applications / edited by Alexandr A. Berlin, DSc, Roman Joswik, PhD, and Nikolai I. Vatin, DSc ; reviewers and advisory members, Gennady E. Zaikov, DSc, and A. K. Haghi, PhD.

Includes bibliographical references and index.
ISBN 978-1-77188-078-7 (bound)
1. Textile industry. 2. Textile chemistry. 3. Textile fibers.
I. Haghi, A. K., author, editor II. Zaikov, G. E. (Gennadiĭ Efremovich), 1935- author, editor
III. Berlin, Alexandr A., editor IV. Joswik, Roman, editor V. Vatin, Nikolai I., editor

TS1445.E55 2015   677'.0283   C2015-905017-0

Library of Congress Cataloging-in-Publication Data

Engineering textiles: research methodologies, concepts, and modern applications / authors, Alexandr A. Berlin, DSc, Roman Joswik, PhD, and Nikolai I. Vatin, DSc ; reviewers and advisory members, Gennady E. Zaikov, DSc, and A. K. Haghi, PhD.

pages cm
Includes bibliographical references and index.
ISBN 978-1-77188-078-7 (hardcover : alk. paper) -- ISBN 978-1-4987-0603-2 (eBook)
1. Textile chemistry--Research. 2. Textile fibers--Technological innovations. 3. Textile fabrics--Technological innovations. I. Berlin, Al. Al., 1940- author. II. Joswik, Roman, author. III. Vatin, Nikolai I. (Nikolai Ivanovich), author.

TS1474.5.E58 2015   677'.028--dc23   2015026613

Apple Academic Press also publishes its books in a variety of electronic formats. Some content that appears in print may not be available in electronic format. For information about Apple Academic Press products, visit our website at **www.appleacademicpress.com** and the CRC Press website at **www.crcpress.com**

ABOUT THE EDITORS

Alexandr A. Berlin, DSc

Professor Alexandr A. Berlin, DSc, is Director of the N. N. Semenov Institute of Chemical Physics at the Russian Academy of Sciences, Moscow, Russia. He is a member of the Russian Academy of Sciences and many national and international associations. Dr. Berlin is world-renowned scientist in the field of chemical kinetics (combustion and flame), chemical physics (thermodynamics), chemistry and physics of oligomers, polymers, and composites and nanocomposites. He is the contributor to 100 books and volumes and has published 1000 original papers and reviews.

Roman Joswik, PhD

Roman Joswik, PhD, is Director of the Military Institute of Chemistry and Radiometry in Warsaw, Poland. He is a specialist in the field of physical chemistry, chemical physics, radiochemistry, organic chemistry, and applied chemistry. He has published several hundred original scientific papers as well as reviews in the field of radiochemistry and applied chemistry.

Nikolai I. Vatin, DSc

Vatin Nikolai Ivanovich, DSc, is Chief Scientific Editor of *Magazine of Civil Engineering* and Editor of *Construction of Unique Buildings and Structures*. He is also a specialist in the field of chemical technology. He has published several hundred scientific papers (original and review) and several volumes and books.

A. K. Haghi, PhD

A. K. Haghi, PhD, holds a BSc in urban and environmental engineering from the University of North Carolina (USA); a MSc in mechanical engineering from North Carolina A&T State University (USA); a DEA in applied mechanics, acoustics and materials from the Université de Technologie de Compiègne (France); and a PhD in engineering sciences from the Université de Franche-Comté (France). He is the author and editor of 165 books as well as 1000 published papers in various journals and conference proceedings. Dr. Haghi has received several grants, consulted for a number of major corporations, and is a frequent speaker to national and international audiences. Since 1983, he served as a professor at several universities. He is currently Editor-in-Chief of the *International Journal of Chemoinformatics and Chemical Engineering* and *Polymers Research Journal* and on the editorial boards of many international journals. He is a member of the Canadian Research and Development Center of Sciences and Cultures (CRDCSC), Montreal, Quebec, Canada.

Gennady E. Zaikov, DSc

Gennady E. Zaikov, DSc, is Head of the Polymer Division at the N. M. Emanuel Institute of Biochemical Physics, Russian Academy of Sciences, Moscow, Russia, and Professor at Moscow State Academy of Fine Chemical Technology, Russia, as well as Professor at Kazan National Research Technological University, Kazan, Russia. He is also a prolific author, researcher, and lecturer. He has received several awards for his work, including the Russian Federation Scholarship for Outstanding Scientists. He has been a member of many professional organizations and on the editorial boards of many international science journals. Dr. Zaikov has recently been honored with tributes in several journals and books on the occasion of his 80th birthday for his long and distinguished career and for his mentorship to many scientists over the years.

CONTENTS

LIST OF CONTRIBUTORS

Arezoo Afzali
University of Guilan, Rasht, Iran

A. K. Haghi
University of Guilan, Rasht, Iran

A. L. Iordanskii
N.N. Semenov Institute of Chemical Physics, RAS, 119991 Moscow, Street Kosygina, 4, Federation of Russia

M. Kanafchian
University of Guilan, Rasht, Iran

A. N. Kazakova
Ufa State Petroleum Technological University, 1 Kosmonavtov Str., 450062 Ufa, Russia; Phone: (347) 2420854, E-mail: nocturne@mail.ru

Shima Maghsoodlou
University of Guilan, Rasht, Iran

N. N. Mikhailova
Ufa State Petroleum Technological University, 1 Kosmonavtov Str., 450062 Ufa, Russia; Phone: (347) 2420854, E-mail: nocturne@mail.ru

V. M. Misin
N. M. Emanuel Institute of Biochemical Physics, Russian Academy of Sciences, Moscow, Russia

T. P. Mudrik
Sterlitamak branch of Bashkir State University, 49 Lenina avenue, 453103 Sterlitamak, Russia

S. S. Nikulin
Voronezh State University of the Engineering Technologies, Voronezh, Russiasss

A. A. Olkhov
Plekhanov Russian University of Economics, Stremyanny per. 36, 117997, Moscow, Federation of Russia; N.N. Semenov Institute of Chemical Physics, RAS, 119991 Moscow, Street Kosygina, 4, Federation of Russia, E-mail: aolkhov72@yandex.ru

I. N. Pugacheva
Voronezh State University of the Engineering Technologies, Voronezh, Russia

G. Z. Raskildina
Ufa State Petroleum Technological University, 1 Kosmonavtov Str., 450062 Ufa, Russia; Phone: (347) 2420854, E-mail: nocturne@mail.ru

O. V. Staroverova
N.N. Semenov Institute of Chemical Physics, RAS, 119991 Moscow, Street Kosygina, 4, Federation of Russia

G. E. Zaikov

L.Ya. Karpov Physicochemical Research Institute, Federation of Russia; N. M. Emanuel Institute of Biochemical Physics, Russian Academy of Sciences, 4 Kosygin str., 119334 Moscow, Russia; E-mail: Chembio@sky.chph.ras.ru

S. S. Zlotsky

Ufa State Petroleum Technological University, 1 Kosmonavtov Str., 450062 Ufa, Russia; Phone: (347) 2420854, E-mail: nocturne@mail.ru

LIST OF ABBREVIATIONS

a-Si	amorphous silicon
AFM	atomic force microscopy
APC	acid powder-like cellulose
BSA	bovine serum albumin
CB	conduction band
CD	cross direction
CdTe	cadmium telluride
CFM	chloroform
CIGS	copper indium selenide
COFs	covalent organic frameworks
CPCs	conductive polymer composites
CRDCSC	Canadian Research and Development Center of Sciences and Cultures
CRE	constant rate of extension
CS	chemical shift
CV	coefficient of variation
CVD	chemical vapor deposition
DMF	N, N-dimethylformamide
DMSO	methyl sulfoxide
DSC	differential scanning calorimetry
DSCs	dye-sensitized cells
DSSC	dye-sensitized solar cell
EAGLE	easily applicable graphical summary layout editor
ECN	Energy Research Centre of the Netherland
EDA	ethylenediamine
EDANA	European Disposables and Nonwovens Association
EDOT	3,4-ethylenedioxythiophene
EG	ethylene glycol
ESCs	embryonic stem cells
ESR	electron spin resonance
FCC	face-centered cubic
FOD	fiber orientation distribution
FT	Fourier transform method
HCP	hexagonal close-packed
HOMO	highest occupied molecular orbital

HT	Hough transform method
INDA	Association of the Nonwovens Fabrics Industry
ITO	indium tin oxide
IUPAC	International Union of Pure and Applied Chemistry
LUMO	lowest unoccupied molecular orbital
LVDT	linear variable differential transformers
MCC	microcrystalline cellulose
MD	machine direction
MePRN	methoxypropionitrile
MOFs	metal organic frameworks
MOPs	microporous organic polymers
MPP	maximum power point
MRA	mechanical rubber articles
MWCO	molecular weight cut-off
OPV	optical photovoltaic
PAN	polyacrylonitrile
PANCMPC	polyacrylonitriles-2-methacryloyloxyethyl phosphoryl choline
PANI	polyaniline
PBI	polybenzimidazole
PBS	poly butylene succinate
PCM	phase change materials
PE	polyethylene
PEK-C	polyetherketone cardo
PET	polyethylene terephthalate
PGD	pores geometry distribution
PHB	polyhydroxybutyrate
PIMs	polymers of intrinsic microporosity
PL	photoluminescence
PMMA	polymethylmethacrylate
PP	polypropylene
PPy	polypyrrole
PRINT	particle repulsion in non wetting templates
PS	polystyrene
PSD	pore size distribution
PSS	polystyrene sulfonic acid
PTh	polythiophenes
PV	photovoltaic devices
PVD	pore volume distributions
PVDE-HFP	polyvinylidene fluoride-hexafluoro-propylene
PVDF	poly(vinylidene fluoride)
PZT	lead-zirconate-titanate

R2R	roll-to-roll processing technology
R2RNIL	roll-to-roll nanoimprint lithography
SEM	scanning electron microscope
SIT	solar integrated technologies
SSCCs	spin-spin coupling constants
TCO	transparent conducting oxide
UPF	UV protective factor
UV	ultraviolet
VB	valence band
WVP	water vapor permeability

LIST OF SYMBOLS

a_1 and a_2	two basis vectors of the graphite cell
$B(i,j)$	the heat exchanged between each node and central blood compartment
c	by the average molecular velocity of the particles
C	convective heat loss
c_v	the specific heat
$C(i,j)$	the heat capacity of each node
C_h	a vector that maps an atom of one end of the tube to the other
C_{pbl}	specific heat at constant pressure of blood
$D(i,j)$	the heat transmitted by conduction to the neighboring layer with the same segment
D_a	the diffusion coefficient of water vapor through the textiles
D_l	the diffusion coefficient of free water in the fibrous batting
$E(i,j)$	the evaporative heat loss at skin surface
e_1, e_2 and e_3	the coordinates in the current configuration
E_{res} and C_{res}	respectively is latent and dry respiration heat loss
F_R and F_L	the total thermal radiation incident traveling to the right and left way
G_1 and G_2	the material coordinates of a point in the initial configuration
h_{mn} and h_{cn}	respectively the mass and heat transfer coefficients of the inner and outer fabric surface
K	thermal conductivity
K_a	conductivity of the air
K_f	conductivity of solid fiber
L	the thickness of the fabric sample
l_{i0} and l_{i1}	defined as the thickness of the left and right gap between neighbored layers
M	the metabolic heat/ shivering metabolic rate
M_d	the moisture flow from the skin
M_t	the heat flow from the skin
N_A	the Avogadro constant
P	pressure
P_0	the vapor pressure of the bulk liquid, ambient pressure
$Q(i,j)$	the sum of the basic metabolic rate,
Q_k	conductive heat flux
Q_w	the moisture transfer rate/weight loss/gain in grams over a period time

R	gas constant
R	radioactive heat loss
R	the radius of the modeled SWCNT
R_1 and R_2	the moisture sorption rates in the first and second stages
R_n	thermal resistance of the waterproof fabric
r_p	pore radius
RSW	sweat rate
S_{cr} and S_{sk}	respectively are the heat storage of core and skin shell
T	temperature
$T(i,j)$	the node temperature
T_{ms} and T_{ml}	respectively are the temperature distributions in a sphere containing solid and liquid PCM
V_{bl}	skin blood flow rate
V_L	molecular volume of the condensate
W	moisture transfer resistance
$W{-}W_f$	the free water content in the fibrous material
W_n	the water vapor resistance
W_n	water vapor resistance
(n, m) CNT	constructed CNT

GREEK SYMBOLS

α	the proportion of uptake occurring during the second stage
β	the radiation absorption constant for textile materials
γ	surface tension
$\Gamma(x,t)$	moisture change
ΔC	the difference of water vapor concentration on the two surfaces of the fabric
ΔC	the vapor concentration gradient of two fabric sides
$\acute{\epsilon}$	the porosity of textile material
$\gamma_{LV}, \gamma_{SV}, \gamma_{SL}$	respectively represents the interfacial tensions that exists between the solid-vapor, solid-liquid and liquid-vapor interfaces
θ	equilibrium contact angle
λ	the mean free path
ρ_b	bulk density
ρ_p	particle density
σ	Boltzmann constant
$\Gamma_{\acute{f}}$	moisture sorption rate
ϕ	volume fraction of voids

PREFACE

This volume provides the textile science community with a forum for critical, authoritative evaluations of advances in many areas of the discipline. This book reports recent advances with significant, up-to-date chapters by internationally recognized researchers. This book highlights applications of chemical physics to subjects that textile engineering students will see in graduate courses.

The book presents biochemical examples and applications focuses on concepts above formal experimental techniques and theoretical methods.

By providing an applied and modern approach, this volume helps students see the value and relevance of studying textile science and technology to all areas of applied engineering, and gives them the depth of coverage they need to develop a solid understanding of the key principles in the field.

The book assumes a working knowledge of calculus, physics, and chemistry, but no prior knowledge of polymers. It is valuable for researchers and for upper-level research students in chemistry, textile engineering and polymers.

CHAPTER 1

UNDERSTANDING NONWOVENS: CONCEPTS AND APPLICATIONS

M. KANAFCHIAN and A. K. HAGHI

University of Guilan, Rasht, Iran

CONTENTS

ABSTRACT

Nonwovens are a distinct class of textile materials made directly from fibers, thus avoiding the intermediate step of yarn production. The nonwoven industry as we know it today has grown from developments in the textile, paper and polymer processing industries. Nonwoven manufacturers operate in a highly competitive environment, where the reduction of production costs is of strategic imperative to stay in business. While the understanding and estimation of the production costs of nonwovens are very important, the literature review reveals a lack of research in this field, in the public domain. To fill the gap, this study is designed to lay the foundation for modeling and analysis of nonwoven structures.

1.1   INTRODUCTION

1.1.1   DEFINITION OF NONWOVEN

Nonwoven is a term used to describe a type of material made from textile fibers, which is not produced on conventional, looms or knitting machines. They combine features from the textile, paper and plastic industries and an early description was 'web textiles' because this term reflects the essential nature of many of these products. The yarn spinning stage is omitted in the nonwoven processing of staple fibers, while bonding of the web by various methods, chemical, mechanical or thermal, replaces the weaving (or knitting) of yarns in traditional textiles. However, even in the early days of the industry, the process of stitch bonding, which originated in Eastern Europe in the 1950s, employed both layered and consolidating yarns, and the parallel developments in the paper and synthetic polymer fields, which have been crucial in shaping today's multibillion dollar nonwovens industry, had only tenuous links with textiles in the first place. But now the precise meaning of the term is somewhat clearer to the experts. According to the experts, Nonwovens is a class of textiles/sheet products, unique in industry, which is defined in the negative; that is, they are defined in what they are not.

Nonwovens are different than paper in that nonwovens usually consist entirely or at least contain a sizeable proportion of long fibers and/or they are bonded intermittently along the length of the fibers. Although paper consists of fiber webs, the fibers are bonded to each other so completely that the entire sheet comprises one unit. In nonwovens we have webs of fibers where fibers are not as rigidly bonded and to a large degree act as individuals [1].

The definitions of the nonwovens most commonly used nowadays are those by the Association of the Nonwovens Fabrics Industry (INDA) and the European Disposables and Nonwovens Association (EDANA).

I. INDA definition:

Nonwovens are a sheet, web, or bat of natural and/or man-made fibers or filaments, excluding paper, that have not been converted into yarns, and that are bonded to each other by any of several means [2].

II. EDANA definition:

Nonwovens are a manufactured sheet, web or bat of directionally or randomly oriented fibers, bonded by friction, and/or cohesion and/or adhesion, excluding paper or products which are woven, knitted, tufted stitch bonded incorporating binding yarns or filaments, or felted by wet milling, whether or not additionally needled. The fibers may be of natural or man-made origin. They may be staple or continuous or be formed in situ [3].

A definition of nonwoven is provided by ISO 9092:1988 which details the fibrous content and other conditions. However, felts, needled fabrics, tufted and stitch-bonded materials are usually grouped under this general heading for convenience, even though they may not strictly be described by the definition. They are made from all types of fiber, natural, regenerated man-made or synthetic or from fiber blends. One of their most significant features is their speed of manufacture, which is usually much faster than all other forms of fabric production; for example, spun bonding can be 2000 times faster than weaving. Therefore, they are very economical but also very versatile materials, which offer the opportunity to blend different fibers or fiber types with different binders in a variety of different physical forms, to produce a wide range of different properties. Nonwoven materials typically lack strength unless reinforced by a backing. They are broadly defined as sheet or web structures bonded together by entangling fiber or filaments mechanically, thermally or chemically. Nonwoven materials are flat or tufted porous sheets that are made directly from separate fibers, molten plastic or plastic film [4].

Thereupon, a nonwoven structure is different from some other textile structures because:

- it principally consists of individual fibers or layers of fibrous webs rather than yarns;
- it is anisotropic both in terms of its structure and properties due to both fiber alignment and the arrangement of the bonding points in its structure;
- it is usually not uniform in fabric weight and/or fabric thickness, or both; and
- it is highly porous and permeable.

1.1.2 APPLICATION OF NONWOVENS

Nonwovens may be a limited-life, single-use fabric or a very durable fabric. They provide specific functions such as absorbency, liquid repellency, resilience, stretch, softness, strength, flame retardancy, washability, cushioning, filtering, bacterial barriers and sterility. These properties are often combined to create fabrics suited for specific jobs while achieving a good balance between product use-life and cost. They can mimic the appearance, texture and strength of a woven fabric, and can be as bulky as the thickest padding. Although it is not possible to list all the applications of nonwovens, some of the important applications are listed in Table 1.1 [5].

TABLE 1.1   Nonwoven Products

Agriculture and Land-scaping	Crop Covers, Turf protection products, Nursery overwintering, Weed control fabrics, Root bags, Containers, Capillary matting
Automotive	Trunk applications, Floor covers, Side liners, Front and back liners, Wheelhouse covers, Rear shelf trim panel covers, Seat applications, Listings, Cover slip sheets, Foam reinforcements, Transmission oil filters, Door trim panel carpets, Door trim panel padding, Vinyl and landau cover backings, Molded headliner substrates, Hood silencer pads, Dash insulators, Carpet tufting fabric and under, Padding
Clothing	Interlinings, Clothing and glove insulation, Bra and shoulder padding, Handbag components, Shoe components
Construction	Roofing and tile underlayment, Acoustical ceilings, Insulation, House wrap, Pipe wrap
Geotextiles	Asphalt overlay, Road and railroad beds, Soil stabilization, Drainage, Dam and stream embankments, Golf and tennis courts, Artificial turf, Sedimentation and erosion, Pond liners
Home Furnishings	Furniture construction sheeting, Insulators, arms, Cushion ticking, Dust covers, Decking, Skirt linings, Pull strips, Bedding construction sheeting, Quilt backing, Dust covers, Flanging, Spring wrap, Quilt backings, Blankets, Wallcovering backings, Acoustical wallcoverings, Upholstery backings, Pillows, pillow cases, Window treatments, Drapery components, Carpet backings, carpets, Pads, Mattress pad components
Health Care	caps, gowns, masks, Shoe covers, Sponges, dressings, wipes, Orthopedic padding, Bandages, tapes, Dental bibs, Drapes, wraps, packs, Sterile packaging, Bed linen, under-pads, Contamination control gowns, Electrodes, Examination gowns, Filters for IV solutions and blood, Oxygenators and kidney, Dialyzers, Transdermal drug delivery
Household	Wipes, wet or dry polishing, Aprons, Scouring pads, Fabric softener sheets, Dust cloths, mops, Tea and coffee bags, Placemats, napkins Ironing board pads, Washcloths, Tablecloths

Industrial or Military	Coated fabrics, Filters, Semiconductor polishing pads, Wipers, Clean room apparel, Air conditioning filters, Military clothing, Abrasives, Cable insulation, Reinforced plastics, Tapes, Protective clothing, lab coats, Sorbents, Lubricating pads, Flame barriers, Packaging, Conveyor belts, Display felts, Papermaker felts, Noise absorbent felt
Leisure, Travel	Sleeping bags, Tarpaulins, tents, Artificial leather, luggage, Airline headrests, pillow cases,
Personal Care and Hygiene	Diapers, Sanitary napkins, tampons, Training pants, Incontinence products, Dry and wet wipes, Cosmetic applicators, removers, Lens tissue, Hand warmers, Vacuum cleaner bags, Tea and coffee bags, Buff pads
School, Office	Book covers, Mailing envelopes, labels, Maps, signs, pennants, Floppy disk liners, Towels, Promotional items, Pen nibs

1.1.3   TECHNOLOGIES FOR NONWOVEN MANUFACTURING [6]

All nonwovens are manufactured by two general steps, which sometimes can even be combined into one. The first is actual web formation, and the second is some method of bonding the web fibers together. Sometimes, but not always a finishing process is required to provide the specific needs of a particular end use, in a similar way to woven or knitted fabrics. Nonwoven materials are produced by a wide variety of processes ranging from hydraulic formation, air assisted processing, direct polymer to web systems, and combinations of these. These materials find application in diverse fields that include filtration, automotive, hygiene products, battery separators, medical, and home furnishings. In the wet-laid process, a mixture of fibers is suspended in a fluid, the fibers are deposited onto a screen or porous surface to remove the fluid and the web is then consolidated mechanically, chemically, or thermally. Wet-laid materials have outstanding uniformity when compared with other nonwovens. Melt-blown materials are produced by melting and extruding molten polymer resin directly into a web. Fine fibers with a narrow fiber distribution can be produced using the melt-blown process. Composites of wet-laid and melt-blown materials can be used to obtain improved performance and strength characteristics in the final application. To make finer fibers, typically in the nanometer range, an electrospinning process is used. During the electrospinning process, electrical charge is applied to polymer solution to form nanometer size fibers on a collector. The structures thus produced have a high surface area to volume ratio, high length to diameter ratio, high porosity, and interconnected pores. In fiber processing it is common to make first a thin layer of fiber called a web and then to lay several webs on top of each other to form a batt, which goes directly to bonding. The words web and batt are cases where it is difficult to decide if a fiber layer is a web or a batt. Nevertheless the first stage of nonwoven processing is normally called batt production.

1.1.3.1   CARDING METHOD

The machines in the nonwoven industry use identical principles and are quite similar but there are some differences. In particular in the process of yarn manufacture there are opportunities for further opening and for improving the levelness of the product after the carding stage, but in nonwoven manufacture there is no further opening at all and very limited improvements in levelness are possible. It therefore follows that the opening and blending stages before carding must be carried out more intensively in a nonwoven plant and the card should be designed to achieve more opening, for instance by including one more cylinder, though it must be admitted that many nonwoven manufacturers do not follow this maxim. Theoretically either short-staple revolving flat cards or long-staple roller cards could be used, the short-staple cards having the advantages of high production and high opening power, especially if this is expressed per unit of floor space occupied. However, the short-staple cards are very narrow, whereas long-staple cards can be many times wider, making them much more suitable for nonwoven manufacture, particularly since nonwoven fabrics are required to be wider and wider for many end-uses. Hence a nonwoven installation of this type will usually consist of automatic fiber blending and opening feeding automatically to one or more wide long staple cards. The cards will usually have some form of autoleveler to control the mass per unit area of the output web.

1.1.3.2   AIR LAYING METHOD

The air-laying method produces the final batt in one stage without first making a lighter weight web. It is also capable of running at high production speeds but is similar to the parallel-lay method in that the width of the final batt is the same as the width of the air-laying machine, usually in the range of 3–4 m. The degree of fiber opening available in an air-lay machine varies from one manufacturer to another, but in all cases it is very much lower than in a card (Fig. 1.1). As a consequence of this more fiber opening should be used prior to air laying and the fibers used should be capable of being more easily opened, otherwise the final batt would show clumps of inadequately opened fiber. In the past the desire for really good fiber opening led to a process consisting of carding, cross laying, then feeding the cross-laid batt to an air-laying machine.

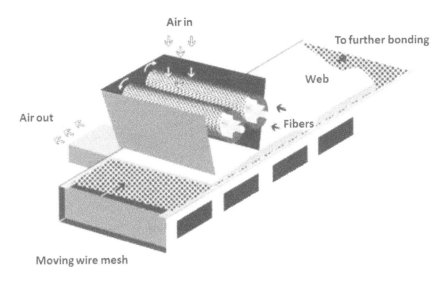

FIGURE 1.1 A schematic of air-laid system.

1.1.3.3 WET LAYING METHOD

The wet-laid process is a development from papermaking that was undertaken because the production speeds of papermaking are very high compared with textile production. Textile fibers are cut very short by textile standards (6–20 mm), but at the same time these are very long in comparison with wood pulp, the usual raw material for paper. The fibers are dispersed into water; the rate of dilution has to be great enough to prevent the fibers aggregating (Fig. 1.2). The required dilution rate turns out to be roughly ten times that required for paper, which means that only specialized forms of paper machines can be used, known as inclined-wire machines. In fact most frequently a blend of textile fibers together with wood pulp is used as the raw material, not only reducing the necessary dilution rate but also leading to a big reduction in the cost of the raw material. It is now possible to appreciate one of the problems of defining nonwoven. It has been agreed that a material containing 50% textile fiber and 50% wood pulp is a nonwoven, but any further increase in the wood pulp content results in a fiber-reinforced paper. A great many products use exactly 50% wood pulp. Wet-laid nonwovens represent about 10% of the total market, but this percentage is tending to decline. They are used widely in disposable products, for example, in hospitals as drapes, gowns, sometimes as sheets, as one-use filters, and as cover stock in disposable nappies.

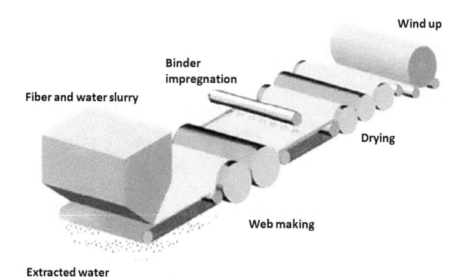

FIGURE 1.2   A schematic of wet-laid system.

1.1.3.4   DRY LAYING METHOD

The paper industry has attempted for many years to develop a dry paper process because of the problems associated with the normal wet process, that is, the removal of very large volumes of water by suction, by squeezing and finally by evaporation (Fig. 1.3). Now a dry process has been developed using wood pulp and either latex binders or thermoplastic fibers or powders to bond the wood pulp together to replace hydrogen bonding which is no longer possible. Owing to the similarity of both the bonding methods and some of the properties to those of nonwovens, these products are being referred to as nonwovens in some areas, although it is clear from the definition above they do not pass the percentage wood pulp criterion. Hence although the dry-laid paper process cannot be regarded as a nonwoven process at present, it is very likely that the process will be modified to accept textile fibers and will become very important in nonwovens in the future.

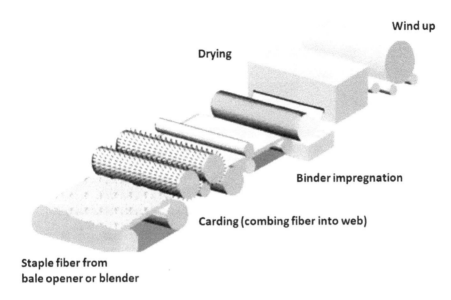

Wind up

Drying

Binder impregnation

Carding (combing fiber into web)

Staple fiber from
bale opener or blender

FIGURE 1.3   A schematic of dry-laid system.

1.1.3.5   SPUN LAYING METHOD

Spun laying includes extrusion of the filaments from the polymer raw material, drawing the filaments and laying them into a batt. As laying and bonding are normally continuous, this process represents the shortest possible textile route from polymer to fabric in one stage (Fig. 1.4). In addition to this the spun-laid process has been made more versatile. When first introduced only large, very expensive machines with large production capabilities were available, but much smaller and relatively inexpensive machines have been developed, permitting the smaller nonwoven producers to use the spun-laid route. Further developments have made it possible to produce microfibers on spun-laid machines giving the advantages of better filament distribution, smaller pores between the fibers for better filtration, softer feel and also the possibility of making lighter-weight fabrics. For these reasons spun-laid production is increasing more rapidly than any other nonwoven process. Spun laying starts with extrusion. Virtually all commercial machines use thermoplastic polymers and melt extrusion. Polyester and polypropylene are by far the most common, but polyamide and polyethylene can also be used. The polymer chips are fed continuously to a screw extruder which delivers the liquid polymer to a metering pump and thence to a bank of spinnerets, or alternatively to a rectangular spinneret plate, each containing a very large number of holes. The liquid polymer pumped through each hole is cooled rapidly to the solid state but some stretching or drawing in the liquid state will inevitably take place.

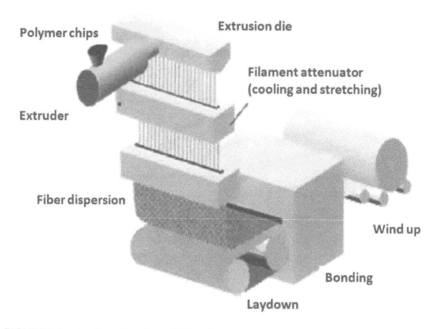

FIGURE 1.4   A schematic of spun-laid system.

In spun laying the most common form of drawing the filaments to obtain the correct modulus is air drawing, in which a high velocity air stream is blown past the filaments moving down a long tube, the conditions of air velocity and tube length being chosen so that sufficient tension is developed in the filaments to cause drawing to take place. In some cases air drawing is not adequate and roller drawing has to be used as in normal textile extrusion, but roller drawing is more complex and tends to slow the process, so that air drawing is preferred. The laying of the drawn filaments must satisfy two criteria; the batt must be as even as possible in mass per unit area at all levels of area, and the distribution of filament orientations must be as desired, which may not be isotropic. Taking the regularity criterion first, the air tubes must direct the filaments onto the conveyor belt in such a way that an even distribution is possible. However, this in itself is not sufficient because the filaments can form agglomerations that make 'strings' or 'ropes' which can be clearly seen in the final fabric. A number of methods have been suggested to prevent this, for instance, charging the spinneret so that the filaments become charged and repel one another or blowing the filaments from the air tubes against a baffle plate, which tends to break up any agglomerations. With regard to the filament orientation, in the absence of any positive traversing of the filaments in the cross direction, only very small random movements would take place. However, the movement of the conveyor makes a very strong machine direction orientation; thus the fabric would have a very strong machine-direction orientation. Cross-direction movement can most easily be applied by oscillating the air tubes backwards and forwards. By controlling the speed and amplitude of this oscillation it is possible to control the degree of cross-direction orientation.

1.1.3.6   FLASH SPINNING METHOD

Flash spinning is a specialized technique for producing very fine fibers without the need to make very fine spinneret holes. In flash spinning the polymer is dissolved in a solvent and is extruded as a sheet at a suitable temperature so that when the pressure falls on leaving the extruder the solvent boils suddenly (Fig. 1.5). This blows the polymer sheet into a mass of bubbles with a large surface area and consequently with very low wall thickness. Subsequent drawing of this sheet, followed by mechanical fibrillation, results in a fiber network of very fine fibers joined together at intervals according to the method of production. This material is then laid to obtain the desired mass per unit area and directional strength. Flash-spun material is only bonded in two ways, so that it seems sensible to discuss both the bonding and the products at this juncture. One method involves melting the fibers under high pressure so that virtually all fibers adhere along the whole of their length and the fabric is almost solid with very little air space. This method of construction makes a very stiff material with high tensile and tear strengths. It is mainly used in competition with paper for making tough waterproof envelopes for mailing important documents and for making banknotes. The fact that the original fibers were very fine means that the material is very smooth and can be used for handwriting or printing. The alternative method of bonding is the same, that is, heat and pressure but is only applied to small areas, say 1 mm square, leaving larger areas, say 4 mm square, completely unbonded. The bonded areas normally form a square or diagonal pattern. This material, known as Tyvek, has a lower tensile strength than the fully bonded fabric, but has good strength and is flexible enough to be used for clothing. Because the fine fibers leave very small pores in the fabric, it is not only waterproof but is also resistant to many other liquids with surface tensions lower than water. The presence of the pores means that the fabric is permeable to water vapor and so is comfortable to wear. Tyvek is used principally for protective clothing in the chemical, nuclear and oil industries, probably as protection for the armed forces and certainly in many industries not requiring such good protection but where it is found to be convenient. The garments can be produced so cheaply that they are usually regarded as disposable.

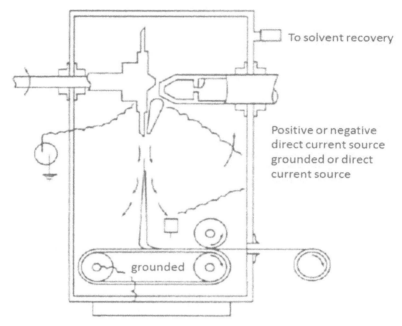

To solvent recovery

Positive or negative
direct current source
grounded or direct
current source

grounded

FIGURE 1.5   A schematic of flash-spinning system.

1.1.3.7   MELT BLOWING METHOD

The process of melt blowing is another method of producing very fine fibers at high production rates without the use of fine spinnerets (Fig. 1.6). As the polymer leaves the extrusion holes it is hit by a high-speed stream of hot air at or even above its melting point, which breaks up the flow and stretches the many filaments until they are very fine. At a later stage, cold air mixes with the hot and the polymer solidifies. Depending on the air velocity after this point a certain amount of aerodynamic drawing may take place but this is by no means as satisfactory as in spun laying and the fibers do not develop much tenacity. At some point the filaments break into staple fibers, but it seems likely that this happens while the polymer is still liquid because if it happened later this would imply that a high tension had been applied to the solid fiber, which would have caused drawing before breaking. The fine staple fibers produced in this way are collected into a batt on a permeable conveyor as in air laying and spun laying. The big difference is that in melt blowing the fibers are extremely fine so that there are many more fiber-to-fiber contacts and the batt has greater integrity. For many end-uses no form of bonding is used and the material is not nonwoven, but simply a batt of loose fibers. Such uses include ultrafine filters for air conditioning and personal facemasks, oil-spill absorbents and personal hygiene products. In other cases the melt-blown batt may be laminated to another nonwoven, especially a spun-laid one or the melt-blown batt itself may be bonded but the method must be chosen carefully to avoid spoiling the openness of the very fine fibers. In the bonded or laminated form the fabric can be used for breathable protective clothing in hospitals, agriculture and industry, as battery separators, industrial wipes and clothing interlinings with good insulation properties. Melt blowing started to develop rapidly in about 1975, although the process was known before 1950. It is continuing to grow at about 10% per annum.

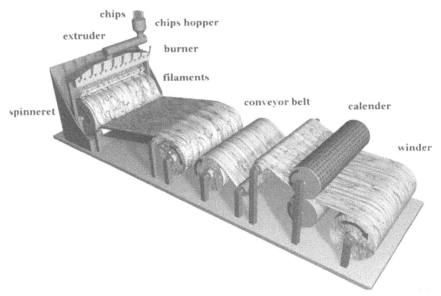

FIGURE 1.6   A schematic of melt-blowing system.

1.1.3.8   BONDING METHOD

The various methods for bonding are:
1. Adding an adhesive.
2. Thermally fusing the fibers or filaments to each other or to the other meltable fibers or powders.
3. Fusing fibers by first dissolving, and then resolidifying their surfaces.
4. Creating physical tangles or tuft among the fibers.
5. Stitching the fibers or filaments in place.

1.1.3.8.1   CHEMICAL BONDING

Chemical bonding involves treating either the complete batt or alternatively isolated portions of the batt with a bonding agent with the intention of sticking the fibers together. Although many different bonding agents could be used, the modern industry uses only synthetic lattices, of which acrylic latex represents at least half and styrene–butadiene latex and vinyl acetate latex roughly a quarter each. When the bonding agent is applied it is essential that it wets the fibers, otherwise poor adhesion will be achieved. Most lattices already contain a surfactant to disperse the polymer particles, but in some cases additional surfactant may be needed to aid wetting. The next stage is to dry the latex by evaporating the aqueous component and leaving the polymer particles together with any additives on and between the fibers. During this stage the surface tension of the water pulls the binder particles together forming a film over the fibers and a rather thicker film over the fiber intersections. Smaller binder particles will form a more effective film than larger particles, other things being equal. The final stage is curing and in this phase the batt is brought up to a higher temperature than for the drying. The purpose of curing is to develop crosslinks both inside and between the polymer particles and so to develop good cohesive strength in the binder film. Typical curing conditions are 120–140°C for 2–4 min.

1.1.3.8.2   THERMAL BONDING

Thermal bonding is increasingly used at the expense of chemical bonding for a number of reasons. Thermal bonding can be run at high speed, whereas the speed of chemical bonding is limited by the drying and curing stage. Thermal bonding takes up little space compared with drying and curing ovens. Also thermal bonding requires less heat compared with the heat required to evaporate water from the binder, so it is more energy efficient. Thermal bonding can use three types of fibrous raw material, each of which may be suitable in some applications but not in others. First, the fibers may be all of the same type, with the same melting point. This is satisfactory if the heat is applied at localized spots, but if overall bonding is used it is possible that all the fibers will melt into a plastic sheet with little or no value. Second, a blend of fusible fiber with either a fiber with a higher melting point or a nonthermoplastic fiber can be used. This is satisfactory in most conditions except where the fusible fiber melts completely, losing its fibrous nature and causing the batt to collapse in thickness. Finally, the fusible fiber may be a bicomponent fiber, that is, a fiber extruded with a core of high melting point polymer surrounded by a sheath of lower melting point polymer (Fig. 1.7). This is an ideal material to process because the core of the fiber does not melt but supports the sheath in its fibrous state. Thermal bonding is used with all the methods of batt production except the wet-laid method, but it is worth pointing out that the spun-laid process and point bonding complement each other so well that they are often thought of as the standard process.

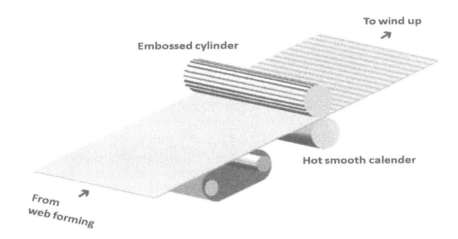

FIGURE 1.7   Thermal bonding method.

1.1.3.8.3   SOLVENT BONDING

This form of bonding is only rarely used but it is interesting from two points of view; first, the solvent can be recycled, so the process is ecologically sound, although whether or not recycling is practical depends on the economics of recovering the solvent. Second, some of the concepts in solvent bonding are both interesting and unique. In one application of the method, a spun-laid polyamide batt is carried through an enclosure containing the solvent gas, NO_2 that softens the skin of the filaments. On leaving the enclosure bonding is completed by cold calendar rolls and the solvent is washed from the fabric with water using traditional textile equipment. This is a suitable method of bonding to follow a spun-laid line because the speeds of production can be matched. The other application uses a so-called latent solvent, by which is meant one that is not a solvent at room temperature but becomes a solvent at higher temperatures. This latent solvent is used in conjunction with carding and cross laying and is applied as a liquid before carding. The action of carding spreads the solvent and at the same time the solvent lubricates the fibers during carding. The batt is passed to a hot air oven, which first activates the solvent and later evaporates it. The product will normally be high loft, but if fewer lofts are required a compression roller could be used.

1.1.3.8.4   MECHANICAL BONDING

I. Needle Felting

All the methods of bonding discussed so far have involved adhesion between the fibers; hence they can be referred to collectively as adhesive bonding. The final three methods, needle felting, hydroentanglement and stitch bonding rely on frictional forces and fiber entanglements, and are known collectively as mechanical bonding. The basic concept of needle felting is apparently simple; the batt is led between two stationary plates, the bed and stripper plates as shown in Fig. 1.8. While between the plates the batt is penetrated by a large number of needles. The needles are usually made triangular and have barbs cut into the three edges. When the needles descend into the batt the barbs catch some fibers and pull them through the other fibers. When the needles return upwards, the loops of fiber formed on the downstroke tend to remain in position, because they are released by the barbs. This downward pressure repeated many times makes the batt much denser, that is, into a needlefelt. The above description illustrates how simple the concept seems to be. Without going into too much detail it may be interesting to look at some of the complications. First, the needles can only form vertical loops or 'pegs' of fiber and increase the density of the batt. This alone does not form a strong fabric unless the vertical pegs pass through loops already present in the horizontal plane of the batt. It follows from this that parallel-laid fabric is not very suitable for needling since there are few fiber loops present, so most needling processes are carried out with cross-laid, air-laid and spun-laid batts. Second, the amount of needling is determined partly by the distance the drawing rollers move between each movement of the needleboard, the 'advance,' and partly by the number of needles per meter across the loom. If the chosen advance happens to be equal to, or even near the distance between needle rows, then the next row of needles will come down in exactly the same position as the previous row, and so on for all the rows of needles. The result will be a severe needle patterning; to avoid this distance between each row of needles must be different. Computer programs have been written to calculate the best set of row spacings. Third, if it is necessary to obtain a higher production from a needleloom, is it better to increase the number of needles in the board or to increase the speed of the needleboard. Finally, in trying to decide how to make a particular felt it is necessary to choose how many needle penetrations there will be per unit area, how deep the needles will penetrate and what type of needles should be used from a total of roughly 5000 different types. The variations possible seem to be infinite, making an optimization process very difficult.

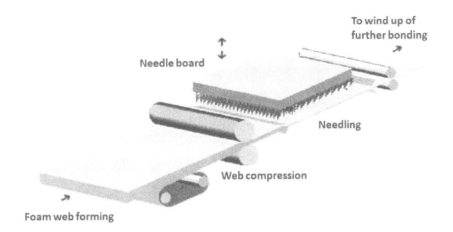

FIGURE 1.8   A schematic of needle-felting system.

II. Stitch Bonding
The idea of stitch bonding was developed almost exclusively in Czechoslovakia, in the former East Germany and to some extent in Russia, though there was a brief development in Britain. The machines have a number of variants, which are best discussed separately; many possible variants have been produced but only a limited number are discussed here for simplicity.

III. Batt Bonding by Threads
Stitch bonding uses mainly cross-laid and air-laid batts. The batt is taken into a mod-ification of a warp-knitting machine and passes between the needles and the guide bar(s). The needles are strengthened and are specially designed to penetrate the batt on each cycle of the machine. The needles are of the compound type having a tongue controlled by a separate bar. After the needles pass through the batt the needle hooks open and the guide bar laps thread into the hooks of the needles. As the needles withdraw again the needle hooks are closed by the tongues, the old loops knock over the needles and new loops are formed. In this way a form of warp knitting action is carried out with the overlaps on one side of the batt and the underlaps on the other. Generally, as in most warp knitting, continuous filament yarns are used to avoid yarn breakages and stoppages on the machine. Two structures are normally knitted on these machines, pillar (or chain) stitch, or tricot. When knitting chain stitch the same guide laps round the same needle continuously, producing a large number of isolated chains of loops. When knitting tricot structure, the guide bar shogs one needle space to the left, then one to the right. Single-guide bar structure is called tri-

cot, whereas the two-guide bar structure is often referred to as full tricot. The nature of this fabric is very textile-like, soft and flexible. At one time it was widely used for curtaining but is now used as a backing fabric for lamination, as covering material for mattresses and beds and as the fabric in training shoes. In deciding whether to use pillar or tricot stitch, both have a similar strength in the machine direction, but in the cross direction the tricot stitch fabric is stronger, owing to the underlaps lying in that direction. A cross-laid web is already stronger in that direction so the advantage is relatively small. The abrasion resistance is the same on the loop or overlap side, but on the underlap side the tricot fabric has significantly better resistance owing to the longer underlaps. However, continuous filament yarn is very expensive relative to the price of the batt, so tricot fabric costs significantly more.

IV. Stitch Bonding Without Threads

In this case the machine is basically the same as in the previous section, but the guide bar(s) are not used. The needle bar moves backwards and forwards as before, pushing the needles through the batt. The main difference is that the timing of the hook closing by the tongues is somewhat delayed, so that the hook of the needle picks up some of the fiber from the batt. These fibers are formed into a loop on the first cycle and on subsequent cycles the newly formed loops are pulled through the previous loops, just as in normal knitting. The final structure is felt-like on one side and like a knitted fabric on the other. The fabric can be used for insulation and as a decorative fabric.

V. Stitch Bonding in Pile Fabric

To form a pile fabric two guide bars are usually used, two types of warp yarns (pile yarn and sewing yarn) and also a set of pile sinkers, which are narrow strips of metal over which the pile yarn is passed and whose height determines the height of the pile. The pile yarn is not fed into the needle's hook so does not form a loop; it is held in place between the underlap of the sewing yarn and the batt itself (Fig. 1.9). It is clear that this is the most efficient way to treat the pile yarn, since any pile yarn in a loop is effectively wasted. This structure has been used for making toweling with single-sided pile and also for making loop-pile carpeting in Eastern Europe. The structure has not been popular in the West owing to competition with double-sided terry toweling and tufted carpets. Equally it has not been used in technical textiles, but it could be a solution in waiting for the correct problem. Strangely a suitable problem has been proposed. Car seating usually has a polyester face with a foam backing, but this material cannot be recycled, because of the laminated foam. It has been suggested that a polyester nonwoven pile fabric could replace the foam and would be 100% recyclable.

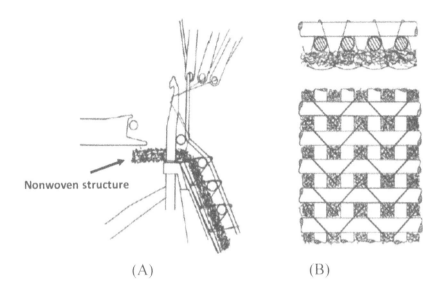

Nonwoven structure

(A) (B)

FIGURE 1.9 Stitch-bonding method (A) and product (B).

VI. Batt Looping

In this technique the needles pass through the supporting fabric and pick up as much fiber from the batt as possible. Special sinkers are used to push fiber into the needle's hook to increase this pick-up. The fiber pulled through the fabric forms a chain of loops, with loose fiber from the batt on the other surface of the fabric. The fabric is finished by raising, not as one might expect on the loose side of the fabric but instead the loops are raised because this gives a thicker pile. This structure was widely used in Eastern Europe, particularly for artificial fur, but in the West it never broke the competition from silver knitting, which gives a fabric with similar properties. The method could be used for making good quality insulating fabrics.

VII. Swing laid yarns

Two distinct types of fabric can be made using the same principle. The first is a simulated woven fabric in which the cross-direction yarns are laid many at a time in a process a bit like cross laying. The machine direction yarns, if any are used, are simply unwound into the machine. These two sets of yarns are sewn together using chain stitch if there are only cross-direction threads and tricot stitch if machine direction threads are present, the underlaps holding the threads down. Although fabric can be made rapidly by this system this turns out to be a situation in which speed is not everything and in fact the system is not usually economically competitive with normal weaving. However, it has one great technical advantage; the machine and cross threads do not interlace but lie straight in the fabric. Consequently the initial modulus of this fabric is very high compared with a woven fabric, which can first extend by straightening out the crimp in the yarn. These fabrics are in demand for making fiber-reinforced plastic using continuous filament glass and similar high modulus fibers or filaments. The alternative system makes a multidirectional fabric. Again sets of yarns are cross-laid but in this case not in the cross direction but at, say, 45° or 60° to the cross direction. Two sets of yarns at, say, +45° and −45° to the cross direction plus another layer of yarns in the machine direction can be sewn together in the usual way. Again high modulus yarns are used, with the advantage that the directional properties of the fabric can be designed to satisfy the stresses in the component being made.

VIII. Hydroentanglement

The process of hydroentanglement was invented as a means of producing an entanglement similar to that made by a needleloom, but using a lighter weight batt. A successful process was developed during the 1960s by Du Pont and was patented. However, Du Pont decided in the mid-1970s to dedicate the patents to the public domain, which resulted in a rush of new development work in the major industrial countries, Japan, USA, France, Germany and Britain. As the name implies the process depends on jets of water working at very high pressures through jet orifices with very small diameters (Fig. 1.10). A very fine jet of this sort is liable to break up into droplets, particularly if there is any turbulence in the water passing through the orifice. If droplets are formed the energy in the jet stream will still be roughly the same, but it will spread over a much larger area of batt so that the energy per unit area will be much less. Consequently the design of the jet to avoid turbulence and to produce a needle-like stream of water is critical. The jets are arranged in banks and the batt is passed continuously under the jets held up by a perforated screen, which removes most of the water. Exactly what happens to the batt underneath the jets is not known, but it is clear that fiber ends become twisted together or entangled by the turbulence in the water after it has hit the batt. It is also known that the supporting screen is vital to the process; changing the screen with all other variables remaining constant will profoundly alter the fabric produced. Although the machines have higher throughputs compared with most bonding systems, and particularly compared with a needleloom, they are still very expensive and require a lot of power, which is also expensive. The other considerable problem lies in supplying clean water to the jets at the correct pH and temperature. Large quantities of water are needed, so recycling is necessary, but the water picks up air bubbles, bits of fiber and fiber lubricant/fiber finish in passing through the process and it is necessary to remove everything before recycling.

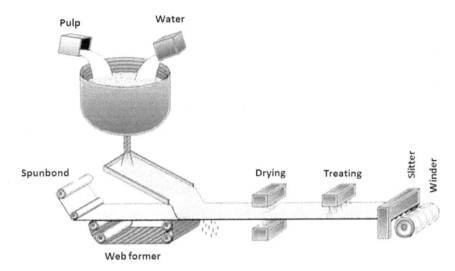

FIGURE 1.10   A schematic of hydroentanglementing process.

It is said that this filtration process is more difficult than running the rest of the machine. Fabric uses include wipes, surgeons' gowns, disposable protective clothing and backing fabrics for coating. The wipes produced by hydroentanglement are guaranteed lint free, because it is argued that if a fiber is loose it will be washed away by the jetting process. It is interesting to note that the hydroentanglement process came into being as a process for entangling batts too light for a needleloom, but that the most recent developments are to use higher water pressures (400 bar) and to process heavier fabrics at the lower end of the needleloom range.

1.2   STRUCTURAL PROPERTIES OF NONWOVEN FABRICS

1.2.1   CLASSIFICATION

The structure and properties of a nonwoven fabric are determined by fiber properties, the type of bonding elements, the bonding interfaces between the fibers and binder elements and the fabric structural architecture. Examples of dimensional and structural parameters may be listed as follows:

1.   Fiber dimensions and properties: fiber diameter, diameter variation (e.g., in melt-blown microfiber and electrospun nanofiber webs), cross-sectional shape, crimp wave frequency and amplitude, length, density; fiber properties (Young's modulus, elasticity, tenacity, bending and torsion rigidity,

compression, friction coefficient), fibrillation propensity, surface chemistry and wetting angle.

2. Fiber alignment: fiber orientation distribution.
3. Fabric dimensions and variation: dimensions (length, width, thickness, and weight per unit area), dimensional stability, density and thickness uniformity.
4. Structural properties of bond points: bonding type, shape, size, bonding area, bonding density, bond strength, bond point distribution, geometrical arrangement, the degree of liberty of fiber movement within the bonding points, interface properties between binder and fiber; surface properties of bond points.
5. Porous structural parameters: fabric porosity, pore size, pore size distribution, pore shape.

Examples of important nonwoven fabric properties are:
• Mechanical properties: tensile properties (Young's modulus, tenacity, strength and elasticity, elastic recovery, work of rupture), compression and compression recovery, bending and shear rigidity, tear resistance, burst strength, crease resistance, abrasion, frictional properties (smoothness, roughness, friction coefficient), energy absorption.
• Fluid handling properties: permeability, liquid absorption (liquid absorbency, penetration time, wicking rate, rewet, bacteria/particle collection, repellency and barrier properties, run-off, strike time), water vapor transport and breathability.
• Physical properties: thermal and acoustic insulation and conductivity, electrostatic properties, dielectric constant and electrical conductivity, opacity and others.
• Chemical properties: surface wetting angle, oleophobicity and hydrophobicity, interface compatibility with binders and resins, chemical resistance and durability to wet treatments, flame resistance, dyeing capability, flammability, soiling resistance.
• Application specific performance: linting (particle generation), esthetics and handle, filtration efficiency, biocompatibility, sterilization compatibility, biodegradability and health and safety status.

1.2.2   DIMENSIONAL PARAMETERS

The structure and dimensions of nonwoven fabrics are frequently characterized in terms of fabric weight per unit area, thickness, density, fabric uniformity, fabric porosity, pore size and pore size distribution, fiber orientation distribution, bonding segment structure and the distribution. Nonwoven fabric weight (or fabric mass) is defined as the mass per unit area of the fabric and is usually measured in g/m^2 (or gsm). Fabric thickness is defined as the distance between the two fabric surfaces under a specified applied pressure, which varies if the fabric is high-loft (or compressible). The fabric weight and thickness determine the fabric packing density, which influences the freedom of movement of the fibers and determines the porosity (the proportion of voids) in a nonwoven structure. The freedom of movement of the fibers plays an important role in nonwoven mechanical properties and the proportion of voids determines the fabric porosity, pore sizes and permeability in a nonwoven structure. Fabric density, or bulk density, is the weight per unit volume of the nonwoven fabric (kg/m³). It equals the measured weight per unit area (kg/m²) divided by the measured thickness of the fabric (m). Fabric bulk density together with fabric porosity is important because they influence how easily fluids, heat and sound transport through a fabric.

1.2.3   WEIGHT UNIFORMITY OF NONWOVEN FABRICS

The fabric weight and thickness usually varies in different locations along and across a nonwoven fabric. The variations are frequently of a periodic nature with a recurring wavelength due to the mechanics of the web formation and/or bonding process. Persistent cross-machine variation in weight is commonly encountered, which is one reason for edge trimming. Variations in either thickness and/or weight per unit area determine variations of local fabric packing density, local fabric porosity and pore size distribution, and therefore influence the appearance, tensile properties, permeability, thermal insulation, sound insulation, filtration, liquid barrier and penetration properties, energy absorption, light opacity and conversion behavior of nonwoven products. Fabric uniformity can be defined in terms of the fabric weight (or fabric density) variation measured directly by sampling different regions of the fabric. The magnitude of the variation depends on the specimen size, for example the variation in fabric weight between smaller fabric samples (e.g., consecutive fabric samples of 1 m² or 10 mm²) will usually be much greater than the variation between bigger fabric samples (e.g., rolls of fabric of hundreds of meters). Commercially, to enable on-line determination of fabric weight variation, the fabric uniformity is measured in terms of the variation in the optical density of fabric images [7], the gray level intensity of fabric images [8] or the amount of electromagnetic rays absorbed by the fabric depending on the measurement techniques used [9, 10]. The basic statistical terms for expressing weight uniformity in the industry are the standard deviation (s) and the coefficient of variation (CV) of measured parameters as follows:

$$\text{Standard deviation:} \quad \sigma^2 = \frac{\sum_{i=1}^{n}(w_i - \overline{w})^2}{n} \tag{1}$$

$$\text{Coefficient of variation: } CV = \frac{\sigma}{w} \tag{2}$$

$$\text{Index of dispersion: } I_{\text{dispersion}} = \frac{\sigma^2}{w} \tag{3}$$

where n is the number of test samples, w is the average of the measured parameter and w_i is the local value of the measured parameter. Usually, the fabric uniformity is referred to as the percentage coefficient of variation (CV%). The fabric uniformity in a nonwoven is normally anisotropic, that is, the uniformity is different in different directions (MD and CD) in the fabric structure. The ratio of the index of dispersion has been used to represent the anisotropy of uniformity [11]. The local anisotropy of mass uniformity in a nonwoven has also been defined by Scharcanski and Dodson in terms of the 'local dominant orientations of fabric weight' [12].

1.2.4   FIBER ORIENTATION IN NONWOVEN

The fibers in a nonwoven fabric are rarely completely randomly orientated, rather, individual fibers are aligned in various directions mostly in-plane. These fiber alignments are inherited from the web formation and bonding processes. The fiber segment orientations in a nonwoven fabric are in two and three dimensions and the orientation angle can be determined (Fig. 1.11). Although the fiber segment orientation in a nonwoven is potentially in any three-dimensional direction, the measurement of fiber alignment in three dimensions is complex and expensive [13]. In certain nonwoven structures, the fibers can be aligned in the fabric plane and nearly vertical to the fabric plane. The structure of a needle punched fabric is frequently simplified in this way. In this case, the structure of a three-dimensional nonwoven may be simplified as a combination of two-dimensional layers connected by fibers orientated perpendicular to the plane (Fig. 1.12). The fiber orientation in such a three-dimensional fabric can be described by measuring the fiber orientation in two dimensions in the fabric plane [14].

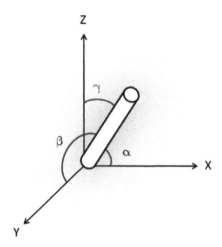

FIGURE 1.11   Fiber orientation angle in 3-D nonwoven fabric.

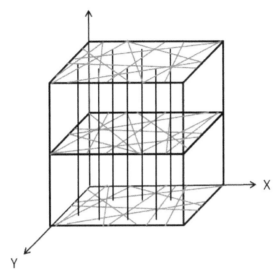

FIGURE 1.12   Simplified 3-D nonwoven structure.

In the two-dimensional fabric plane, fiber orientation is measured by the fiber orientation angle, which is defined as the relative directional position of individual fibers in the structure relative to the machine direction as shown in Fig. 1.13. The orientation angles of individual fibers or fiber segments can be determined by evaluating photomicrographs of the fabric or directly by means of microscopy and image analysis.

The frequency distribution (or statistical function) of the fiber orientation angles in a nonwoven fabric is called fiber orientation distribution (FOD) or ODF (orientation distribution function). Frequency distributions are obtained by determining the fraction of the total number of fibers (fiber segments) falling within a series of pre-defined ranges of orientation angle. Discrete frequency distributions are used to estimate continuous probability density functions. The following general relationship is proposed for the fiber orientation distribution in a two-dimensional web or fabric:

$$\int_0^\pi \Omega(\alpha)\,d\alpha = 1 \qquad \left(\Omega(\alpha) \geq 0\right)$$

(4)

or

$$\sum_{\alpha=0}^\pi \Omega(\alpha)\,\Delta\alpha = 1 \qquad \left(\Omega(\alpha) \geq 0\right)$$

(5)

where α is the fiber orientation angle, and $\Omega(\alpha)$ is the fiber orientation distribution function in the examined area. The numerical value of the orientation distribution indicates the number of observations that fall in the direction α, which is the angle relative to the examined area. Attempts have been made to fit the fiber orientation distribution frequency with mathematical functions including uniform, normal and exponential distribution density functions. The following two functions in combination with the constrain in the equation $\int_0^\pi \Omega(\alpha)d\alpha = 1$ have been suggested by Petterson [15] and Hansen [16] respectively.

Petterson: $\Omega(\alpha) = A + B\cos\alpha + C\cos^3\alpha + D\cos^8\alpha + E\cos^{16}\alpha$ (6)

Hansen: $\Omega(\alpha) = A + B\cos^2(2\alpha)$ (7)

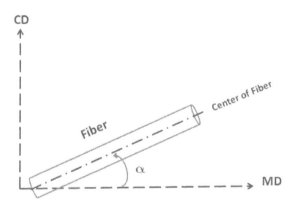

FIGURE 1.13   Fiber orientation and orientation angle.

Fiber alignments in nonwoven fabrics are usually anisotropic, that is, the number of fibers in each direction in a non-woven fabric is not equal. The differences between the fiber orientation in the fabric plane and in the direction perpendicular to the fabric plane (i.e., transverse direction or fabric thickness direction) are particularly important. In most nonwovens except some air- laid structures, most of the fibers are preferentially aligned in the fabric plane rather than in the fabric thickness. Significant in-plane differences in fiber orientation are also found in the machine direction and in the fabric cross direction in nonwovens. Preferential fiber (either staple fiber or continuous filament) orientation in one or multiple directions is introduced during web formation and to some extent during mechanical bonding processes. A simplified example of an anisotropic nonwoven structure is a unidirectional fibrous bundle in which fibers are aligned in one direction only. Parallel-laid or cross-laid carded webs are usually anisotropic with a highly preferential direction of fiber orientation. Fiber orientation in airlaid structures is usually more isotropic than in other dry-laid fabrics both in two and three dimensions. In perpendicular-laid webs, fibers are orientated in the direction of the fabric thickness. Spunlaid non-wovens composed of filaments are less anisotropic in the fabric plane than layered carded webs [17], however, the anisotropy of continuous filament webs depends on the way in which the webs are collected and tensioned. This structural anisotropy can be characterized in terms of the fiber orientation distribution functions. This anisotropy is important because of its influence on the anisotropy of fabric mechanical and physical properties including tensile, bending, thermal insulation, acoustic absorption, dielectric behavior and permeability. The ratio of physical properties obtained in different directions in the fabric, usually the MD/CD, is a well established means of expressing the anisotropy. The MD/CD ratio of tensile strength is most commonly encountered, although the same approach may be used to express directional in-plane differences in elongation, liquid wicking distance, liquid transport rate, dielectric constant and permeability. However, these anisotropy terms use indirect experimental methods to characterize the nonwoven structure, and they are just ratios in two specific directions in the fabric plane, which can misrepresent the true anisotropy of a nonwoven structure.

1.2.5 FABRIC POROSITY

The pore structure in a nonwoven may be characterized in terms of the total pore volume (or porosity), the pore size, pore size distribution and the pore connectivity. Porosity provides information on the overall pore volume of a porous material and is defined as the ratio of the nonsolid volume (voids) to the total volume of the non-woven fabric. The volume fraction of solid material is defined as the ratio of solid fiber material to the total volume of the fabric. While the fiber density is the weight of a given volume of the solid component only (i.e., not containing other materials), the porosity can be calculated as follows using the fabric bulk density and the fiber density:

$$\varnothing(\%) = \frac{\rho_{fibric}}{\rho_{fibre}} \times 100\% \tag{8}$$

$$\varepsilon(\%) = (1-\varnothing) \times 100\% \tag{9}$$

where ε is the fabric porosity (%), ϕ is the volume fraction of solid material (%), ρ_{fabric} (kg/m^3) is the fabric bulk density and ρ_{fiber} (kg/m^3) is the fiber density.

In resin coated, impregnated or laminated nonwoven composites, a small proportion of the pores in the fabric is not accessible (i.e., they are not connected to the fabric surface). The definition of porosity as shown above refers to the so-called total porosity of the fabric. Thus, the open porosity (or effective porosity) is defined as the ratio of accessible pore volume to total fabric volume, which is a component part of the total fabric porosity. The majority of nonwoven fabrics have porosities >50% and usually above 80%. A fabric with a porosity of 100% is a totally open fabric and there is no such fabric, while a fabric with a porosity of 0% is a solid polymer without any pore volume; there is no such fabric either. High-loft nonwoven fabrics usually have a low bulk density because they have more pore space than a heavily compacted nonwoven fabric; the porosity of high-loft nonwovens can reach >98%. Pore connectivity, which gives the geometric pathway between pores cannot be readily quantified and described. If the total pore area responsible for liquid transport across any distance along the direction of liquid transport is known, its magnitude and change in magnitude are believed to indicate the combined characteristics of the pore structure and connectivity.

1.3   CHARACTERIZATION OF NONWOVEN FABRICS

Fabric thickness Testing [18, 19] of nonwoven fabric thickness and fabric weight is similar to other textile fabrics but due to the greater compressibility and unevenness a different sampling procedure is adopted. The thickness of a nonwoven fabric is defined as the distance between the face and back of the fabric and is measured as the distance between a reference plate on which the nonwoven rests and a parallel presser-foot that applies a pressure to the fabric (See BS EN ISO 9703–2:1995, ITS 10.1).

Nonwoven fabrics with a high specific volume, i.e., bulky fabrics, require a special procedure. In this context, bulky fabrics are defined as those that are compressible by 20% or more when the pressure applied changes from 0.1 kPa to 0.5 kPa. Three procedures are defined in the test standard (BS EN ISO 9703–2:1995). Three test methods, (i.e., ASTM D5729–97 (ITS 120.1), ASTM D5736–01 (ITS 120.2), ASTM D6571–01 (ITS 120.3)) are defined for the measurement of the thickness, compression and recovery of conventional nonwovens and high-loft nonwovens (it is defined, in the ASTM, as a low density fiber network structure characterized by a high ratio of thickness to mass per unit area. High-loft batts have no more than a 10% solid volume and are greater than 3 mm in thickness). Two more test standards (ITS 120.4, ITS 120.5) are defined for rapid measurement of the compression and recovery of high-loft nonwovens.

1.3.1   *FABRIC* MASS PER UNIT AREA

The measurement of a nonwoven weight per unit area requires a specific sampling procedure, specific dimensions for the test samples, and a greater balance accuracy than for conventional textiles. According to the ISO standards (BS EN 29073–1:1992, ISO 9073–1:1989, ITS10.1), the measurement of nonwoven fabric mass per unit area of nonwovens requires each piece of fabric sample to be at least 50,000

mm^2. The mean value of fabric weight is calculated in grams per square meter and the coefficient of variation is expressed as a percentage.

1.3.2 FABRIC WEIGHT UNIFORMITY

Nonwoven fabric uniformity refers to the variations in local fabric structures, which include thickness and density, but is usually expressed as the variation of the weight per unit area. Both subjective and objective techniques are used to evaluate the fabric uniformity. In subjective assessment, visual inspection can distinguish non-uniform areas as small as about 10 mm^2 from a distance of about 30 cm. Qualitative assessments of this type can be used to produce ratings of nonwoven fabric samples by a group of experts against benchmark standards. The consensual benchmark standards are usually established by an observer panel using paired comparison, graduated scales or similar voting techniques; these standard samples are then used to grade future samples. Indirect objective measurements of the web weight uniformity have been developed based on variations in other properties that vary with fabric weight including the transmission and reflection of beta rays, gamma rays (CO60), lasers, optical and infra-red light [20], and variation in tensile strength. With optical light scanning methods, the fabrics are evaluated for uniformity using an optical electronic method, which screens the nonwoven to register 32 different shades of gray [21]. The intensity of the points in the different shades of gray provides a measure of the uniformity. A statistical analysis of the optical transparency and the fabric uniformity is then produced. This method is suitable for lightweight nonwovens of 10–50 g/m^2. Optical light measurements are commonly coupled with image analysis to determine the coefficient of variation of gray level intensities from scanned images of nonwoven fabrics [22]. In practice, nonwoven fabric uniformity depends on fiber properties, fabric weight and manufacturing conditions. It is usually true that the variation in fabric thickness and fabric weight decreases as the mean fabric weight increases. Wet-laid nonwovens are usually more uniform in terms of thickness than dry-laid fabrics. Short fiber airlaid fabrics are commonly more uniform than carded and crosslaid and parallel-laid fabrics, and spunbond and melt-blown fabrics are often more uniform than fabrics produced from staple fibers.

1.3.3 MEASURING FIBER ORIENTATION DISTRIBUTION

In modeling the properties of nonwoven fabrics and particularly in any quantitative analysis of the anisotropic properties of nonwoven fabrics, it is important to obtain an accurate measurement of FOD. A number of measuring techniques have been developed. A direct visual and manual method of measurement was first described by Petterson [23]. Hearle and co-workers [24, 25] found that visual methods produce accurate measurements and it is the most reliable way to evaluate the fiber orientation. Manual measurements of fiber segment angles relative to a given direction were conducted and the lengths of segment curves were obtained within a given range. Chuleigh [26] developed an optical processing method in which an opaque mask was used in a light microscope to highlight fiber segments that are orientated in a known direction. However, the application of this method is limited by the tedious and time-consuming work required in visual examinations.

To increase the speed of assessment, various indirect-measuring techniques have been introduced including both the zero span [27, 28] and short span [29] tensile analysis for predicting the fiber orientation distribution. Stenemur [30] devised a computer system to monitor fiber orientation on running webs based on the light diffraction phenomenon. Methods that employ X-ray diffraction analysis and X-ray diffraction patterns of fiber webs have also been studied [31, 32]. In this method the distribution of the diffraction peak of the fiber to X-ray is directly related to the distribution of the fiber orientation. Other methods include the use of microwaves [33], ultrasound [34], light diffraction methods [35], light reflection and light refraction [36], electrical measurements [37, 38] and liquid-migration-pattern analysis [39, 40] In the last few decades, image analysis has been employed to identify fibers and their orientation [41–44] and computer simulation techniques have come into use for the creation of computer models of various nonwoven fabrics [45–48]. Huang and Bressee89 developed a random sampling algorithm and software to analyze fiber orientation in thin webs. In this method, fibers are randomly selected and traced to estimate the orientation angles; test results showed excellent agreement with results from visual measurements. Pourdeyhimi et al. [49–52] completed a series of studies on the fiber orientation of nonwovens by using an image analyzer to determine the fiber orientation in which image processing techniques such as computer simulation, fiber tracking, Fourier transforms and flow field techniques were employed. In contrast to two-dimensional imaging techniques suitable only for thin nonwoven fabrics, the theory of Hilliard-Komori-Makishima [53] and the visualizations made by Gilmore et al. [54] using X-ray tomographic techniques have provided a means of analyzing the three-dimensional orientation.

Image analysis is a computer-based means of converting the visual qualitative features of a particular image into quantitative data. The measurement of the fiber orientation distribution in nonwoven fabrics using image analysis is based on the assumption that in thin materials a two-dimensional structure can be assumed, although in reality the fibers in a nonwoven are arranged in three dimensions. However, there is currently no generally accepted way of characterizing the fabric structure in terms of the three-dimensional geometry. The fabric geometry is reduced to two dimensions by evaluation of the planar projections of the fibers within the fabric. The assumption of a two-dimensional fabric structure is adequate to describe thin fabrics. The image analysis system in the measurement of the fiber orientation distribution is based on a computerized image capture system operating with an integrated image analysis software package in which numerous functions can be performed [55]. A series of sequential operations is required to perform image analysis and, in a simple system, the following procedures are carried out: production of a gray image of the sample fabric, processing the gray image, detection of the gray image and conversion into binary form, storage and processing of the binary image, measurement of the fiber orientation and output of results.

1.3.4 MEASURING POROSITY

Porosity can be obtained from the ratio of the fabric density and the fiber density. In addition to the direct method of determination for resin impregnated dense non-woven composites, the fabric porosity can be determined by measuring densities using liquid buoyancy or gas expansion porosimetry [56]. Other methods include small angle neutron, small angle X-ray scattering and quantitative image analysis for total porosity. Open porosity may be obtained from xylene and water impregnation techniques [57], liquid metal (mercury) impregnation, nitrogen adsorption and air or helium penetration. Existing definitions of pore geometry and the size of pores in a nonwoven are based on various physical models of fabrics for specific applications. In general, cylindrical, spherical or convex shaped pores are assumed with a distribution of pore diameters. Three groups of pore size are defined: (i) the near-largest pore size (known as apparent opening pore size, or opening pore size), (ii) the constriction pore size (known as the pore-throat size), and (iii) the pore volume size. Pore size and the pore size distribution of nonwoven fabrics can be measured using optical methods, density methods, gas expansion and adsorption, electrical resistance, image analysis, porosimetry and porometry. The apparent pore opening (or opening pore) size is determined by the passage of spherical solid glass beads of different sizes (50 μm to 500 μm) through the largest pore size of the fabric under specified conditions. The pore size can be measured using sieving test methods (dry sieving, wet sieving and hydrodynamic sieving). The opening pore sizes are important for determining the filtration and clogging performance of nonwoven geotextiles and it enables the determination of the absolute rating of filter fabrics. The constriction pore size, or porethroat size, is different from the apparent pore opening size. The constriction pore size is the dimension of the smallest part of the flow channel in a pore and it is important for fluid flow transport in nonwoven fabrics. The largest pore-throat size is called the bubble point pore size, which is related to the degree of clogging of geotextiles and the performance of filter fabrics. The pore-throat size distribution and the bubble point pore size can be obtained by liquid expulsion methods. However, it is found that wetting fluid, air pressure and equipment type affects the measured constriction pore size [58, 59].

1.3.4.1 LIQUID POROSIMETRY

Liquid porosimetry, also referred to as liquid porometry, is a general term to describe procedures for the evaluation of the distribution of pore dimensions in a porous material based on the use of liquids. Both pore volumes and pore throat dimensions are important quantities in connection with the use of fiber networks as absorption and barrier media. Pore volumes determine the capacity of a network to absorb liquid, that is, the total liquid uptake. Pore throat dimensions, on the other hand, are related to the rate of liquid uptake and to the barrier characteristics of a network. Liquid porosimetry evaluates pore volume distributions (PVD) by measuring the volume of liquid located in different size pores of a porous structure. Each pore is sized according to its effective radius, and the contribution of each pore size to the total free volume of the porous network is determined. The effective radius R of any pore is defined by the Laplace equation:

$$R = \frac{2\gamma cos\theta}{\Delta p} \tag{10}$$

where γ = liquid surface tension, θ = advancing or receding contact angle of the liquid, ΔP = pressure difference across the liquid/air meniscus For liquid to enter or drain from a pore, an external gas pressure must be applied that is just enough to overcome the Laplace pressure ΔP.

In the case of a dry heteroporous network, as the external gas pressure is decreased, either continuously or in steps, pores that have capillary pressures lower than the given gas pressure ΔP will fill with liquid. This is referred to as liquid intrusion porosimetry and requires knowledge of the advancing liquid contact angle. In the case of a liquid-saturated heteroporous network, as the external gas pressure is increased, liquid will drain from those pores whose capillary pressure corresponds to the given gas pressure ΔP. This is referred to as liquid extrusion porosimetry and requires knowledge of the receding liquid contact angle. In both cases, the distribution of pore volumes is based on measuring the incremental volume of liquid that either enters a dry network or drains from a saturated network at each increment of pressure.

1.3.4.2   INSTRUMENTATION

Until recently, the only version of this type of analysis to evaluate PVDs in general use was mercury porosimetry [60]. Mercury was chosen as the liquid because of its very high surface tension so that it would not be able to penetrate any pore without the imposition of considerable external pressure. For example, to force mercury into a pore 5 μm in radius requires a pressure increase of about 2 atm. While this might not be a problem with hard and rigid networks, such as stone, sand structures, and ceramics, it makes the procedure unsuitable for use with fiber materials that would be distorted by such compressive loading. Furthermore, mercury intrusion porosimetry is best suited for pore dimensions less than 5 μm, while important pores in typical textile structures may be as large as 1000 μm. Some of the other limitations of mercury porosimetry have been discussed by Winslow [61] and by Good [62].

A more general version of liquid porosimetry for PVD analysis, particularly well suited for textiles and other compressible planar materials, has been developed by Miller and Tyomkin [63]. The underlying concept was earlier demonstrated for low-density webs and pads by Burgeni and Kapur [64]. Any stable liquid of relatively low viscosity that has a known $\cos \theta > 0$ can be used. In the extrusion mode, the receding contact angle is the appropriate term in the Laplace equation, while in the intrusion mode the advancing contact angle must be used. There are many advantages to using different liquids with a given material, not the least of which is the fact that liquids can be chosen that relate to a particular end use of a material. The basic arrangement for liquid extrusion porosimetry is shown in Fig. 1.14. In the case of liquid extrusion, a presaturated specimen is placed on a microporous membrane, which is itself supported by a rigid porous plate. The gas pressure within the closed chamber is increased in steps, causing liquid to flow out of some of the pores, largest ones first. The amount of liquid removed at each pressure level is monitored by the top-loading recording balance. In this way, each incremental change in pressure (corresponding to a pore size according to the Laplace equation) is related to an increment of liquid mass. To induce stepwise drainage from large pores requires very small increases in pressure over a narrow range that are only slightly above atmospheric pressure, whereas to analyze for small pores the pressure changes must be quite large. In early versions of instrumentation for liquid extrusion porosimetry, pressurization of the specimen chamber was accomplished either by hydrostatic head changes or by means of a single-stroke pump that injected discrete drops of liquid into a free volume space that included the chamber [63]. In the most recent instrumentation developed by Miller and Tyomkin [65], the chamber is pressurized by means of a computer-controlled, reversible, motor-driven piston/cylinder arrangement that can produce the required changes in pressure to cover a pore radius range from 1 to 1000 μm.

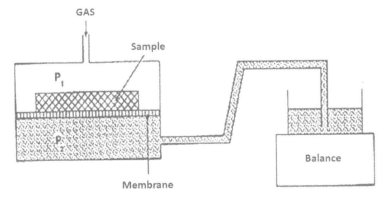

FIGURE 1.14 Basic arrangement for liquid porosimetry to quantify pore volume distribution.

DATA ANALYSIS AND APPLICATIONS

Prototype data output for a single cycle incremental liquid extrusion run is shown in Fig. 1.15. The experiment starts from the right as the pressure is increased, draining liquid first from the largest pores. The cumulative curve represents the amount of liquid remaining in the pores of the material at any given level of pressure. The first derivative of this cumulative curve as a function of pore size becomes the pore volume distribution, showing the fraction of the free volume of the material made up of pores of each indicated size. PVD curves for two typical fabrics woven from spun yarns are shown in Fig. 1.15. PVD curves for several nonwoven materials are shown in Fig. 1.16. These materials normally have unimodal PVD curves, but generally the pores are larger than those associated with typical woven fabrics. Several interesting points can be noted. First, the pore volumes become smaller with increasing compression (decreasing mat thickness), and at the same time the structures appear to have become less heteroporous; that is, the breadth of the PVD curves decreases with increasing compression. Also, the total pore volume (area under each curve), and therefore the sorptive capacity, decreases with compression. Since in many end-use applications fibrous materials are used under some level of compression, it is particularly important to evaluate pore structure under appropriate compression conditions. The PVD instrumentation described here, referred to as the TRI Autoporosimeter, is extremely versatile and can be used with just about any porous material, including textiles, paper products, membrane filters, particulates, and rigid foams. It also allows quantification of interlayer pores, surface pores, absorption/desorption hysteresis, uptake and retention capillary pressures, and effective contact angles in porous networks [65]. The technique has also been used to quantify pore volume dimensions and sorptive capacity of artificial skin in relation to processing conditions [66].

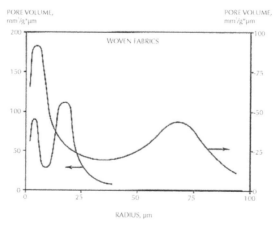

FIGURE 1.15   PVD curves for two typical spun yarn woven fabrics [65].

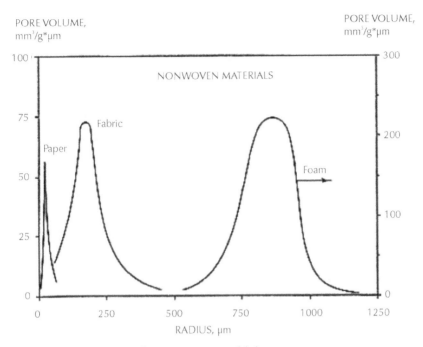

FIGURE 1.16 PVD curves for some nonwoven fabrics.

1.3.4.4   IMAGE ANALYSIS [67]

The dynamic development of computer techniques creates broad possibilities for their application, including identifying and measuring the geometrical dimensions of very small objects including textile objects. Using digital image analysis permits a more detailed analysis of such basic structural parameters of nonwoven products. The digital analysis of two-dimensional images is based on processing the image acquirement, with the use of a computer. The image is described by a two-dimensional matrix of real or imaginary numbers presented by a definite number of bytes. Image modeling is based on digitizing the real image. This process consists of sampling and quantifying the image. The digital image can be described in the form of a two-dimensional matrix, whose elements include quantified values of the intensity function, referred to as gray levels. The digital image is defined by the spatial image resolution and the gray level resolution. The smallest element of the digital image is called the pixel. The number of pixels and the number of brightness levels may be unlimited, although while presenting computer technique data it is customary to use values which are multiplications of the number 2, for example 512×512 pixels and 256 gray levels.

1.3.5 *IMAGE* QUALITY IMPROVEMENT

Image quality improvement and highlighting its distinguished features are the most often used application techniques for image processing. The process of image quality improvement does not increase the essential information represented by the image data, but increases the dynamic range of selected features of the acquired object, which facilitates their detection.

The following operations are carried out during image quality improvement:
- changes of the gray level system and contrast improvement;
- edge exposition;
- pseudo-colorization;
- improvement of sharpness, decreasing the noise level;
- space filtration;
- interpolation and magnification;
- compensation of the influence of interference factors, for example, possible under-exposure.

1.3.5.1   REINSTATING DESIRED IMAGE FEATURES

Reinstating desired image features is connected with eliminating and minimizing any image features which lower its quality. Acquiring images by optical, opto-electronic or electronic methods involves the unavoidable degradation of some image features during the detection process. Aberrations, internal noise in the image sensor, image blurring caused by camera defocusing, as well as turbulence and air pollution in the surrounding atmosphere may cause a worsening of quality. Reinstating the desired image features differs from image improvement, whose procedure is related to highlighting or bringing to light the distinctive features of the existing image. Reinstating the desired image features mainly includes the following corrections:
- reinstating the sharpness lowered as the result of disadvantageous features of the image;
- sensor or its surrounding;
- noise filtration;
- distortion correction;
- correction of sensors' nonlinearity.

1.3.5.2   IMAGE DATA COMPRESSION

Image data compression is based on minimizing the number of bytes demanded for image representation. The compression effect is achieved by transforming the given digital image to a different number table in such a manner that the preliminary information amount is packed into a smaller number of samples.

1.3.5.3   MEASURING FIBER ORIENTATION IN NONWOVEN USING IMAGE PROCESSING

Fiber orientation is an important characteristic of nonwoven fabrics, since it directly influences their mechanical properties and performances. Since visual assessments of fiber orientation are not reliable, image processing techniques are used to assess them automatically. The image of a fabric is taken and digitized. The digitized image of the surface of a fabric is progressively simplified by a line operator and an edge-smoothing and thinning operation to produce an image in which fibers are represented by curves of one pixel in thickness. The fiber-orientation distribution of a fabric is then obtained by tracing and measuring the length and orientation of the curves. The image-analysis method for automatically measuring the fiber-orientation distributions is validated by manual measurements made by experts. Spatial uniformity of fibrous structures can been described statistically using an index of dispersion. The spatial uniformity of web mass is described by dispersion of its surface relief distribution. Surface relief is representative of mass density gradients in local regions. In principle, this concept is similar to topographical surface relief that represents the variations in elevation of the earth's surface: the higher is the elevation, the greater the mass present on a surface. The surface relief in a local region is representative of the average mass density gradient in that local region. Surface relief of a web can be measured from its gray-scale images taken in transmitted or reflected light or any other radiation. Each pixel of a gray-scale image (with 256 gray-levels or 8-bit) is considered a column of cubes with each cube having area (1 pixel) X (1 pixel) and height equal to one intensity level. Figure 1.14 shows the translation of square pixels of Fig. 1.15 into columns of cubes. Surface relief of a pixel is the difference in height (or graylevel intensity, G) of two adjacent pixels sharing a side. The surface relief area was calculated as the total number of exposed faces of cube columns in a "quadrat," which is defined as a rectangular region of interest. Only lateral surfaces were considered and flat tops of columns were ignored in the calculation of relief area.

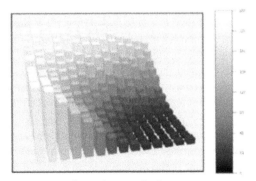

FIGURE 1.17   Pixels represented as columns of cubes.

FIGURE 1.18   Detailed image pixels.

Various techniques are used to estimate fiber orientation distribution in non-wovens, such as direct tracking, flow field analysis and the Fourier transform (FT) and Hough transform (HT) methods. In direct tracking, the actual pixels of the lines are tracked. The algorithm used is time consuming and therefore is mostly used for research and development. In denser structures, as the number of crossover points increases, this method becomes less efficient. Flow field analysis is based on the assumption that edges in an image are representative of orientation fields in the image. This method is mainly used to obtain the mean orientation distribution angle, but not the orientation distribution function. The FT method is widely used in many image processing and measurement operations. It transforms the gray scale intensity domain to a frequency spectrum. Compression of the image, leveling of the image brightness and contrast and enhancement can more conveniently be performed on the frequency spectrum. In most cases the spatial domain needs to be recovered from the frequency spectrum. Measuring the orientation is one of the applications of the FT method. The algorithm is fast and does not need high computational power for the calculations. The HT method is used in estimating fiber orientation distribution directly. The advantage of this method compared with other indirect methods is that the actual orientation of the lines is included directly in the computation of the transform. The capability of the HT method in object recognition is used to measure the length of the straight lines in the image. However, more computational time and resources are needed to run the algorithm for larger-scale images with higher accuracy. The images acquired using backlighting and optical microscopes were reported to be of low quality having poor contrast. Consequently, they limited the image acquisition technique to very thin webs for which they could obtain images that could be successfully processed. When images are digitized at high magnification, the depth of field is poor, resulting in fuzzy fiber segments and inferior image quality. To overcome this problem, in the present work a scanning electron microscope (SEM) has been used to produce high quality images. The SEM provides high depth of field even at high magnifications. This technique is not restricted to thin webs and can also be used for fabrics with very high density.

1.3.6 *PORE* VOLUME DISTRIBUTION AND MERCURY POROSIMETRY

Unlike porometry where the measurement of the pore-throat size distribution is based on measurement of the airflow rate through a fabric sample, the pore volume distribution is determined by liquid porosimetry, which is based on the liquid uptake concept proposed by Haines [68]. A fabric sample (either dry or saturated) is placed on a perforated plate and connected to a liquid reservoir. The liquid having a known surface tension and contact angle is gradually forced into or out of the pores in the fabric by an external applied pressure. Porosimetry is grouped into two categories based on the liquid used, which is either nonwetting (e.g., mercury) or wetting (e.g., water). Each is used for intrusion porosimetry and extrusion porosimetry where the advancing contact angle and receding contact angle are applied in liquid intrusion and extrusion porosimetry respectively.

Mercury has a high surface tension and is strongly nonwetting on most fabrics at room temperature. In a typical mercury porosimetry measurement, a nonwoven fabric is evacuated to remove moisture and impurities and then immersed in mercury. A gradually increasing pressure is applied to the sample forcing mercury into increasingly smaller 'pores' in the fabric. The pressure P required to force a nonwetting fluid into a circular cross-section capillary of diameter d is given by:

$$P = \frac{4\sigma_{Hg} \cos \gamma_{Hg}}{d} \tag{11}$$

where σ_{Hg} is the surface tension of the mercury (0.47 N/m), 130 and γ_{Hg} is the contact angle of the mercury on the material being intruded (the contact angle ranges from $135°\sim180°$), and d is the diameter of a cylindrical pore.

The incremental volume of mercury is recorded as a function of the applied pressure to obtain a mercury intrusion curve. The pore size distribution of the sample can be estimated in terms of the volume of the pores intruded for a given cylindrical pore diameter d. The pressure can be increased incrementally or continuously (scanning porosimetry). The process is reversed by lowering the pressure to allow the mercury to extrude from the pores in the fabric to generate a mercury extrusion curve. Analysis of the data is based on a model that assumes the pores in the fabric are a series of parallel nonintersecting cylindrical capillaries of random diameters [69]. However, as a consequence of the nonwetting behavior of mercury in mercury intrusion porosimetry, relatively high pressure is needed to force mercury into the smaller pores therefore compressible nonwoven fabrics are not suitable for testing using the mercury porosimetry method.

Liquids other than mercury find use in porosimetry [70–72] and have been commercialized [73]. Test procedure is similar to that of mercury porosimetry but any liquid that wets the sample, such as water, organic liquids, or solutions may be utilized. The cumulative and differential pore volume distribution, total pore volume, porosity, average, main, effective and equivalent pore size can be obtained.

1.3.7 MEASUREMENT OF SPECIFIC SURFACE AREA BY USING GAS ADSORPTION

The number of gas molecules adsorbed on the surface of nonwoven materials depends on both the gas pressure and the temperature. An experimental adsorption isotherm plot of the incremental increases in weight of the fabric due to absorption against the gas pressure can be obtained in isothermal conditions. Prior to measurement, the sample needs to be pretreated at an elevated temperature in a vacuum or flowing gas to remove contaminants. In physical gas adsorption, when an inert gas (such as nitrogen or argon) is used as an absorbent gas, the adsorption isotherm indicates the surface area and/ or the pore size distribution of the objective material by applying experimental data to the theoretical adsorption isotherm for gas adsorption on the polymer surface. In chemical gas adsorption, the chemical properties of a polymeric surface are revealed if the absorbent is acidic or basic. In some experiments, a liquid absorbent such as water is used in the same manner [74–76].

1.3.8 MEASURING TENSILE PROPERTIES

Some of the most important fabric properties governing the functionality of nonwoven materials include mechanical properties (tensile, compression, bending and stiffness), gaseous and liquid permeability, water vapor transmission, liquid barrier properties, sound absorption properties and dielectric properties.

Mechanical properties of nonwoven fabrics are usually tested in both machine direction (MD) and cross-direction (CD), and may be tested in other bias directions if required. Several test methods are available for tensile testing of nonwovens, chief among these are the strip and grab test methods. In the grab test, the central section across the fabric width is clamped by jaws a fixed distance apart. The edges of the sample therefore extend beyond the width of the jaws. In the standard grab tests for nonwoven fabrics, the width of the nonwoven fabric strip is 100 mm, and the clamping width in the central section of the fabric is 25 mm. The fabric is stretched at a rate of 100 mm/min (according to the ISO standards) or 300 mm/min (according to the ASTM standards) and the separation distance of the two clamps is 200 mm (ISO standards) or 75 mm (ASTM standards). Nonwoven fabrics usually give a maximum force before rupture. In the strip test, the full width of the fabric specimen is gripped between the two clamps. The width of the fabric strip is 50 mm (ISO standard) or either 25 mm or 50 mm (ASTM standards). Both the stretch rate and the separation distance of the two clamps in a strip test are the same as they are in the grab test. The separation distance of the two clamps is 200 mm (ISO standards) or 75 mm (ASTM standards). The observed force for a 50 mm specimen is not necessarily double the observed force for a 25 mm specimen [76–77].

1.3.9 MEASURING WATER VAPOR TRANSMISSION

The water vapor transmission rate through a nonwoven refers to the mass of the water vapor (or moisture) at a steady state flow through a thickness of unit area per unit time. This is taken at a unit differential pressure across the fabric thickness under specific conditions of temperature and humidity ($g/Pa.s.m^2$). It can be tested by two standard methods, the desiccant method and water methods. In the desiccant method, the specimen is sealed to an open mouth of a test dish containing a desiccant, and the assembly is placed in a controlled atmosphere. Periodic weighing's determine the rate of water vapor movement through the specimen into the desiccant. In the water method, the dish contains distilled water, and the weighing's determine the rate of water vapor movement through the specimen to the controlled atmosphere. The vapor pressure difference is nominally the same in both methods except when testing conditions are with the extremes of humidity on opposite sides [78].

1.3.10 MEASURING WETTING AND LIQUID ABSORPTION

There are two main types of liquid transport in nonwovens. One is the liquid absorption which is driven by the capillary pressure in a porous fabric and the liquid is taken up by a fabric through a negative capillary pressure gradient. The other type of liquid transport is forced flow in which liquid is driven through the fabric by an external pressure gradient. The liquid absorption that takes place when one edge of a fabric is dipped in a liquid so that it is absorbed primarily in the fabric plane is referred to as wicking. When the liquid front enters into the fabric from one face to the other face of the fabric, it is referred to as demand absorbency or spontaneous uptake [79, 80].

1.3.11 MEASURING THERMAL CONDUCTIVITY AND INSULATION [81, 82]

The thermal resistance and the thermal conductivity of flat nonwoven fabrics, fibrous slabs and mats can be measured with a guarded hot plate apparatus according to BS 4745: 2005, ISO 5085–1:1989, ISO 5085–2:1990. For testing the thermal resistance of quilt, the testing standard is defined in BS 5335 Part 1:1991. The heat transfer in the measurement of thermal resistance and thermal conductivity in current standard methods is the overall heat transfer by conduction, radiation, and by convection where applicable.

The core components of the guarded hot plate apparatus consist of one cold plate and a guarded hot plate. A sample of the fabric or insulating wadding to be tested, 330 mm in diameter and disc shaped, is placed over the heated hot metal plate. The sample is heated by the hot plate and the temperature on both sides of the sample is recorded using thermocouples. The apparatus is encased in a fan-assisted cabinet and the fan ensures enough air movement to prevent heat build up around the sample and also isolates the test sample from external influences. The test takes approximately eight hours including warm-up time. The thermal resistance is calculated based on the surface area of the plate and the difference in temperature between the inside and outside surfaces. When the hot and cold plates of the apparatus are in contact and a steady state has been established, the contact resistance, R_c $(m^2 KW^{-1})$, is given by the equation:

$$\frac{R_c}{R_s} = \frac{\theta_2 - \theta_3}{\theta_1 - \theta_2} \tag{12}$$

R_s is the thermal resistance of the 'standard', θ_1 is the temperature registered by thermocouple T_1, θ_2 is the temperature registered by T_2 and θ_3 is the temperature registered by T_3. Thus, the thermal resistance of the test specimen, R_f $(m^2 KW^{-1})$, is given by the equation:

$$\frac{R_f}{R_s} = \frac{\theta_2' - \theta_3'}{\theta_1' - \theta_2'} - \frac{\theta_2 - \theta_3}{\theta_1 - \theta_2} \tag{13}$$

where θ_1' is the temperature registered by T_1, θ_2' is the temperature registered by T_2 and θ_3' is the temperature registered by T_3. Since R_s $(m^2 KW^{-1})$ is a known constant and can be calibrated for each specific apparatus, R_f $(m^2 KW^{-1})$ can thus be calculated. Then the thermal conductivity of the specimen, $k (Wm^{-1} K^{-1})$ can be calculated from the equation:

$$k = \frac{d(mm)^* 10^{-3}}{R_f (m^2 KW^{-1})} \tag{14}$$

The conditioning and testing atmosphere shall be one of the standard atmospheres for testing textiles defined in ISO139, that is, a relative humidity of 65%+/– 2% R.H. and a temperature of 20°C+/– 2°C.

1.4   MODELLING

Several theoretical models have been proposed to predict nonwoven fabric properties that can be used or adapted for spunbonded filament networks:

Backer and Petterson [83]	– The filaments are assumed to be straight and oriented in the machine direction. – Filament properties and orientation are assumed to be uniform from point to point in the fabric.
Hearle et al. [84, 85]	– The model accounts for the local fiber curvature (curl). – The fiber orientation distribution, fiber stress-strain relationships and the fabric's Poisson ratio must be determined in advance.
Komori and Makishima [86]	– Estimation of fiber orientation and length. – Assumed that the fibers are straight-line segments of the same length and are uniformly suspended in a unit volume of the assembly.
Britton et al. [8]	– Demonstrated the feasibility of computer simulation of nonwovens. – The model is not based on real fabrics and is designed for mathematical convenience.
Grindstaff and Hansen [88]	– Stress-strain curve simulation of point-bonded fabrics. – Fiber orientation is not considered.
Mi and Batra [89]	– A model to predict the stress-strain behavior of certain point-bonded geometries. – Incorporated fiber stress-strain properties and the bond geometry into the model.
Kim and Pourdeyhimi [90]	– Image simulation and data acquisition. – Prediction of stress-strain curves from fiber stress-strain properties, network orientation, and bond geometry. – Simulated fibers are represented as straight lines.

1.4.1   MODELING BENDING RIGIDITY

The bending rigidity (or flexural rigidity) of adhesive bonded nonwovens was evaluated by Freeston and Platt [91]. A nonwoven fabric is assumed to be composed of unit cells and the bending rigidity of the fabric is the sum of the bending rigidities of all the unit cells in the fabric, defined as the bending moment times the radius of curvature of a unit cell. The analytical equations for bending rigidity were established in the two cases of 'no freedom' and 'complete freedom' of relative motion of the fibers inside a fabric. The following assumptions about the nonwoven structure are made for modeling the bending rigidity.

1. The fiber cross-section is cylindrical and constant along the fiber length.
2. The shear stresses in the fiber are negligible.

3. The fibers are initially straight and the axes of the fibers in the bent cell follow a cylindrical helical path.
4. The fiber diameter and fabric thickness are small compared to the radius of curvature; the neutral axis of bending is in the geometric centerline of the fiber.
5. The fabric density is high enough that the fiber orientation distribution density function is continuous.
6. The fabric is homogeneous in the fabric plane and in the fabric thickness.

The general unit cell bending rigidity, $(EI)_{cell}$, is therefore as follows:

$$(EI)_{cell} = N_f \int_{-\pi/2}^{\pi/2} [E_f I_f \cos^4 \theta + G I_p \sin^2 \theta \cos^2 \theta] \Omega(\theta) \, d\theta \qquad (15)$$

where N_f = number of fibers in the unit cell, $E_f I_f$ = fiber bending rigidity around the fiber axis, G = shear modulus of the fiber, I_p = polar moment of the inertia of the fiber cross section, which is a torsion term and $f(\theta)$ = the fiber orientation distribution in the direction, θ.

The bending rigidities of a nonwoven fabric in two specific cases of fiber mobility are as follows:

1. 'Complete freedom' of relative fiber motion. If the fibers are free to twist during fabric bending (e.g., in a needle punched fabric), the torsion term $(G I_p \sin^2\theta \cos^2\theta)$ will be zero. Therefore,

$$(EI)_{cell} = \pi \, d^4_f N_f E_f / 64 \int_{-\pi/2}^{\pi/2} f(\theta) \cos^4\theta \, d\theta \qquad (16)$$

where d_f = fiber diameter, E_f = Young's modulus of the fiber.

2. 'No freedom' of relative fiber motion. In chemically bonded nonwovens, the freedom of relative fiber motion is severely restricted. It is assumed in this case that there is no freedom of relative fiber motion and the unit cell bending rigidity $(EI)_{cell}$, is therefore:

$$(EI)_{cell} = \pi \, N_f E_f \, d^2_f h / 48 \int_{-\pi/2}^{\pi/2} f(\theta) \cos^4\theta \, d\theta \qquad (17)$$

where h = fabric thickness and d_f = fiber diameter.

1.4.2 MODELING SPECIFIC PERMEABILITY

The specific permeability of a nonwoven fabric is solely determined by nonwoven fabric structure and is defined based on D'Arcy's Law [92] which may be written as follows:

$$Q = - \frac{k}{\eta} \frac{\Delta p}{h} \qquad (18)$$

where Q is the volumetric flow rate of the fluid flow through a unit cross-sectional area in the porous structure (m/s), η is the viscosity of the fluid ($Pa.s$), Δp is the pressure drop (Pa) along the conduit length of the fluid flow $h(m)$ and k is the specific permeability of the porous material (m^2). Numerous theoretical models describing laminar flow through porous media have been proposed to predict permeability. The existing theoretical models of permeability applied in nonwoven fabrics can be grouped into two main categories based on:

1. Capillary channel theory (e.g., Kozeny [104], Carman [93], Davies [94], Piekaar and Clarenburg [95] and Dent [96]).
2. Drag force theory (e.g., Emersleben [97], Brinkman [98], Iberall [99], Happel [100], Kuwabara [101], Cox [102], and Sangani and Acrivos [103]).

Many permeability models established for textile fabrics are based on capillary channel theory or the hydraulic radius model, which is based on the work of Kozeny [104] and Carman [93]. The flow through a nonwoven fabric is treated as a conduit flow between cylindrical capillary tubes. The Hagen–Poiseuille equation for fluid flow through such a cylindrical capillary tube structure is as follows:

$$q = \frac{\pi r^4}{8} \frac{\Delta p}{h} \qquad (19)$$

where r is the radius of the hydraulic cylindrical tube. However, it has been argued that models based on capillary channel theory are suitable only for materials having a low porosity and are unsuitable for highly porous media where the porosity is greater than 0.8, for example, Carman [93].

1.4.3 MODELING PORE SIZE AND PORE SIZE DISTRIBUTION

In the design and engineering of nonwoven fabrics to meet the performance requirements of industrial applications, it is desirable to make predictions based on the fabric components and the structural parameters of the fabric. Although work has been conducted to simulate isotropic nonwoven structures in terms of, for example, the fiber contact point numbers [105]. and intercross distances [106], only the models concerned with predicting the pore size are summarized in this section.

1.4.3.1 MODELS OF PORE SIZE

Although it is arguable if the term 'pore' accurately describes the voids in a highly connective, low density nonwoven fabric, it is still helpful to use this term in quantifying a porous nonwoven structure. The pore size in simplified,

$$r = \left(0.075737 \sqrt{\frac{Tex}{\rho_{fibric}}} \right) - \frac{d_f}{2} \qquad (20)$$

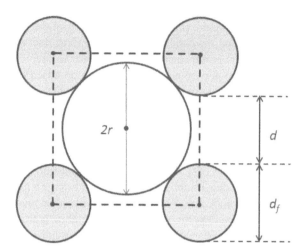

FIGURE 1.19   Wrotnowski's model for pore size in a bundle of parallel cylindrical fibers arranged in a square pattern.

where Tex = fiber linear density (tex), ρ_{fabric} is the fabric density (g/cm³) and d_f is the fiber diameter (m). Several other models relating pore size and fiber size by earlier researchers can also be used in nonwoven materials. For example, both the largest pore size and the mean pore size can be predicted as follows, by using, Goeminne's equation. The porosity is defined as nonwoven structures can be approximately estimated by Wrotnowski's model [107, 109] although the assumptions that are made for the fabric structure are based on fibers that are circular in cross-section, straight, parallel, equidistant and arranged in a square pattern (Fig. 1.19). The radius of a pore in Wrotnowski's model is shown as follows,

$$\text{Largest pore size } (2r_{max}): r_{max} = \frac{d_f}{2(1-\varepsilon)} \tag{21}$$

$$\text{Mean pore size } (2r) \text{ (porosity} < 0.9): r = \frac{d_f}{4(1-\varepsilon)} \tag{22}$$

In addition, pore size ($2r$) can also be obtained based on Hagen-Poiseuille's law in a cylindrical tube,

$$r = \sqrt[4]{\frac{8k}{\pi}} \tag{23}$$

where k is the specific permeability (m²) in Darcy's law.

1.4.3.2 MODELS OF PORE SIZE DISTRIBUTION

If it is assumed that the fibers are randomly aligned in a nonwoven fabric following Poisson's law, then the probability, $P(r)$, of a circular pore of known radius, r, is distributed as follows [110],

$$P(r) = -\left(2\pi v'\right)exp(-\pi r^2 v') \tag{24}$$

where $v = \dfrac{0.36}{r^2}$ is defined as the number of fibers per unit area.

Giroud [111] proposed a theoretical equation for calculating the filtration pore size of nonwoven geotextiles. The equation is based on the fabric porosity, fabric thickness, and fiber diameter in a nonwoven geotextile fabric.

$$O_f = \left[\frac{1}{\sqrt{1-\varepsilon}} - 1 + \frac{\xi \in d_f}{(1-\varepsilon)^h}\right] d_f \tag{25}$$

where d_f = fiber diameter, ε = porosity, h = fabric thickness, ξ = an unknown dimensionless parameter to be obtained by calibration with test data to account for the further influence of geotextile porosity and ξ = 10 for particular experimental results, and O_f = filtration opening size, usually given by the nearly largest constriction size of a geotextile.

Lambard [112] and Faure [113] applied Poissonian line network theory to establish a theoretical model of the 'opening sizing' of nonwoven fabrics. In this model, the fabric thickness is assumed to consist of randomly stacked elementary layers, each layer has a thickness T_e and is simulated by two- dimensional straight lines, (a Poissonian line network). Faure et al. [114] and Gourc and Faure [115] also presented a theoretical technique for determining constriction size based on the Poissonian polyhedral model. In Faure's approach, epoxy-impregnated nonwoven geotextile specimens were sliced and the nonwoven geotextile was modeled as a pile of elementary layers, in which fibers were randomly distributed in planar images of the fabric. The cross-sectional images were obtained by slicing at a thickness of fiber diameter d_f and the statistical distribution of pores was modeled by inscribing a circle into each polygon defined by the fibers (Fig. 1.20). The pore size distribution, which is obtained from the probability of passage of different spherical particles through the layers forming the geotextile, can thus be determined theoretically using the following equation:

$$Q(d) = (1-\phi)\left(\frac{2+\lambda\left(d+d_f\right)}{2+\lambda d_f}\right)^{2N} e^{-\lambda Nd} \tag{26}$$

$$\lambda = \frac{4}{\pi}\frac{(1-\phi)}{d_f}, \quad and \quad N = \frac{T}{d_f} \tag{27}$$

where $Q(d)$ = probability of a particle with a diameter d passing through a pore channel in the geotextile, φ = fraction of solid fiber materials in the fabric, λ = total length of straight lines per unit area in a planar surface (also termed specific length) and N = number of slices in a cross-sectional image. Because of the assumption in Faure's approach that the constriction size in geotextiles of relatively great thicknesses tends to approach zero, Faure's model generally produces lower values [111]. The use of this method is thus not recommended for geotextiles with a porosity of 50% or less [116].

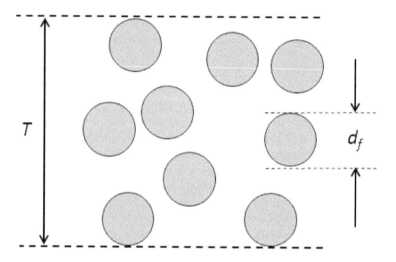

FIGURE 1.20   Model for constriction pore size in a nonwoven fabric consisting of randomly stacked elementary layers of fibers.

1.4.3.3   MODELING OF THE PORE NETWORK BY IMAGE PROCESSING

Industry to extract the influence of the structural properties on the functional properties. It is also widely used in theoretical modeling or specific applications to infer nonwoven hydraulic and other properties of nonwoven behavior. Specific devices, like porometer, have been developed to provide indirect measurements of the PSD in small textile structure. Other devices have been developed to provide direct measurements: X-ray, Electronic Scan Microscopy, optical profiling system. The use of porometer is based on a variation of the pressure of the liquid through the sample. It saves instantly the quantity of liquid passed through the pores beginning by the largest pore and finishing by the smallest ones. It gives so a PSD based on the equivalent radius. However this distribution is not a real distribution. In fact, the number of smallest pores is vague because it requires a big air pressure to be measured. Moreover the geometry and the shape of pores are neglected. So, it is important, to take into account the morphological aspect of the pores. In this approach, we want to study the PSD without neglecting the geometry and the shape of pores. Moreover, all of those devices are not able to measure the dynamic aspect of nonwovens and so it is difficult to understand hydraulic properties with such kind of devices. For all those reasons, we set up a test bed that allows us to study more easily the PSD and the hydraulic properties of pores through time.

1.4.3.4 THE PORES SIZE AND GEOMETRIC DISTRIBUTION FUNCTION

The nonwoven materials are made up of pores in which gas effluents or liquids can run out. Fatt [117] in 1956 introduced the model of porous media in the rock science. A lot of works are done also in the same field [118–126]. For nonwoven research, some papers have been written [127–135]. A good description of this structure permits to model the fluid transport in the medium. It permits also to extract relevant parameters characterizing the pore network. It allows also a classification of porous materials according to their hydraulic properties [136–138]. We take a black and white image of the sample. To determine the PSD, our approach is based on the detection of the 8-connected pixels belonging to the background of the image [139]. The 8-connected pixels are connected if their edges or corners touch. This means that if two adjoining pixels are on, they are part of the same object, regardless of whether they are connected along the horizontal, vertical, or diagonal direction. The algorithm consists on many steps. First of all, we convert the gray image into a binary image by using Otsu threshold. This method consists in calculating the white pixels [140]. In the second step of the algorithm, we label the resulting image by using the following general procedure:

- Scan all image pixels, assigning preliminary labels to nonzero pixels and recording label equivalences in a union-find table.
- Solve the equivalence classes using the union-find algorithm [141].
- Relabel the pixels based on the resolved equivalence classes.

The resulting matrix contains positive integer elements that correspond to different regions. For example, the set of elements equal to 1 corresponds to region 1; the set of elements equal to 2 corresponds to region 2; and so on. We can extract from the labeled matrix all the geometric parameters concerning every pore (equivalent radius, orientation, area, eccentricity etc.). We can model a pore with an ellipse or a rectangle or a convex hull. In this paper, we choose to characterize a pore with his equivalent radius r, which represents the size of the pore, and his eccentricity e, which represents the geometry of the pore. The equivalent radius corresponds to the radius of a circle with the same area as the region:

$$r = (Area/\pi)^{0.5} \tag{28}$$

Where *Area* is the actual number of pixels in the region. Another approach is based on the extraction of the biggest disk contained in a region [142], but some pixels will be ignored. A such PSD is a histogram describing the:

$$frequency\ (\%) = \frac{Number\ of\ pores\ having\ radius\ r}{Total\ number\ of\ pores} \tag{29}$$

By using the same algorithm, we extract the pores geometry distribution (PGD). More exactly; we determine the eccentricity of each pore. The eccentricity is the ratio of the distance between the foci of the ellipse and its major axis length. The value varies between 0 and 1 (an ellipse whose eccentricity is 0 is actually a circle, while an ellipse whose eccentricity is 1 is a line segment). The resulting PGD is a histogram describing the:

$$frequency\ (\%) = \frac{Number\ of\ pores\ having\ eccentricity\ e}{Total\ number\ of\ pores} \tag{30}$$

1.4.4 MODELING ABSORBENCY AND LIQUID RETENTION

Liquid absorbency (or liquid absorption capacity), C, is defined as the weight of the liquid absorbed at equilibrium by a unit weight of nonwoven fabric. Thus, liquid absorbency is based on determining the total interstitial space available for holding fluid per unit dry mass of fiber. The equation is shown as follows [143]:

$$C = A\frac{T}{W_f} - \frac{1}{\rho_f} + (1 - \alpha)\frac{V_d}{W_f} \tag{31}$$

where A is the area of the fabric, T is the thickness of the fabric, W_f is the mass of the dry fabric, ρ_f is the density of the dry fiber, V_d is the amount of fluid diffused into the structure of the fibers and a is the ratio of increase in volume of a fiber upon wetting to the volume of fluid diffused into the fiber. In the above equation, the second term is negligible compared to the first term, and the third term is nearly zero if a fiber is assumed to swell strictly by replacement of fiber volume with fluid volume [144]. Thus, the dominant factor that controls the fabric absorbent capacity is the fabric thickness per unit mass on a dry basis (T/W_f).

1.5 CONCLUSION

Nonwovens are porous fiber assemblies that are engineered to meet the technical requirements of numerous industrial, medical and consumer products. In this review, we briefly summarized the properties of nonwoven fabrics that govern their suitability for use in various applications. Several techniques were employed to characterize the structure and properties of nonwoven fabrics. Also, we reviewed models of nonwoven fabric structure and introduce examples of analytical modeless that link nonwoven fabric structure and properties. We concluded that analytical models are helpful in providing insights in to proposed mechanisms or interactions; they can show whether a mechanism is at least theoretically feasible and help to suggest experiments that might further elucidate and discriminate the influence of individual variables on fabric properties. Therefore, understanding and estimation of nonwoven behavior are very important for manufacturers where the reduction of production costs is of strategic imperative to stay in business.

KEYWORDS

- **characterization**
- **modeling**
- **nonwoven structure**
- **yarn production**

NOMENCLATURE

d	Diameter of a cylindrical pore
d_f	Fiber diameter
E_f	Young's modulus of the fiber
$E_f I_f$	Fiber bending rigidity
G	Shear modulus of the fiber
h	Fabric thickness
I_p	Polar moment of the inertia of the fiber cross section
k	Specific permeability in Darcy's law
N_f	Number of fibers in the unit cell
O_f	Filtration opening size
ΔP	Pressure difference
Q	Volumetric flow rate
$Q(d)$	Probability of a particle with a diameter d passing through a pore
r	Radius of the hydraulic cylindrical tube
R_c	Contact resistance $(m^2 KW^{-1})$
R_f	Thermal resistance of the test specimen $(m^2 KW^{-1})$
R_s	Thermal resistance of the 'standard' $(m^2 KW^{-1})$

GREEK SYMBOLS

α	Fiber orientation angle
ε	Fabric porosity (%)
ϕ	Volume fraction of solid material (%)
γ	Liquid surface tension
γ_{Hg}	Contact angle of the mercury on the material
λ	Specific length
θ	Contact angle of the liquid
ρ_{fabric}	Fabric bulk density (kg/m^3)
ρ_{fiber}	Fiber density (kg/m^3)
σ_{Hg}	Surface tension of the mercury
$\Omega(\alpha)$	Fiber orientation distribution function
$\Omega(\theta)$	Fiber orientation distribution in the direction θ

REFERENCES

1. Arthur Drelich. A Nonwoven classification: A simple system, Nonwovens Industry, 54–55, Oct. 1998.
2. http://www.inda.org/about/nonwovens.html.
3. EDANA 2004 Nonwoven Statistics.
4. ISO 9092:1988; BS EN 29092:1992.
5. The Nonwoven Fabrics Handbook, Association of the Nonwoven Fabrics Industry, Cary, NC, 2007.
6. Handbook of nonwovens, S. J. Russell, CRC Press, Boca Raton, USA (2007).
7. Pound W.H. Real world uniformity measurement in nonwoven coverstock, Int. Nonwovens J., 10 (1), 2001, pp. 35–39.
8. Huang. X., Bresee R.R., Characterizing nonwoven web structure using image analysis techniques, Part III: Web Uniformity Analysis, Int. Nonwovens J., 5 (3), 1993, pp. 28–38.
9. Aggarwal R.K., Kennon W.R. Porat I., A Scanned-laser Technique for Monitoring Fibrous Webs and Nonwoven Fabrics, J. Text. Inst., 83 (3), 1992, pp. 386–398.
10. Boeckerman P.A., Meeting the Special Requirements for On-line Basis Weight Measurement of Lightweight Nonwoven Fabrics, Tappi J., 75 (12), 1992, pp. 166–172.
11. Chhabra R., Nonwoven Uniformity – Measurements Using Image Analysis, Intl. Nonwovens J., 12(1), 2003, pp. 43–50.
12. Scharcanski J., Dodson C.T., Texture analysis for estimating spatial variability and anisotropy in planar stochastic structures, Optical Engineering, 35(8), pp. 2302–2309, 1996.
13. Gilmore T., Davis H., Mi Z., Tomographic approaches to nonwovens structure. definition, National Textile Center Annual Report, USA, Sept., 1993.
14. Mao N., Russell S.J., 2003, Modeling of permeability in homogeneous three-dimensional nonwoven fabrics, Text. Res. J., 91, pp. 243–258.
15. Petterson D.R., The mechanics of nonwoven fabrics, Sc D., Thesis, MIT, Cambridge, MA, 1958.
16. Hansen S.M., Nonwoven Engineering Principles, in (ed. by Turbak AF) Nonwovens Theory, Process, Performance and Testing, Tappi Press, Atlanta, 1993.
17. Groitzsch D., Ultrafine Microfiber Spunbond for Hygiene and Medical Application, http://www.technica.net/NT/NT2/eedana.htm.

18. ASTM D5729–97 Standard Test Method for Thickness of Nonwoven Fabrics.
19. ASTM D5736–95 Standard Test Method for Thickness of Highloft Nonwoven Fabrics.
20. Chen H.J., Huang D.K., Online measurement of nonwoven weight evenness using optical methods, ACT paper, 1999.
21. Hunter Lab Color Scale, http://www.hunterlab.com/appnotes/an08_96a.pdf.
22. Chhabra R., Nonwoven Uniformity – Measurements Using Image Analysis, Int. Nonwovens J., 12(1), pp. 43–50, 2003.
23. Petterson D.R., The mechanics of nonwoven fabrics, Sc D., Thesis, MIT, Cambridge,MA, 1958.
24. Hearle J.W.S., Stevenson P.J., Nonwoven fabric studies, part 3: The anisotropy of nonwoven fabrics, Text. Res. J., 33, pp. 877–888, 1963.
25. Hearle J.W.S., Ozsanlav V., Nonwoven fabric studies, part 5: Studies of adhesive-bonded non-woven fabrics, J. Text. Inst., 70, pp. 487–497, 1979.
26. Chuleigh P.W., Image formation by fibers and fiber assemblies, Text. Res. J., 54, p. 813, 1983.
27. Kallmes O.J., Techniques for determining the fiber orientation distribution throughout the thickness of a sheet, TAPPI, No. 52, pp. 482–485, 1969.
28. Votava A., Practical method-measuring paper asymmetry regarding fiber orientation. Tappi J., 65, p. 67, 1982.
29. Cowan W.F., Cowdrey E.J.K., Evaluation of paper strength components by short span tensile analysis, Tappi J., 57(2), p. 90, 1973.
30. Stenemur B., Method and device for monitoring fiber orientation distributions based on light diffraction phenomenon, Int. Nonwovens J., 4, pp. 42–45, 1992.
31. Comparative degree of preferred orientation in 19 wood pulps as evaluated from X-ray diffraction patterns, Tappi J., 33, p. 384, 1950.
32. Prud'homme B., et al., determination of fiber orientation of cellulosic samples by X-ray diffraction, J. Polym. Sci., 19, p. 2609, 1975.
33. Osaki S., Dielectric anisotropy of nonwoven fabrics by using the microwave method, Tappi J., 72, p. 171, 1989.
34. Lee S., Effect of fiber orientation on thermal radiation in fibrous media, J. Heat Mass Transfer, 32(2), p. 311, 1989.
35. McGee S.H., McCullough R.L., Characterization of fiber orientation in short-fiber composites, J. Appl. Phys., 55(1), p. 1394, 1983.
36. Orchard G.A., The measurement of fiber orientation in card webs, J. Text. Inst., 44, T380, 1953.
37. Tsai PP and Bresse RR, Fiber orientation distribution from electrical measurements. Part 1, theory, Int. Nonwovens J., 3(3), p. 36, 1991.
38. Tsai P.P., Bresse RR, Fiber orientation distribution from electrical measurements. Part 2, instrument and experimental measurements, Int. Nonwovens J., 3(4), p. 32, 1991.
39. Chaudhray M.M., MSc Dissertation, University of Manchester, 1972.
40. Judge S.M., MSc Dissertation, University of Manchester, 1973.
41. Huang X.C., Bressee R.R., Characteristizing nonwoven web structure using image analyzing techniques, Part 2: Fiber orientation analysis in thin webs, Int. Nonwovens J., No. 2, pp. 14–21, 1993.
42. Pourdeyhimi B., Nayernouri A., Assessing fiber orientation in nonwoven fabrics, INDA J. Nonw. Res., 5, pp. 29–36, 1993.
43. Pouredyhimi B., Xu B., Characterizing pore size in nonwoven fabrics: Shape considerations, Int. Nonwoven J., 6(1), pp. 26–30, 1993.
44. Gong R.H., Newton A., Image analysis techniques Part II: The measurement of fiber orientation in nonwoven fabrics, Text. Res. J., 87, p. 371, 1996.

45. Britton P.N., Sampson A.J., Jr., Elliot C.F., Grabben H.W., Gettys W.E., Computer simulation of the technical properties of nonwoven fabrics, part 1: The method, Text. Res. J., 53, pp. 363–368, 1983.

46. Grindstaff T.H., Hansen S.M., Computer model for predicting point-bonded nonwoven fabric strength, Part 1: Text. Res. J., 56, pp. 383–388, 1986.

47. Jirsak O., Lukas D., Charrat R., A two-dimensional model of mechanical properties of textiles, J. Text. Inst., 84, pp. 1–14, 1993.

48. Xu B., Ting Y., Measuring structural characteristics of fiber segments in nonwoven fabrics, Text. Res. J., 65, pp. 41–48, 1995.

49. Pourdeyhimi B., Dent R., Davis H., Measuring fiber orientation in nonwovens. Part 3: Fourier transform, Text. Res. J., 67, pp. 143–151, 1997.

50. Pourdeyhimi B., Ramanathan R., Dent R., Measuring fiber orientation in nonwovens. Part 2: Direct tracking, Text. Res. J., 66, pp. 747–753, 1996.

51. Pourdeyhimi B., Ramanathan R., Dent R., Measuring fiber orientation in nonwovens. Part 1: Simulation, Text. Res. J., 66, pp. 713–722, 1996.

52. Pourdeyhimi B., Dent R., Measuring fiber orientation in nonwovens. Part 4: Flow field analysis, Text. Res. J., 67, pp. 181–187, 1997.

53. Komori T., Makishima K., Number of fiber-to-fiber contacts in general fiber assemblies, Text. Res. J., 47, pp. 13–17, 1977.

54. Gilmore T., Davis H., Mi Z., Tomographic approaches to nonwovens structure definition, National Textile Center Annual Report, USA, Sept., 1993.

55. Manual of Quantimet 570, Leica Microsystems Imaging Solutions, Cambridge, UK, 1993.

56. BS 1902–3.8, Determination of bulk density, true porosity and apparent porosity of dense shaped products (method 1902–308).

57. BS EN 993–1:1995, BS 1902–3.8:1995 Methods of test for dense shaped refractory products. Determination of bulk density, apparent porosity and true porosity.

58. Bhatia S.K., Smith J.L., 1995, Application of the bubble point method to the characterization of the pore size distribution of geotextile, Geotech. Test. J., 18(1), pp. 94–105.

59. Bhatia S.K., Smith J.L., 1996, Geotextile characterization and pore size distribution, Part II: A review of test methods and results, Geosynthet. Int., 3(2), pp. 155–180.

60. R. J. Good, Contact angle, wetting, and adhesion: A critical review, J. Adhesion Sci. Technol. 6 (12):1269–1302 (1992).

61. Olderman, G. M. Liquid repellency and surgical fabric barrier properties, Eng. Med 13(1):35–43 (1984).

62. S. S. Block, Disinfection, Sterilization, and Preservation, 4th ed., Lea & Febiger, Philadelphia, PA, 1996.

63. Weast, R. C., Astle, M. J., Beyer, W. H. eds., CRC Handbook of Chemistry and Physics, 69th ed., CRC Press, Boca Raton, FL, 1988.

64. Lentner, C. ed., Geigy Scientific Tables, Vol. 1, Units of Measurement, Body Fluids, Composition of the Body, Nutrition, Medical Education Division, Ciba-Geigy Corporation, West Caldwell, NJ, 1981.

65. Lentner, C. ed., Geigy Scientific Tables, Vol. 3, Physical Chemistry, Composition of Blood, Hematology, Somatometric Data, Medical Education Division, Ciba-Geigy Corporation, West Caldwell, NJ, 1984.

66. Internal Research, W. L. Gore & Associates, Inc., Elkton, MD.

67. Behera, B. K. "Image Processing in Textiles" Textile Institutes.

68. Haines W.B., J. Agriculture Sci. 20, pp. 97–116, 1930.

69. Washburn E., The Dynamics of Capillary Flow, Physics Review, 17(3), pp. 273–283, 1921.

70. ASTM E. 1294–89 Standard Test Method for Pore Size Characteristics of Membrane Filters Using Automated Liquid Porosimeter.

71. Miller B., Tyomkin I., Wehner J.A., 1986, Quantifying the Porous Structure of Fabrics for Filtration Applications, Fluid Filtration: Gas, 1, Raber RR (editor), ASTM Special Technical Publication 975, proceedings of a symposium held in Philadelphia, Pennsylvania, USA, pp. 97–109.
72. Miller B., Tyomkin I., 1994, An Extended Range Liquid Extrusion Method for Determining Pore Size Distributions, Textile Research Journal, 56(1), pp. 35–40.
73. http://www.triprinceton.org/instrument_sales/autoporosimeter.html.
74. ISO/DIS 15901–2, Pore size distribution and porosimetry of materials. Evaluation by mercury posimetry and gas adsorption, Part 2: Analysis of meso-pores and macropores by gas adsorption.
75. ISO/DIS 15901–3, Pore size distribution and porosity of solid materials by mercury porosimetry and gas adsorption, Part 3: Analysis of micropores by gas adsorption.
76. BS 7591–2:1992 Porosity and pore size distribution of materials. Method of evaluation by gas adsorption.
77. Hearle, J.W.S., Stevenson P.J., Studies in Nonwoven Fabrics: Prediction of Tensile Properties, Text. Res. J., 34, pp. 181–191, 1964.
78. WSP 70.4: WSP70.5; WSP70.6.
79. Kissa E., Wetting and Wicking, Text. Res. J. 66, pp. 660, 1996.
80. Harnett P.R., Mehta P.N., A survey and comparison of laboratory test methods for measuring wicking, Text. Res. J., 54, pp. 471–478, 1984.
81. Grewal R.S., Banks-Lee P., Development of Thermal Insulation For Textile Wet Processing Machinery Using Needle-punched Nonwoven Fabrics, Int. Nonwovens J., 2, pp. 121–129, 1999.
82. Baxter S., The thermal conductivity of textiles, Proceedings of the Physical Society.58, pp. 105–118, 1946.
83. Backer, S., Patterson, D. R.: Textile Research Journal, 30, 704–711, 1960.
84. Hearle, J. W. S., Stevenson, P. J.: Textile Research Journal, 38, 343–351, 1968.
85. Hearle, J. W. S., Stevenson, P. J.: Textile Research Journal, 34, 181–191, 1964.
86. Komori, T., Makishima, K.: Textile Research Journal, 47, 13–17. 1977.
87. Britton, P., Simpson, A. J.: Textile Research Journal, 53, pp 1–5 and pp 363– 368, 1983.
88. Grindstaff, T. H., Hansen, S. M.: Textile Research Journal, 56(6) 383. 1986.
89. Gilmore, T. F., Mi Z., Batra, S. K.: Proceedings of the TAPPI Conference May 1993.
90. Kim, H. S., Pourdeyhimi, B.: Journal of Textile and Apparel Technology and Management, Vol. 1 (4) 1–7, 2001.
91. Freeston W.D., Platt M.M., Mechanics of Elastic Performance of Textile Materials, Part XVI: Bending Rigidity of Nonwoven Fabrics, Text. Res. J. 35(1), pp. 48–57, 1965.
92. Darcy H., 1856, Les Fontaines Publiques de la Ville de Dijon (Paris: Victor Valmont).
93. Carman P.C., Flow of Gases through Porous Media, Academic Press, New York, 1956.
94. Davies C.N., 1952, The separation of airborne dust and particles. Proc. Instn mech. Engrs IB, 185–213.
95. Piekaar H.W., Clarenburg L.A., Aerosol filters: Pore size distribution in fibrous filters, Chem. Eng. Sci., 22, pp. 1399, 1967.
96. Dent R.W., The air permeability of nonwoven fabrics, J. Text. Inst., 67, pp. 220–223, 1976.
97. Emersleben V.O., Das darcysche Filtergesetz, Physikalische Zeitschrift, 26, p. 601, 1925.
98. Brinkman H.C., On the permeability of media consisting of closely packed porous particles, Applied Scientific Research, A1, p. 81, 1948.
99. Iberall A.S., Permeability of glass wool and other highly porous media, J. Res. Natl. Bureau Standards 45: 398, 1950.
100. Happel J., 1959, Viscous flow relative to arrays of cylinders, AIChE J., 5, pp. 174–177.

101. Kuwabara SJ, 1959, The forces experienced by randomly distributed parallel circular cylinder or spheres in a viscous flow at small Reynolds numbers, J. Phys. Soc. Japan, 14, p. 527.
102. Cox RG, The motion of long slender bodies in a viscous fluid, Part 1, J. Fluid Mechanics, 44, pp. 791–810, 1970.
103. Sangani A.S., Acrivos A., Slow flow past periodic arrays of cylinders with applications to heat transfer, Int. J. Multiphase Flow, 8, pp. 193–206, 1982.
104. Kozeny J., Royal Academy of Science, Vienna, Proc. Class 1, 136, p. 271, 1927.
105. Dodson C.T.J., Fiber crowding, fiber contacts and fiber flocculation, Tappi J. 79(9), pp. 211–216, 1996.
106. Dodson C.T.J., Sampson W.W., Spatial statistics of stochastic fiber networks, J. Stat. Phys, 96 (1–2), pp. 447–458, 1999.
107. Wrotnowski A.C., 1962, nonwoven filter media, chemical engineering progress, 58(12), pp. 61–67.
108. Wrotnowski A.C., Felt filter media, Filtration and Separation, Sept./Oct., pp. 426–431, 1968.
109. Goeminne, H., The Geometrical and Filtration Characteristics of Metal-Fiber Filters, Filtration and Separation 1974 (August), pp. 350–355.
110. Rollin A.L., Denis R., Estaque L., Masounave J., 1982, Hydraulic behavior of synthetic nonwoven filter fabrics, The Canadian Journal of Chemical Engineering, 60, pp. 226–234.
111. Giroud J.P., Granular filters and geotextile filters, Proc., Geo-filters'96, Montréal, 565–680, 1996.
112. Lambard G., et al., Theoretical and Experimental opening size of heat-bonded geotextiles, Text. Res. J., April, pp. 208–217, 1988.
113. Faure Y.H., et al., Theoretical and Experimental determination of the filtration opening size of geotextiles, 3rd International Conference on Geotextiles, Vienna, Austria, pp. 1275–1280, 1989.
114. Faure Y.H., Gourc J.P., Gendrin P., 1990, Structural study of porometry and filtration opening size of geotextiles, Geosynthetics: microstructure and performance, ASTM STP 1076, Peggs ID (ed.), Philadelphia, pp. 102–119.
115. Gourc J.P., Faure Y.H., Soil particle, water and fiber – A fruitful interaction non controlled, Proc., 4th Int. Conf. on Geotextiles, Geomembranes and Related Products, The Hague, The Netherlands, pp. 949–971, 1990.
116. Aydilek A.H., Oguz S.H., Edil T.B., Constriction Size of Geotextile Filters, Journal of Geotechnical and Geoenvironmental Engineering, 131(1), pp. 28–38, 2005.
117. Fatt, I. "The network model of porous media I. capillary pressure characteristics," AIME Petroleum Transactions, vol. 207 (1956) p. 144.
118. Bakke S., Oren P. E. 3-d pore-scale modeling of heterogeneous sandstone reservoir rocks and quantitative analysis of the architecture, geometry and spatial continuity of the pore network. In European 3-D Reservoir Modelling Conference, Stavanger, Norway, SPE (1996) 35–45.
119. Blunt M. J., King. Relative permeabilities from two- and three-dimensional porescale network modeling. Transport in Porous Media, 6 (1991) 407–433.
120. Blunt M. J. Sher H. Pore network modeling of wetting. Physical Review E, 52(December 1995) 63–87.
121. Feyen and K. Wiyo, editors, Modelling of transport processes in soils, Wageningen Pers, The Netherlands, (1999) 153–163.
122. Jean-Fran, Ois Delerue and Edith Perrier. DX Soil, a library for 3D image analysis in soil science. Computer & Geosciences, 28 (2002) 1041–1050.
123. Denesuk, M., Smith, G.L., Zelinski, B.J.J., Kreidl, N.J., and Uhlmann, D.R., "Capillary Penetration of Liquid Droplets into a Porous Materials" Journal of Colloid and Interface Science, 158 (1993) 114–120.

124. Hidajat, A., Rastogi, M. Singh, and K. Nohanty Transport properties of porous media reconstructed from thinsections. SPE Journal, (March 2002) 40–48.
125. Lindquist W. B., Venkatarangan A. Investigating 3D geometry of porous media from high-resolution images. Phys. Chem. Earth (A), 25(7):(1999) 593–599.
126. Oren, P. E., Bakke, S., Arntzen O. J. Extending predictive capabilities to network models. SPE Journal, (December 1998) 324–336. Yarn torsion," Polymer testing, 20 (2001) 553–561.
127. Pourdeyhimi, B., Dent, R., and Davis, H., "Measuring Fiber Orientation in nonwovens Part III: Fourier Transform," Textile Research Journal, 2 (1997) 143–151.
128. Pourdeyhimi, B., Ramanathan, R., "Measuring Fiber Orientation in Nonwovens Part II: Direct Tracking," Textile Research Journal, 12 (1996), 747–753.
129. Pourdeyhimi, B. Reply to "Comments on Measuring Fiber Orientation in Nonwovens," Textile Research Journal, 4 (1998) 307–308. 593–599.
130. Marmur, A., "Penetration and Displacement in Capillary Systems" Advances in Colloid and Interface Science, 39 (1992) 13–33.
131. Marmur, A., Cohen, R.D., "Characterization of Porous Media by the Kinetics of Liquid Penetration: The Vertical Capillaries Model," Journal and Colloid and Interface Science, 189 (1997) 299–304.
132. MeBratney A.B. Moran C.J. Soil pore structure modeling using fuzzy random pseudo fractal sets, International working meeting on soil micromorphology, pp.495–506, 1994.
133. Rebenfeld, L., Miller, B., "Using Liquid Flow to Quantify the Pore Structure of Fibrous Materials," J. Text. Inst., 2 (1995) 241–251.
134. Sedgewick, R. Algorithms in C, 3rd ed., Addison-Wesley, (1998) 11–20.
135. Zeng, X., Vasseur, C., Fayala, F. Modeling micro geometric structures of porous media with a predominant axis for predicting diffusive flow in capillaries, Applied Mathematical Modeling, vol. 24(2000) pp 969–986.
136. Anderson A.N, McBratney, A.B., FitzPatrick, E.A. Soil Mass, Surface, and Spectral fractal Dimensions Estimated from Thin Section Photographs, Soil Sci. Soc. Am. J, vol 60 (1996) pp 962–969.
137. Perwelz, A., Mondon, P., Caze C. "Experimental study of capillary flow in yarns," Textile Research Journal, 70(4), (2000) 333–339.
138. Perwelz, A., Cassetta, M., Caze C. "Liquid organization during capillary rise in yarns-influence of yarn torsion," Polymer testing, 20 (2001) 553–561.
139. Serra J. Image Analysis and Mathematical Morphology. Academic Press, New York, 1982.
140. N.Otsu, "A threshold Selection Method from Grey-level Histograms," IEEE Transactions Systems, Man, and Cybernetics, 9(1) (1979) 62–66.
141. Sedgewick R., Algorithms in C, 3rd ed., Addison-Wesley, (1998) 11–20.
142. Delerue J. F., Perrier E., Timmerman A., Rieu M., Leuven K. U. New computer tools to quantify 3D porous structures in relation with hydraulic properties. In J. Feyen and K. Wiyo, editors, Modelling of transport processes in soils, Wageningen Pers, The Netherlands, (1999) 153–163.
143. Gupta B.S., Smith D.K., Nonwovens in Absorbent Materials, Textile Sci. and Technol., 13, pp. 349–388, 2002.
144. Gupta B.S., The Effect of Structural Factors on Absorbent Characteristics of Nonwovens, Tappi J. 71, pp. 147–152, 1988.

CHAPTER 2

CELLULOSE-BASED TEXTILE WASTE TREATMENT INTO POWDER-LIKE FILLERS FOR EMULSION RUBBERS

V. M. MISIN,[1] S. S. NIKULIN,[2] and I. N. PUGACHEVA[2]

[1]N. M. Emanuel Institute of Biochemical Physics, Russian Academy of Sciences, Moscow, Russia

[2]Voronezh State University of the Engineering Technologies, Voronezh, Russia

CONTENTS

2.1   INTRODUCTION

Fiberfills have a wide diverse raw materials base that is practically unlimited. A great amount of various fiberfill wastes is formed at the textile enterprises, garment workshops, etc. Therefore, an important and actual practical task is a search for most perspective directions of their usage [1].

In some of the published works it was shown that fiberfills can be applied in the composite structures of different intended purposes. Special attention is paid to the use of fiberfills in the polymer composites. One of the ways of their usage is the production of mechanical-rubber articles (MRA). Incorporation of the fiberfills and additives into MRA in the industrial production is performed with rolling mills in the process of producing of the rubber compounds. This way of incorporation does not allow attaining of the uniform distribution in the bulk of rubber compound that will further have a negative effect on the properties of the obtained vulcanizates. A uniform distribution of fiberfills in the bulk of polymer matrix can be obtained due to the change of way of their incorporation. For example, incorporation of fiberfills into the latex of butadiene-styrene rubber before its supply to coagulation allows to attain a uniform distribution of the fiber in the obtained rubber crumb and this results in an increase of such quality factors of vulcanizates as their immunity to thermal-oxidation effect, multiple deformations and so on [2, 3].

Results of the investigations on the influence of small doses of fiberfills (up to 1 mass % in a rubber) on the process of the rubber extraction from latex and the properties of the obtained composites are presented in Refs. [2, 3]. In Ref. [4] a technological difficulty was noted that was related with the incorporation of the fiberfills into the process in the dosage of more than 1 mass % in the rubber.

In this situation it would be interesting to transform fiberfills into the powder state. This will make possible to incorporate a greater amount of the filler into the rubber just at the stage of its production and thus to attain its uniform distribution in the rubber matrix.

Powder-like fillers get rather wide application in producing of tires and in mechanical rubber industry [5]. The overwhelming amounts of the used powder-like fillers are of inorganic nature and they are incorporated into rubber compounds with the use of rolling mills during their production. This way of incorporation just as in the case of fibers does not make it possible to attain a uniform distribution of the filler in the rubber compound that is further affected on the properties of produced articles. Therefore the elaboration of the new ways of fillers incorporation into the polymer composites in order to obtain the articles having a set of new properties seems to be very actual both from scientific and practical viewpoints.

The aim of this work is to study the possibility of filling butadiene-styrene rubber of SKS-30 ARK grade with powder-like fillers made of the cotton fiber. Another aim was the choice of the way of incorporation of the rubber at the stage of latex as well as the estimation of the fillers effect on the process of coagulation and the properties of the composites.

2.2   EXPERIMENTAL PART

At first we have elaborated the technique of obtaining the powder-like cellulose fillers from the fibrous from cellulose-containing textile waste products.

To make this cotton fiber was subjected to a rough crumbling and treated with sulphuric acid according to the following technique. Fibers of 0.5–3.0 cm in size were treated with 1.5–2.0 parts of sulfuric acid under stirring (acid concentration was of 20–30 mass %). Next the pasty mass (fibers + sulfuric acid solution) were filtrated. The obtained powder-like filler was dried for 1–2 h. After the final drying the powder-like mass was subjected to the additional crumbling up to more highly dispersed state. Thus, acid powder-like cellulose (APC) filler was obtained. To obtain neutral powder-like cellulose filler (NPC) APC was neutralized with the aqueous solution of sodium hydroxide with the concentration of 1.0–2.0 mass %.

At the second stage of investigations some certain ways of incorporation of the powder-like cellulose fillers into butadiene-styrene rubber were estimated and proposed for the use.

Thereto APC, NPC and microcrystalline cellulose (MCC) were applied with the dosage for every sample 1, 3, 5, and 10 mass % in the rubber. For all of the studied ways of incorporation the powder-like fillers were incorporated into the latex of butadiene-styrene rubber just before its supply to coagulation. Aqueous solution of NaCl (24 mass %), $MgCl_2$ (12 mass %) or $AlCl_3$ (10 mass %) were applied as a coagulant and aqueous solution of sulfuric acid with a concentration of 1–2 mass % was used as an acidifying agent. Powder-like cellulose fillers were incorporated in the following ways:

- in the dry form just into latex immediately before its supply for coagulation;
- in the dry form into latex, involving coagulant;
- simultaneously with aqueous solution of coagulant in latex;
- with serum at the completion phase of extraction rubber from the latex.

At the third stage we studied the influence of the powder-like cellulose fillers on the process of coagulation.

Coagulation process was performed in the following way. 20 mL of latex SKS-30 ARK (dry residue of ~18%) was loaded into coagulator which was made in the form of capacitance provided with a stirring device and then it was thermally stabilized for 15–20 min at the temperature of 60 °C. After that 24% aqueous solution of coagulant (NaCl, $MgCl_2$, $AlCl_3$) supplied. Coagulation process was completed by addition of 1–2% aqueous solution of sulfuric acid to the mixture up to pH value ≈ 2.0–2.5. The formed rubber crump was separated from aqueous phase by filtration, washed up with water, dried in desiccator at the temperature of 80–85 °C and then weighed. Mass of the formed rubber crump was calculated basing from the dry residue of the original latex. After sedimentation of filtrate (or as it is named in the industry of synthetic rubber – "serum") a possible presence of high-dispersive rubber crump in the sample was determined visually. Powder-like cellulose fillers were incorporated into the latex of butadiene-styrene rubber using all of the above-named ways.

At the fourth stage an estimation of the influence of powder-like cellulose fillers on the properties of obtained rubbers, rubber compounds and vulcanizates. To make the estimations first rubber compounds were prepared with the use of conventional ingredients with their further vulcanization at the standard facilities [6, 7]. Produced vulcanizates were subjected to physical-mechanical tests and simultaneous study of vulcanization kinetics and swelling ability kinetics.

Swelling kinetics ability of vulcanizates filled with different fibers was studied in solvents of different polarity according to the following technique. Samples of vulcanizates were cut-of in the form of squares of 1x1 cm in sized and weighed. Number of samples for each series of measurements was 5. The samples were put into solvents for eight hours. Every hour they were extracted from the solvents, then their sizes were measured and they were weighed. The last point to measure the sizes and mass of a sample was in 24 h. After that we processed the obtained data:

- in order to determine swelling degree α (mass %) we subtracted mass of the original sample from that one of the swollen sample: the obtained solvent mass was divided by mass of original sample and multiplied the result by 100%; from the five obtained results for each of the samples the greatest (equilibrium) value α_{max} was chosen;
- swelling constant rate was determined as:

$k = (1/\tau) \times (\ln [\alpha_{max}/(\alpha_{max} - \alpha_{\tau})])$, where

τ – is time (hr.); α_{τ} – is the value of the current swelling degree at the time τ.

2.3   RESULTS AND DISCUSSION

The obtained APC involved the rest of sulfuric acid. However, this disadvantage is transformed into an advantage in case of the use of this filler in the production of emulsion rubbers where acidifying of the system takes place at the stage of rubber extraction from a latex.

One can expect that the use of APC fillers in the technological process of buta-diene-styrene rubber production should reduce a total consumption of sulfuric acid and stabilize coagulation stage. It should be noted that the process performed in the real industrial scale the separation stage of the obtained powder-like filler from sulfuric acid solution and its drying can be eliminated since extraction of butadiene-styrene rubber from the latex is accompanied by acidifying of the system by a solution of sulfuric acid. Therefore the obtained pasty mixture composed of a sulfuric acid solution and a powder-like filler on the basis of cellulose is reasonable to dilute with water in order to decrease sulfuric acid concentration up to 1–2 mass % and to perform incorporation of the obtained dispersed mass into coagulated latex instead of the "pure" solution of sulfuric acid. In order to make more overall estimate of the influence of powder-like cellulose fillers on the coagulation process and the properties of obtained composites comparative tests with the samples of MCC were performed.

Fractional composition of APC, NPC and MCC fillers is presented in Fig. 2.1. On the basis of the fractional composition of the powder-like fillers weight-average size of their particles was determined: APC ≈ 0.57 mm; NPC ≈ 0.14 mm; MCC ≈ 0.15 mm. The calculated specific surface of the particles in these powder-like fillers was of 70, 286 and 267 cm^2/g, respectively accounting for cellulose density ρ =1.5 g/cm^3.

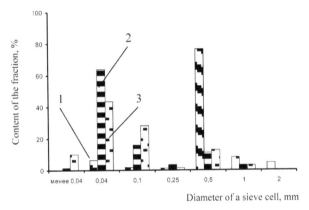

FIGURE 2.1 Fractional composition of the powder-like cellulose fillers: APC (1), NPC (2), MCC (3).

Images of the powder-like cellulose fillers obtained with the use of scanning electron microscopy JSM-6380 LV (magnification by 220–250 times) are presented in Fig. 2.2.

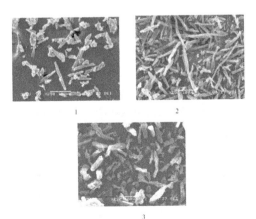

FIGURE 2.2 Electron microphotographs of APC (1), NPC (2), MCC (3).

These images were obtained in secondary-electron emission mode. In order to prevent thermal destruction of the samples and their electrical charging under the impact of the electron beam gold layer with a thickness of 10 nm was deposited on the samples. The particles of cellulose powders were mainly presented by the crystals with the shape factor (the ratio of length l to the diameter d) l/d varying within the interval of 1–9 for APC, 1–25 for NPC and 1–8 for MCC. This was also confirmed indirectly by the value of packed density for the fillers: MCC = 0.79, NPC = 0.44 and APC = 0.68 g/cm^3. In the presence of such needle-like fillers it is possible to observe anisotropy effect of elastic-strengthening factors for vulcanizates.

In turn, analysis of the particles elemental composition demonstrated the presence of sulfate groups in APC and their absence in NPC and MCC (Table 2.1).

TABLE 2.1   Elemental Composition of Cellulose Powder-Like Fillers

Name of the element	Content of the elements, mass %		
	NPC	**APC**	**MCC**
C	43.76	39.02	44.53
O	53.80	53.47	55.47
Na	0.62	0	0
S	0	5.49	0
Other	1.82	2.02	0

Comparing the possible ways of incorporation of the powder-like cellulose fillers into butadiene-styrene rubber with the account of their properties one can make a conclusion that incorporation of APC is appropriate to perform with a coagulant while MCC and NPC are reasonable to incorporate in the form of dry powders into latex just before its supply to coagulation [8, 9]. However, since incorporation of the fillers in a dry form is connected with certain technological problems, so in what follows, NPC and MCC were incorporated jointly with a coagulant.

Tables 2.2–2.4 represent the results of the study of influence of the powder-like cellulose fillers on the yield of rubber crump obtained from latex in the presence of different coagulants: NaCl, MgCl$_2$, AlCl$_3$ Analysis of the obtained results demonstrated that incorporation of all the above-named powder-like cellulose fillers resulted in an increase of the yield of the formed rubber crump.

TABLE 2.2 Influence of Coagulant (NaCl) Consumption on the Yield of the Formed Rubber Crump for Different Amounts of the Investigated Powder-Like Fillers

NaCl consumption, kg/ton of rubber	Yield of the formed rubber crump, mass %.												
	Without filler	Amount of APC, mass % per rubber				Amount of NPC, mass % per rubber				Amount of MCC, mass % per rubber			
		1	3	5	10	1	3	5	10	1	3	5	10
1	35.2	52.5	48.2	52.7	49.2	45.4	44.7	50.1	48.0	41.3	48.1	49.2	45.1
5	45.9	59.8	58.4	61.4	61.7	56.9	59.0	58.5	68.6	58.6	60.9	56.5	57.9
10	56.8	69.8	71.1	79.6	77.5	78.1	78.6	76.3	79.0	76.4	77.5	78.2	72.6
25	80.0	87.2	85.6	92.3	90.5	90.7	91.8	94.9	90.8	92.5	93.6	91.3	89.2
50	92.8	93.1	92.9	95.7	92.7	95.5	95.8	97.9	92.3	95.9	96.5	94.9	92.7
75	95.3	98.7	98.6	98.1	96.4	97.3	97.9	98.1	96.0	98.0	98.4	98.5	95.6
100	97.8	98.9	99.0	99.2	97.5	97.9	99.2	99.2	98.2	99.1	99.0	99.2	98.2

Table 2.3. Influence of Coagulant (MgCl$_2$) Consumption on the Yield of the Formed Rubber Crump For Different Amounts of the Investigated Powder-Like Fillers

MgCl$_2$ consumption, kg/ton of rubber	Yield of the formed rubber crump, mass %.												
	Without filler	Amount of APC, mass % per rubber				Amount of NPC, mass % per rubber				Amount of MCC, mass % per rubber			
		1	3	5	10	1	3	5	10	1	3	5	10
1	37.9	50.1	60.7	62.8	60.3	44.2	45.7	48.8	49.1	45.2	44.5	43.2	47.6
2	48.9	54.5	70.2	70.1	73.3	50.7	54.9	51.7	52.3	52.4	50.3	51.4	59.5
3	61.7	62.5	76.7	85.6	83.8	66.4	61.1	64.9	65.3	60.8	61.3	60.4	61.4
6	75.2	88.5	87.6	90.9	97.1	79.4	77.9	80.1	83.1	80.5	81.5	83.9	83.0
9	80.4	91.2	97.5	98.2	97.5	84.1	82.3	86.3	92.9	91.6	91.0	89.9	91.7
10	89.6	95.6	98.5	98.6	98.3	90.3	95.2	95.8	95.9	98.6	95.9	96.4	97.5
15	92.8	98.2	99.1	99.0	98.9	96.1	98.2	97.8	98.2	98.9	97.4	98.5	98.6
20	95.7	99.2	99.2	99.2	99.1	96.2	98.9	98.9	98.9	99.0	98.6	98.9	99.2

TABLE 2.4 Influence of Coagulant (AlCl$_3$) Consumption on the Yield of the Formed Rubber Crump For Different Amounts of the Investigated Powder-Like Fillers

AlCl$_3$ consumption, kg/ ton of rubber	Yield of the formed rubber crump, mass %.												
	Without filler	Amount of APC, mass % per rubber				Amount of NPC, mass % per rubber				Amount of MCC, mass % per rubber			
		1	3	5	10	1	3	5	10	1	3	5	10
0.3	43.0	57.9	68.9	66.0	74.3	56.7	54.6	56.1	55.3	57.1	56.8	59.7	55.0
0.7	52.3	79.9	82.7	82.8	83.4	74.0	70.9	72.0	75.1	70.0	69.3	69.3	65.1
1	71.2	92.3	93.7	92.7	91.1	83.0	82.2	80.8	81.9	81.4	80.1	81.2	81.7
2	90.4	95.0	97.1	96.3	95.3	94.1	92.1	93.3	90.9	92.8	91.8	91.2	91.5
3	93.5	98.6	98.2	97.5	96.2	97.5	98.1	97.2	97.3	97.7	94.0	95.7	93.7
4	96.5	99.1	98.6	99.2	99.1	98.2	99.2	98.2	99.2	98.2	98.3	97.3	95.9

This can be connected with a decrease of rubber losses in the form of highly dispersed crump with serum and rinsing water. Incorporation of the powder-like cellulose fillers allowed to reduce consumption of the coagulant and acidifying reagent required for a complete separation of the rubber from latex. In case of application of APC filler with a dosage of more than 7 mass % in the rubber complete coagulation of latex can be attained without additional incorporation of the acidifying reagent – sulfuric acid solution – into the process. At the same time powder-like cellulose fillers can absorb surface active materials on their surface as well as coagulant and the components of the emulsion system thus facilitating a decrease of environmental pollution by sewage water, for example APC (Table 2.5). Similar data were obtained both for NPC and MCC.

TABLE 2.5   Elemental Composition of APC Before and After Its Application in the Process of Rubber Separation From Latex

Name of the element	Content, mass %			
	APC composition before coagulation process	APC composition treated with the components of emulsion system in the presence of different electrolytes		
		NaCl	**MgCl$_2$**	**AlCl$_3$**
C	39.02	54.98	50.28	52.09
O	53.47	34.44	44.63	44.34
S	5.49	0.48	0.45	0.41
Cl	0	5.75	3.37	2.16
Na	0	4.06	0	0
Mg	0	0	0.97	0
Al	0	0	0	0.84
K	0	0.16	0.29	0.15
Additives	2.02	0.13	0.01	0.01

Proposals on the changes in any technological parameters of the rubber coagulation process should not have any negative effects on the properties of the rubber and its vulcanizates. Therefore, we investigated the process of vulcanization for the rubber samples separated with the application of different amounts of the studied fillers (3; 5; 10 mass % in the rubber). Results of these investigations are presented in Table 2.6. Properties of the rubber compounds and physical-mechanical factors for the filled vulcanizates are given in Table 2.7.

TABLE 2.6 Characteristics of Vulcanization Process of the Rubber Compounds Involving Fillers

Factor	Reference sample (without a filler)	Value of the factor for the rubbers with different fillers								
		Content of APC, mass % in a rubber			Content of NPC, mass % in a rubber			Content of MCC, mass % in a rubber		
		3	5	10	3	5	10	3	5	10
M_L, dN·m	7.5	6.5	7.0	7.5	7.0	7.3	7.7	7.0	6.9	7.5
M_H, dN·m	32.8	31.5	32.9	33.0	34.0	34.3	37.5	33.0	34.8	36.7
t_S, min	3.0	4.3	3.8	3.0	4.0	3.9	2.3	4.4	4.4	4.0
$t_{C(25)}$, min	9.9	8.3	8.9	8.7	10.0	10.0	8.7	8.8	8.1	8.3
$t_{C(50)}$, min	12.6	10.7	11.4	11.4	12.6	12.7	11.4	11.9	10.8	10.8
$t_{C(90)}$, min	22.0	22.0	22.5	21.8	22.8	23.1	21.9	22.8	22.1	21.4
R_v, min^{-1}	5.3	5.6	5.3	5.3	5.3	5.2	5.1	5.4	5.6	5.7

Note: M_L – minimal torsion moment; M_H – conditional maximal torsion moment; t_S – time of vulcanization start; $t_{C(25)}$ – time of attaining 25% vulcanization, min.; $t_{C(50)}$ – time of attaining 50% vulcanization; $t_{C(90)}$ – time of attaining 90% vulcanization; R_v – vulcanization rate.

TABLE 2.7 Properties of the Rubber Compounds and Vulcanizates Involving Fillers

Factor	Reference sample (without a filler)	Content of APC, mass % in a rubber				Content of NPC, mass % in a rubber				Content of MCC, mass % in a rubber			
		1	3	5	10	1	3	5	10	1	3	5	10
Mooney viscosity MB 1+4 (100°C), a.u.	57.0	52.0	53.0	54.0	57.0	54.0	54.0	52.0	58.6	54.0	54.0	54.0	55.0
Plasticity, a.u.	0.40	0.40	0.40	0.40	0.40	0.35	0.35	0.32	0.35	0.28	0.29	0.28	0.29
Elastic recovery, mm	1.10	0.80	0.70	0.70	1.10	1.50	1.40	1.60	1.29	1.88	1.97	1.86	1.82
M_{300}, MPa	8.1	8.0	8.1	8.0	7.8	8.0	7.9	8.5	9.2	8.8	8.6	8.8	8.6
f_p, MPa	22.8	20.5	22.0	24.4	21.9	24.0	23.6	22.7	21.7	20.0	20.0	20.3	22.3
ε_p, %	620	620	620	670	623	630	637	620	570	500	520	544	562
ε_{oct}, %	14	14	14	16	14	14	14	15	17	12	13	14	16
Ball drop resilience, %	38	42	42	41	38	40	40	40	38	40	42	38	39
Shore hardness A, a.u.	57	57	57	55	57	58	57	57	62	58	59	60	61

Note: M_{300} — elongation stress 300%, MPa; f_p — conditional disruption strength; ε_p — relative elongation under disruption; ε_{oct} — relative residual deformation after disruption.

With an increase of content of all cellulose fillers from 3 to 10 mass % in a rubber the rise of the minimal (M_L) and conditional maximal (M_n) torsion moments was observed in the rubber compound under vulcanization. The presence of powder-like cellulose fillers increased the time of the vulcanization start for the rubber compounds but actually did not have an effect on the time of vulcanization ending (Table 2.6).

The presence of NPC and MCC in the amount of 1–10 mass % in the rubber increased vulcanizates hardness due to the effect of the fillers presence as a result of their large shape factor (1–25 for NPC), as well as their large specific surface (83 cm³/100 g for MCC) [10]. However, the presence of NPC and MCC did not influence on the rubber ball drop resilience.

Reinforcing effect in the polymer matrix provided a linear increase of M_{300} for the rubbers with an increase of content of NPC with a large shape factor (1–25). In turn, the increase of content of the defect centers in polymer matrix due to the accumulation of the coarse grains of particles (Fig. 2.1) reduced the strength and relative elongation of vulcanizates under disruption. An increase of MCC content in vulcanizates provided the rise of stress under 300% elongation (M_{300}), with a simultaneous decrease of the swelling rate of vulcanizates in solvents (Table 2.8). Moreover, an increase of MCC content multiplied concentration of the transverse bonds and increased vulcanization rate along with a decrease of the strength and relative elongation at the disruption due to accumulation of the defect centers in polymer matrix, that can be explained by the absence of wetting for MCC particles with a rubber (Table 2.7).

Swelling process of the obtained vulcanizates was investigated in Ref. [11] in different environments allowing to simulate real service conditions of the vulcanizates. Aromatic toluene and aliphatic nefrac (benzine) were taken as solvents as mostly wide-spread coupling media.

Analysis of the reference literature sources [12] showed that for all of the solvents swelling kinetics of vulcanizates in the presence of the fillers could be described in semilogarithmic coordinates of descending direct line (Fig. 2.3) of the form:

$$lg(Q_{max} - Q_\tau) = lgQ_{max} - b\tau,$$

By changing $lg(Q_{max} - Q_\tau) = Y$, and $lgQ_{max} = a$, the original equation can be written as

$$Y = a - b\tau,$$

where b – is swelling rate, hr⁻¹; τ – duration of swelling, hr; Q_{max} – is an equilibrium swelling degree, %; Q_τ – current swelling rate, %.

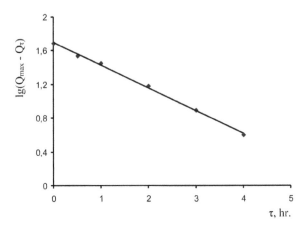

FIGURE 2.3   Swelling kinetics of vulcanizates in organic solvents.

Equilibrium swelling degree of vulcanizates, involving 3–10 mass %, in nefrac ($Q^{nf}_{mcc,\,max}$) was reduced up to 46–52% as compared with 84% for vulcanizate without MCC (Fig. 2.4, straight line 2). Swelling rate of vulcanizates in nefrac in the presence of 3–10 mass % of MCC was of $b^{nf}_{mcc} = -0.27$ hr^{-1}, did not depend on the value of MCC content but it was less than the swelling rate of vulcanizate without MCC (-0.38 hr^{-1}) (Table 2.8).

In turn, swelling rate of vulcanizates in toluene b^{tl}_{mcc} involving 3–10 mass % of MCC was of – (0.30–0.34) hr^{-1} and it was by 2 times lower than for vulcanizate without the filler (-0.62 hr^{-1}) (Table 2.8). An increase of MCC content within the investigated interval reduced as the equilibrium swelling degree $Q^{tl}_{mcc,\,max}$ from 250 to 209% (Fig. 2.4, straight line 1), as swelling rate from -0.34 to -0.30 hr^{-1}.

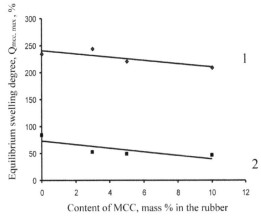

FIGURE 2.4   Influence of MCC content (mass %) and the nature of a solvent on the equilibrium swelling degree of vulcanizates ($Q_{MCC,\,max}$, %) in toluene (1) and nefrac (2).

Table 2.8 Influence of Amount of the Fillers and Nature of a Solution on the Ability to Vulcanizates Swelling

Solvent	Swelling rate of vulcanizates in toluene involving 3–10 mass % of MCC							
	Without fillers		3%		5%		10%	
	b	a	b	a	b	a	b	a
				APC				
Toluene (1)	−0.62	2.4	−0.54	2.3	−0.61	2.4	−0.58	2.4
Nefrac (2)	−0.38	1.9	−0.67	1.9	−0.89	1.9	−0.84	1.9
$1/Q_{max}^{tl} \times 10^3$	4.3		4.1		4.5		4.8	
				NPC				
Toluene (1)	−0.62	2.4	−0.82	2.4	−0.80	2.4	−0.58	2.4
Nefrac (2)	−0.38	1.9	−0.77	1.8	−0.64	1.9	−0.54	2.0
$1/Q_{max}^{tl} \times 10^3$	4.3		4.6		4.0		3.9	
				MCC				
Toluene (1)	−0.62	2.4	−0.34	2.4	−0.31	2.3	−0.30	2.3
Nefrac (2)	−0.38	1.9	−0.27	1.8	−0.27	1.7	−0.27	1.7
$1/Q_{max}^{tl} \times 10^3$	4.3		4.2		4.0		3.9	

Note: b – swelling rate (hr^{-1}); $a = lgQ_{max}$ – equilibrium swelling rate, %; $1/Q_{max}^{tl} \times 10^3$ – concentration of transverse bonds.

Similar regularity is characteristic of vulcanizates involving active fillers (e.g., technical carbon) [10].

The value of equilibrium swelling degree of vulcanizates involving 1–10 mass % of NPC ($Q_{npc, max}^{nf}$) in nefrac, was of 80–96% and actually did not differ (84%) from that one in vulcanizates without the filler (Fig. 2.5, straight line 2).

Swelling rate of vulcanizates in nefrac (b_{npc}^{nf}) in the presence of 1–10 mass % of NPC increased up to the values of (0.48–0.77) hr^{-1} as compared with the value of (−0.38 hr^{-1}) for vulcanizate without the filler (Table 2.7). Dependence of swelling rate b_{npc}^{tl} on the NPC content within the interval of 3–5 mass % was of −0.80 to −0.82 hr^{-1}. These values were higher than the swelling rate of vulcanizate without the filler – 0.62 hr^{-1}.

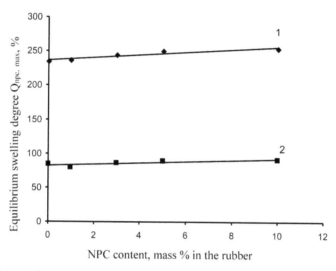

FIGURE 2.5   Influence of NPC content (mass %) and the nature of solvent on the equilibrium swelling degree of vulcanizate ($Q_{npc, max}$, %). 1 – toluene; 2 – nefrac.

An increase of $Q^{tl}_{npc, max}$ in toluene was observed with the rise of NPC content within the interval of 1–10 mass % (Fig. 2.5, straight line 1). Equilibrium swelling degree of vulcanizates in toluene $Q^{tl}_{npc, max}$ with an increase of NPC content was grown up to 262% as compared with a vulcanizate without the filler (235%). Increase of $Q^{tl}_{npc, max}$ can be explained by decrease of the density of transverse bonds in vulcanizate at the boundary between the phases of "polymer-filler" and the greater value of the shape factor for NPC.

For vulcanizates involving 1–10 mass % of APC in nefrac the value of $Q^{nf}_{apc, max}$ was of 79–85% as compared with 84% for vulcanizates without the additives (Fig. 2.6, straight line 1), that is, it did not actually change.

An increase of b^{nf}_{apc} values for vulcanizates involving APC up to – 0.89 hr^{-1} was observed as compared with the value of -0.38 hr^{-1} for vulcanizates without the filler (Table 2.7) by the reason of a bad wettability of the large APC particles with the rubber and occurrence of the tunnel effect. Equilibrium swelling degree of vulcanizates with 3–10 mass % of APC in toluene $Q^{tl}_{apc, max}$ = 240–252% was greater than that one for vulcanizates without the additives (235%) (Fig. 2.6, straight line 1). This can be explained by a decrease of the density of transverse bonds in vulcanizate ($1/Q^{tl}_{apc, max}$) at the boundary of "polymer-filler" due to a bad wettability of the particles of the acid powder-like filler.

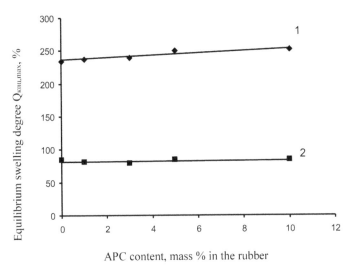

APC content, mass % in the rubber

FIGURE 2.6   Influence of APC content (mass %) and the nature of solvent on the equilibrium swelling degree of vulcanizate ($Q_{apc, max}$, %). 1 – toluene; 2 – nefrac.

2.4 CONCLUSIONS

1. Textile waste products can be used not only as fiberfills but also as powder-like fillers.
2. Differences in sizes, shape factor, specific surface, fractional and chemical composition of the particles of powder-like cellulose fillers obtained in different ways were found in the work. Powders of the neutral and microcrystalline cellulose are characterized by the higher specific surface (267–286 cm^2/g). The largest scattering interval of the shape factor was characteristic of the neutral cellulose ($l/d = 1$–25). Greater content of sulfate groups was determined in the particles of the acid powder-like cellulose (5.49 mass % accounting for the bound sulfur).
3. Incorporation of the powder-like cellulose fillers into SKS-30 ARK rubber does not have a negative effect on the physical-mechanical quality factors of vulcanizates. Thus it is possible to get a considerable improvement of elastic-strength quality factors of butadiene-styrene rubber in the presence of these additives due to the choice of the required reactants for interphase combination of cellulose with the rubber matrix.
4. Linear dependence of the equilibrium swelling degree for vulcanizates in the aliphatic and aromatic solvents on the content of powder-like cellulose fillers was found as a result of the work. With an increase of MCC content

equilibrium swelling degree reduced and for NPC and APC it was on the contrary enhanced.

5. Certain difference were found in swelling kinetics of vulcanizates in nefrac and toluene in the presence of MCC, APC and NPC. In the presence of 3–10 mass % of MCC swelling rate of vulcanizates in solvents reduced by 1.25–2.0 times that is characteristic for the rubbers with active fillers. Incorporation of 1–10 mass % of NPC increased the swelling rate of vulcanizate in nefrac by 1.25–2.0 times.

6. Diverse ways of incorporation of the powder-like cellulose fillers into rubber emulsion before its coagulation with the use of different coagulants makes it possible to improve the distribution of the additives in the rubber matrix, to reduce coagulant consumption, and in case of APC application – to decrease the consumption of the acidifying reactant up to its complete elimination from the coagulation process.

KEYWORDS

- **composite material**
- **fibrous filling material**
- **physicomechanical properties**
- **rubber latex**
- **wastes**

REFERENCES

1. Nikulin S. S., Pugacheva I. N., Chernykh O. N. Composite materials on the basis of butadiene-styrene rubbers. M.: "Academy of Natural Sciences," 2008. 145 p.
2. Nikulin S. S., Akatova I. N. Influence of capron fiber on coagulation, rubber properties, rubber compounds and vulcanizates. Zhurnal prikladnoi khimii, 2004. V. 77. Iss. 4. P. 696–698.
3. Nikulin S. S., Akatova I. N. Influence of the flax and wiscose fiber on the process of separation of butadiene-styrene rubber from the latex. Advances in modern natural sciences, 2003. № 6. P. 10–13.
4. Akatova I. N., Nikulin S. S. Influence of fiberfill with high dosages on the process of separation of butadiene-styrene rubber from the latex. Production and usage of elastomers, 2003. № 6. P. 13–16
5. Mikhailin Yu. A. Constructional polymer composite materials. SPb.: Scientific foundations and technologies, 2008. 822 p.
6. Laboratory practice in rubber technology. Ed. by Zakharov N. D. M.: Khimia, 1988. 237 p.
7. GOST 15627–79. Synthetic rubbers – butadiene-methylstyrene SKMS-30 ARK and butadiene-styrene SKS-30 ARK.
8. Pugacheva I. N., Nikulin S. S. Application of powder-like filler on the basis of cellulose in the production of emulsion rubbers. Modern high-end technologies, 2010. № 5. 52–56.

9. Pugacheva I. N., Nikulin S. S. Application of powder-like fillers in the production emulsion rubbers. Industrial production and use of elastomers, 2010. Iss. 1. 25–28.
10. Koshelev F. F., Korneev A. F., Bukanov A. M. General rubber technology. M.: Khimia, 1978. 528 p.
11. Zakharov N. D., Usachov S. V., Zakharkin O. A., Drovenikova M. P., Bolotov V. S. Laboratory practice in rubber technology. Basic processes of rubber production and methods of their control. M.: Khimia, 1977. 168 p.
12. Tager A. A. Physics and chemistry of polymers. M.: Scientific world, 2007. 573 p.

CHAPTER 3

STRUCTURE AND PARAMETERS OF POLYHYDROXYBUTYRATE NANOFIBERS

A. A. OLKHOV,[1, 2] O. V. STAROVEROVA,[2] A. L. IORDANSKII,[2] and G. E. ZAIKOV[3, 4]

[1]Plekhanov Russian University of Economics, Stremyanny per. 36, 117997, Moscow, Federation of Russia, E-mail: aolkhov72@yandex.ru

[2]N.N. Semenov Institute of Chemical Physics, RAS, 119991 Moscow, Street Kosygina, 4, Federation of Russia

[3]L.Ya. Karpov Physicochemical Research Institute, Federation of Russia

[4]N.M. Emanuel Institute of Biochemical Physics, 119991 Moscow, Street Kosygina, 4, Federation of Russia

CONTENTS

3.1 INTRODUCTION

This research work focuses on process characteristics of polymer solutions, such as viscosity and electrical conductivity, as well as the parameters of electrospinning using poly 3-hydroxybutyrate modified by titanium dioxide nanoparticles, which have been optimized. Both physical-mechanical characteristics and photooxidation stability of materials have been improved. The structure of materials has been examined by means of X-ray diffraction, differential scanning calorimetry (DSC), IR-spectroscopy, and physical-mechanical testing. The fibrous materials obtained can find a wide application in medicine and filtration techniques as scaffolds for cell growth, filters for body fluids and gas-air media, and sorbents.

Development of materials with revolutionary characteristics is closely connected with obtaining nanosized systems. Of greatest interest today are compositions derived from polymers and nanosized objects which show a unique set of characteristics, have no counterparts, and drastically change present ideas about a polymer material.

Titanium dioxide nanoparticles are the most attractive because of the developed surface of titanium dioxide, the formation of surface hydroxyl groups with high reactivity resulted from reacting with electrolytes as crystallite sizes decrease down to 100Å and lower, and a high efficiency of oxidation of virtually any organic substance or many biological objects.

Many modern applications of TiO_2 are based on using its anatase modification which shows the minimum surface energy and greater concentration of OH-groups on sample surface compared with other modifications. According to Dadachov's patents [1, 2], the nanosized η-modification of TiO_2 is considerably superior to anatase in the above mentioned properties. The main characteristics of titanium dioxide modifications in use are sample composition, TiO_2 modification, nanoparticle size and crystallite size, specific surface area, pore size and pore volume.

Polyhydroxybutyrate (PHB) is the most common type of a new class of biodegradable termoplasts, namely polyoxyalkanoates. It demonstrates a high strength and the ability to biodegrade under natural environmental conditions, as well as a moderate hydrophilicity and nontoxicity (biodegrades to CO_2 and water) [3]. PHB shows a wide range of useful performance characteristics [4]; it is superior to polyesters which are the standard materials for implants, can find application in different branches of medicine, and is of great importance for cell engineering due to its biocompatibility. [5].

PHB is a unique sample of a moderate hydrophobic polymer being biocompatible and biodegradable at both high melting and crystallization temperatures. However, its strength and other characteristics, such as thermal stability, gas permeability, and both reduced solubility and fire resistance, are insufficient for its large-scale application.

The objective of the research was to prepare ultra-fine polymer composition fibers based on polyhydroxybutyrate and titanium dioxide and to determine the role played by nanosized titanium dioxide modifications in achieving special properties of the compositions.

3.2   EXPERIMENTAL PART

The nanosized η-TiO_2 and anatase (S12 and S30) were prepared by sulfate process from the two starting reagents, $(TiO)SO_4 \cdot xH_2SO_4 \cdot yH_2O$ (I) and $(TiO)SO_4 \cdot 2H_2O$ (II) correspondingly [6].

Samples of titanium dioxide and its compositions with polymers were analyzed by X-ray diffraction technique, using HZG-4 (Ni filter) and (plane graphite mono-chromator) diffractometers, CuK_α radiation, diffracted beam, in the range of 2θ 2–80°, rotating sample, stepwise mode (the impulse accumulation time is 10s, by step of 0.02°). Experimental data array was processed with PROFILE FITTING V 4.0 software. Qualitative phase analysis of samples was carried out by using JCPDS PDF-2 database, ICSD structure data bank, and original papers.

Particle sizes (coherent scattering region) of TiO_2 samples were calculated by the Selyakov-Scherrer equation $L = \dfrac{k\lambda}{\beta \cdot \cos\theta_{hkl}}$, where is physical peak width for the phase under study (diffraction reflections were approximated by Gaussian function), B is integral peak width, b is instrument error correction ($b \sim 0.14°$ for α-Al_2O_3 as a reference), $k \sim 0.9$ is empirical coefficient, λ is wavelength. Calculations were based on a strongest reflection at $2\theta \sim 25°$. Standard deviation was ±5%.

Starting PHB with molecular weight of 450 kDa was prepared through microbiological synthesis by BIOMER (Germany). Chloroform (CFM) was used as solvent for preparing polymer solution. Both HCOOH (FA) and $[CH_3(CH_2)_3]_4N$ (TBAI) were used as special additives.

Electrostatic spinning of fibers based on PHB and titanium dioxide was carried out with original laboratory installation [7].

The dynamic viscosity of polymer solutions of various compositions was measured as a function of PHB concentration with Heppler and Brookfield viscosimeter. The electrical conductivity of polymer solutions was calculated by the equation $\lambda = \alpha/R$, Ohm^{-1} cm^{-1}; the electrical resistance was measured with E7–15 instrument.

The fiber diameter distribution was studied by microscopy (optical microscope, Hitachi TM-1000 scanning electron microscope). Fiber orientation was studied by using birefringence and polarization IR-spectroscopy (SPECORD M 80 IR-spectrometer). Crystalline phase of polymer was studied by differential scanning calorimetry (DSC) (differential scanning calorimeter). The packing density of fibrous materials was calculated as a function of airflow resistance variation with a special manometric pressure unit [8].

Physical-mechanical characteristics of fibrous materials were determined with PM-3–1 tensile testing machine according to TU 25.061065–72. Kinetics of UV-aging was studied with Feutron 1001 environmental test chamber (Germany). Irradiation of samples was carried out with a 375 W high pressure Hg-lamp, at a distance of 30 cm.

For in vitro biodegradation studies, the materials were incubated in test tubes filled with 10 mL of 0.025 M phosphate buffer solution (pH = 7.4) at 70°C for 21 days. At regular time intervals the materials were removed from the buffer and rinsed with distilled water, then placed into incubator at 70°C for 3 h, and finally weighed within 0.001 g.

3.3   RESULTS AND DISCUSSION

Diffraction patterns of the nanosized anatase and η-TiO$_2$ prepared are presented in Figs. 3.1a and 3.1b correspondingly. The sample S30 referring to anatase contains trace amounts of β-TiO$_2$ (JCPDS 46–1238) (Fig. 3.1a).

Analysis of the obtained size values of coherent scattering regions (*L*-values) of the η-TiO$_2$ and anatase samples showed that L = 50 (2) Å and L =100 (5) Å, that is, crystallite sizes for the η-TiO$_2$ samples are substantially less.

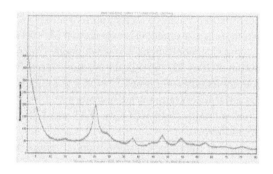

a. S30

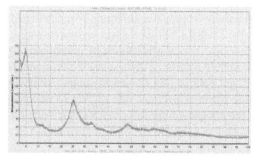

b. S12

FIGURE 3.1   Diffraction patterns of anatase (a) and η-TiO$_2$ (b).

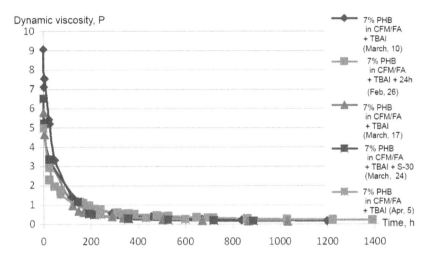

FIGURE 3.2   Dynamic viscosity of PHB solutions as a function of time.

Time dependencies of the dynamic viscosity of PHB solutions are shown in Fig. 3.2. Analysis of the dynamic viscosity of polymer solutions showed how the solution viscosity varies with time as the (HCOOH) – (S1) is added. Most probably, the decrease in viscosity results from the decrease in the molecular weight of polymer.

Examination of the electrical conductivity of polymer solutions and choosing solvent mixture allowed the $([CH_3 (CH_2)_3]_4 N)$ – (TBAI) concentration in solution to be decreased from 5 down to 1 g/L due to adding the S1. The increase in PHB concentration up to 7 wt. % in a new solvent mixture was gained.

The fiber diameter distribution was studied by using fibers from 5% PHB solution in chloroform/formic acid mixture and 7% PHB solution as well. It was found that 550–750 nm diameter fibers are produced from 5% solution, while 850–1250 nm diameter fibers are produced from 7% solution, that is, the fiber diameter increases as the solution concentration rises (Fig.3). The increase in the process velocity leads to the fiber diameter virtually unchanged.

Examination of fibrous materials by electron microscopy showed that the fibrous material uncovered by the current-conducting layer decomposes in 1–2 min, when exposed by electron beam.

Examination of polymer orientation in fibers by using birefringence showed that elementary fibers in nonwoven fabric are well oriented along fiber direction. However, it is impossible to determine the degree of orientation quantitatively because the fibers are packed randomly.

Analysis of fibrous materials by differential scanning calorimetry showed that at low scanning rates a small endothermic peak appears at 190÷200°C which indicates the presence of maximum straightened polymer chains (orientation takes place). Upon polymer remelting this peak disappears.

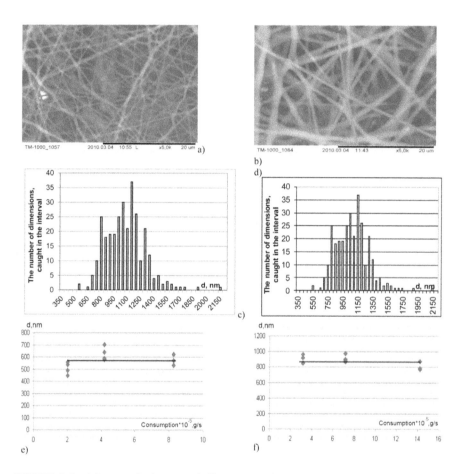

FIGURE 3.3 Microscopic images of fibrous material as a function of PHB solution concentration: (a) 5%; (b) 7%; fiber diameter distributions: (c) 5%; (d) 7%; the fiber diameter as a function of the process velocity (flow rate); (e) 5%; (f) 7%.

IR-examination showed that this method can be also applied only for qualitative estimation of orientation occurrence.

Measurement of the packing density of fibrous materials showed that the formulation used a day after preparing the solution has the greatest density of fiber packing. Probably, it is concerned with a great fiber diameter spread when smaller fibers are spread between bigger ones. The TiO_2-containing fibers are also characterized by the increased packing density.

The results of powder diffraction analysis of PHB powder (a) and PHB fibers obtained from 7% solution with modifying additives (b) and the nanosized TiO_2 modifications (c, d) are presented in Fig. 3.4.

The results of physical-mechanical testing of fibrous materials of different formulations are presented in Table 3.1.

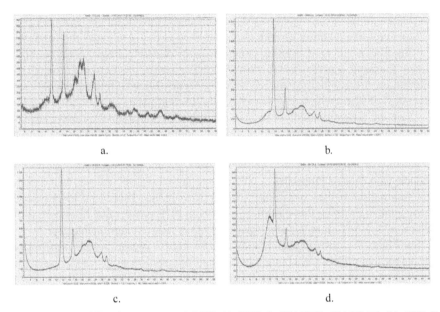

a.   b.

c.   d.

FIGURE 3.4 Diffraction patterns of: PHB (a), PHB-B (7% PHB+1% TBAI) (b), PHB-B +S 30 (c), PHB-B + S12 (d).

TABLE 3.1 Physical-Mechanical Characteristics of Fibrous Materials Prepared From Polymer Compositions As a Function of Formulation

Formulation	L, m	ε, %
PHB + TBAI	724.43	20.83
PHB + TBAI (after a day)	1001.78	12.48
PHB + TBAI + S-12	1430.45	53.37
PHB + TBAI + S-30	1235.91	62.02

L is breaking length, ε is breaking elongation.

Physical-mechanical tests showed that introducing nanosized TiO_2 into solution substantially alters the properties of the resulting fibrous material.

Non-woven fibrous material contains randomly packed fiber layers. Actually, its deformation can be considered as a "creeping," and more and more fibers while straightening during deformation contribute to the breaking stress growing up to its maximum. In other words, the two processes take place simultaneously, namely fiber straightening and deformation of the fibers straightened.

Obviously, as the TiO_2 content increases the fiber flexibility rises due to the decreased crystallinity. This results in a greater amount of fibers simultaneously contributing to the breaking stress and, consequently, the greater slope of the curve.

Moreover, the addition of both S-12 and S-30 is likely to cause the formation of a firm fiber bonding at crossing points, possibly due to hydrogen bonding. Figuratively speaking, a network is formed. At the beginning, this network takes all the deformation stress. Finally, network breaking occurs followed by straightening and "creeping" of fibers which does not affect the stress growth.

There are different additives used in solutions. By purpose, all additives can be divided into process additives and production additives. The first ones are used to control both viscosity and electrical conductivity of spinning solutions as well as the velocity of fiber formation. The second ones are intended for obtaining fibrous products with desired properties.

Examination of the crystallinity of fibrous materials, films, and PHB powder showed that main crystallite modification of both PHB powder and fibers and films melts at 175–177°C, that is, the morphology of crystals remains unchanged.

For fibers, the low-melting shoulder at 160–163°C appears which confirms either the presence of smaller crystallites or their imperfection. For the S-12 based formulation, the low-melting peak disappears which is the evidence of more uniform distribution of the additive within the material.

The broadened crystallization peak is observed for the S-12 based samples. The increased friction and decreased chain mobility lead to the obstructed crystallization and the decreased crystallization rate.

The narrowed crystallization peak is observed for the S-30 based samples. Crystallization proceeds only where no contact with this additive is. Hence, the S-12 and S-30 have different energies of intermolecular interaction of PHB chains and the additive surface.

The crystallinity was calculated by the following equation:

$$\alpha_{\kappa p} = \frac{H_{n\pi}}{146},$$

where $H_{n\pi}$ is melting heat calculated from melting peak area, J/g;
146 is melting heat of monocrystal, J [9]

The most loosely packed structure, that is, the most imperfect, is observed for the formulation prepared a day after.

The S-12 based samples demonstrate excessive fiber bonding; the TiO_2 particles themselves obstruct crystallization. In the case of the S-12, the distribution is good.

The S-30 obstructs crystallization to a lesser extent; the fibers crystallize worse, and the chains are not extended.

Crystallization in fibers occurs upon orientation. The additive is not considered as a nucleating agent.

TABLE 3.2 Widths of Crystallization Peaks At the Scanning Rate of 20°C/min

Material	Heating No.	Tm, °C	Crystallization peak width (at the onset), °C
PHB (starting)	1	174.9	18.61
	2	173.97	18.02
PHB fibers 1 g/L TBAI	1	176.83	20.9
	2	169.27–159.74	20.93
PHB fibers 1 g/L TBAI a day after	1	176.94	26.24
	2	171.98–163.14	22.8
PHB fibers 1 g/L TBAI based on S-12	1	177.49	34.11
	2	170.33	27.38
PHB film from chloroform solution	1	177.21	18.27
	2	171.95–163.1	18.57

UV-aging tests showed that TiO_2-modified fibers demonstrate greater UV-aging resistance. Although the induction period for S-30 based samples is less than that for other formulations, the UV-degradation rate for the S-30 based sample is comparable to that for the S-12 based one.

For TiO_2-containing samples the increased thermal degradation heat after UV-aging is characteristic since the UV-treated TiO_2 acts as the initiator of both thermal and thermooxidation degradation due to OH-groups traveling to powder granule surface.

The onset temperature of both thermal and thermooxidation degradation decreases due to UV-degradation proceeded not only within amorphous phase but also within crystalline phase, as said above.

The TiO_2 acts as the initiator in UV-aging processes.

3.4   CONCLUSIONS

1. Physical-mechanical characteristics of fibrous materials increase as the TiO_2 is introduced.
2. The morphology of main PHB crystallites in powder and fibers is kept unchanged. However, the fibers show the low-melting shoulder (small and imperfect crystals). The TiO_2 obstructs crystallization.
3. PHB fibers are characterized by strongly pronounced molecular anisotropy.
4. S-12 based samples show the best thermal- and thermooxidation degradation stability as well as UV-aging resistance.

The results obtained can be considered as the ground for designing new biocompatible materials, such as self-sterilizing packing material for medical tools or a support for cell growth.

KEYWORDS

- electrical conductivity
- nanofibers
- nanoparticles
- physical-mechanical characteristics
- polyhydroxybutyrate
- viscosity

REFERENCES

1. Dadachov M. U.S. Patent Application Publication. US 2006/0171877.
2. Dadachov M. U.S. Patent Application Publication. US 2006/0144793.
3. Fomin V.A., Guzeev V.V. Plasticheskie Massy. 2001. No 2. 42–46.
4. Boskhomdgiev A.P. Author's abstract of thesis for Ph.D. in Biology. Moscow, 2010.
5. Bonartsev A.P. et al. Journal of Balkan Tribological Association. 2008. No 14. 359–395.
6. Kuzmicheva, G.M., Savinkina, E.V., Obolenskaya, L.N., Belogorokhova, L.I., Mavrin, B.N., Chernobrovkin, M.G., Belogorokhov, A.I. Kristallografiya. 2010. V.55. No 5. 913–918.
7. Filatov, Yu.N. Electrospinning of Fibrous Materials (ES-process). Moscow: Khimiya, 2001,231 p.
8. Barham, P.J., Keller, A., Otum, E.L., Holms, P.A. J. Master Sci. 19(27), 1984, 81–279.

CHAPTER 4

PREPARATION OF AMINES INCLUDING CYCLOACETALIC AND GEM-DICHLOROCYCLOPROPANE FRAGMENTS

A. N. KAZAKOVA,[1] G. Z. RASKILDINA,[1] N. N. MIKHAILOVA,[1] T. P. MUDRIK,[2] S. S. ZLOTSKY,[1] and G. E. ZAIKOV[3]

[1]Ufa State Petroleum Technological University, 1 Kosmonavtov Str., 450062 Ufa, Russia; Phone (347) 2420854, E-mail: nocturne@mail.ru

[2]Sterlitamak branch of Bashkir State University, 49 Lenina avenue, 453103 Sterlitamak, Russia

[3]N. M. Emanuel Institute of Biochemical Physics, Russian Academy of Sciences, 4 Kosygin str., 119334 Moscow, Russia; E-mail: Chembio@sky.chph.ras.ru

CONTENTS

ABSTRACT

This chapter discloses the synthesis of secondary and tertiary amines including *gem*-dichlorocyclopropane and cycloacetal fragments. The method of formation of the cyanuric chloride derivatives, containing same and different amino groups in the side chain, has been proposed.

4.1   INTRODUCTION

Amines of different structure are widely used in various fields of chemistry and chemical technology. So, aromatic and heterocyclic amines are central among additives to oil and polymers. According to this we carried out obtaining of types of amines, including *gem*-dichlorocyclopropane fragments.

4.2   RESULTS AND DISCUSSION

Monochlorinated butenes and pentenes, easily prepared from the industrial dienes (isoprene and butadiene), are widely used as (co)monomers as well as in fine organic synthesis [1]. In particular, their dichlorocarbenylation by the Makosza method affords the corresponding *gem*-substituted dichlorocyclopropanes with quantitative yield [2, 3]. The so prepared monochloroalkyl- and *cis*-2,3-dichloromethyl-*gem*-dichlorocyclopropanes Ia–Ic can be used for *N*-alkylation of the primary and secondary amines IIa–IIc.

2-Chloroethyl-*gem*-dichlorocyclopropane Ia and 2,2-dimethyl-3-chloromethyl-*gem*-dichlorocyclopropane Ib reacted with *n*-butylamine IIa to give the corresponding secondary amines IIIa and IVa with 93–95% yield.

The reactions with diethylamine IIc were slower, giving tertiary amines IIIb and IVb (see table). The yield of the IVb tertiary amine from Ib was below 50% even after 20h.

Under conditions of microwave irradiation, IIc alkylation led to the same yields of IIIb and IVb within only 1 h (see Table 4.1).

cis-2,3-Dichloromethyl-*gem*-dichlorocyclopropane Ic reacted with the primary amines IIa and IIc to yield the bicyclic tertiary amines Va and Vb (see Table 4.1).

Under conditions of thermal heating, the benzyl-amine derivative Vb was produced with yield of 60% within 18 h, whereas the use of microwave radiation gave the same yield within 1 h.

Comparison the chlorinated derivatives Ia and Ib reactivity towards primary amine IIa showed that the presence of geminal dimethyl groups in cyclopropane ring reduced the rate of the exocyclic chlorine substitution by 2.5 times. In all the cases, the endocyclic chlorine atoms did not participate in the reaction with the amines.

TABLE 4.1   *N*-Alkylation of Amines with Monochloroalkyl- and *cis*-2,3-dichloromethyl-*gem*-dichlorocyclopropanes Ia–Ic[a]

Reactants		Product	Reaction time, h	Yield, %
Ia	IIa	IIIa	4	95
	IIc	IIIb	15	93
			1[b]	95
Ib	IIa	IVa	12	93
	IIc	IVb	20	40
			1[b]	42
Ic	IIa	Va	8	90
	IIb	Vb	18	55
			1[b]	58

[a] Molar ratio II: Ia, Ib = 3:1, DMSO, T = 75–80°C; II: Ic; KOH = 2:1:0.3, DMSO, phase transfer catalyst: triethylbenzylammonium chloride, T = 75–80°C.

[b] Microwave irradiation (240 W).

Since derivatives, including *sym*-triazine fragments, are considerable interest as biological active compounds [1, 2], we synthesized their by using of secondary amines, containing 1,3-dioxolane 2a and *gem*-dichlorocyclopropane 2b fragments. In the triazine cycle, as we know, two atoms of chlorine are substituted consecutively whereas the third chlorine atom remains inert [3]. The derivatives of mono-3a,b and disubstituted 4a,b have been obtained consistently in the reaction of cyanuric chloride 1 with amines 2a,b.

R^1=Bu, R^2=H (IIa, IIIa, IVa, Va); R^1=Bn, R^2=H (IIb, Vb);
R^1=R^2=C$_2$H$_5$ (IIc, IIIb, IVb)

At an equimolar ratio between the reagents (secondary amine: triazine = 1:1), the corresponding mono amino derivatives are formed in anhydrous 1,4-dioxane. As the molar ratio between a secondary amine and triazine increases from 1:1 to 2:1, disubstituted triazines IXa, IXb become the main reaction products.

It is noteworthy that, judging from the yields of the triazines formed, secondary amines VIIa and VIIb are close in activity.

The structure of the synthesized and isolated amino substituted triazines was confirmed by ^{1}H and ^{13}C NMR spectroscopic and chromato-mass-spectrometric data.

An analysis of the ^{1}H NMR spectrum of compound VIIIa demonstrated that $H_b^{1'}$ and H_b^5 protons in the *trans* position with respect to the proton of the C^4H group of the dioxolane ring have a chemical shift (CS) at 3.50 and 3.56 ppm in the form of doublet-doublets with spin-spin coupling constants (SSCCs) (Hz): $^2J_{1'} = 14.4$, $^2J_5 = 8.4$, $^3J_{1'b-4} = 8.1$, and $^3J_{5b-4} = 5.7$. The protons H_a^5 and $H_a^{1'}$ in the *cis*-position to the C^4H proton are observed with CS 3.80 and 3.88 ppm (SSCC, Hz: $^2J_{1'} = 14.4$, $^2J_4 = 8.4$, $^3J_{1'a-4} = 3.1$, $^3J_{5a-4} = 6.2$). The proton of the CH group of the cyclic acetal appears as four doublets at 4.36 ppm, with SSCCs (Hz): $^3J_{1'b-4} = 8.1$, $^3J_{1'a-4}$, $3J_{5a-4} = 6.2$, $^3J_{5b-4} = 5.7$. Atoms of the CH_2 group of the second atom have CS = 4.85 and 5.06 ppm in the form of two high-intensity singlets. Signals of methylene protons of the benzyl group form two doublets with CS = 4.7 and 5.53 ppm (SSCC $2J = 15.4$ Hz) and a multiplet in the range 7.2–7.4 ppm from protons of the aromatic ring. The structure of compound IXa was determined in a similar way.

The ^{1}H NMR spectrum of compound VIIIb demonstrated that the proton of the CH group of the cyclopropane ring appears at CS = 1.9 ppm as four doublets with SSCCs (Hz): $^3J_{2a-3} = 10.7$, $^3J_{2b-3} = 7.6$, $3J_{1'a-3} = 7.7$, $^3J_{1'b-1} = 5.8$. Protons of the CH_2 groups bonded to the nitrogen atom are characterized by the presence of a doublet-doublet signal with CS = 3.5 ppm (SSCC, Hz: $^2J = 14.6$, $^3J_{1'a-1} = 7.7$) for C_1H_a and 4.00 ppm ($^2J = 14.7$ Hz, Hz) for C_1H_b. As also in compound VIIIa, to the benzyl group belong two doublets with CS = 4.8 and 5.3 ppm (SSCC $^2J = 15.4$ Hz), which correspond to signals from the CH_2 group, and a multiplet at 7.2–7.4 ppm from protons of the aromatic ring.

A characteristic feature of ^{13}C NMR spectra of the synthesized compounds VIIIa, VIIIb and IXa, IXb is the presence of signals in the weak-field part of the spectrum, which correspond to carbon atoms of the triazine ring. The chemical shifts of 165.49 (VIIIa), 165.42 (VIIIb), 165.09 (IXb), 165.39 and 165.61 ppm (IXa) correspond to carbons C–N bonded to the amine, and those of 170.59 (VIIIb), 167.7 (IXa), 170.16 (IXb), 170.32 and 170.49 ppm (VIIIa), to carbon atoms of the triazine ring with a C–Cl bond.

In the mass spectra of the synthesized compounds VIIIa, VIIIb and IXa, IXb, molecular ions are characterized by low-intensity peaks (<<1%). In all the spectra, the maximum intensity is observed for the $[ArCH_2]^+$ ion with $m/z = 91$ (100%). Under the electron impact, molecular ions of amino derivatives of triazine form high-intensity ions of *N*-benzyl-*N*-(1,3-dioxolan-4-yl)methylamine or *N*-benzyl-4,6-dichloro-*N*-[(2,2-dichlorocyclopropyl)methylamine substituents, which further decompose.

The content of ions corresponding to the presence of a triazine ring in the spectrum does not exceed 10% for VIIIa, VIIIb and 5% for IXa, IXb. Destruction of the triazine ring in the molecular ion M$^+$ of the compounds occurs with loss of HNCCl and formation of $[C_4H_7O_2NHCNCH_2Cl]^+$ moiety (m/z = 176/178) for VIIIa and IXa and $[ArCH_2C_4H_7O_2NCNCCl]^+$ (m/z = 266/268) for (IXa).

4.3 EXPERIMENTAL PART

^{1}H NMR and ^{13}C NMR spectra of the solutions in CDCl$_3$ were recorded with Bruker AM-300 spectrometer (500 and 75.47 MHz, respectively) relative to internal Me$_4$Si standard. GC-MS traces were recorded with Focus instrument equipped with Finnigan DSQ II MS detector (ion source temperature of 200°C, direct injection temperature of 50–270°C, heating rate of 10 deg min^{-1}, column: Thermo TR-5 MS 50×2.5×10^{-4} m, helium flow rate 0.7 mL min^{-1}). GLC analysis was performed with Kristal-2000 M chromatograph equipped with thermal conductivity detector (helium as carrier gas at flow rate of 1.5 L h^{-1}, column: h = 2 m, filled with 5% SE-30 on Chromaton N-AW).

Dihalocarbenylation of chloroalkenes Ia–Ic. 320 g of NaOH aqueous solution (50 wt %) was added drop-wise to a vigorously stirred mixture of 0.1 mol of the corresponding chloroalkene Ia–Ic and 0.2 g of phase transfer catalyst catamine AB in 300 mL of chloroform during 2 h under cooling to 10°C (or under heating up to 40°C in the case of Ic). Then the mixture was stirred during another 1–6 h at 20°C (or 40°C in the case of Ic) and was washed with water. The extract was dried over calcined MgSO$_4$. After the solvent evaporation, the residue was distilled in vacuum.

1,1-DICHLORO-2-CHLOROETHYLCYCLOPROPANE (IA).

Yield 90%, colorless liquid, bp 56°C (3 mm Hg). ^{1}H NMR spectrum, δ, ppm (*J*, Hz): 1.32 d (3H, C^{2}H$_3$), 1.47 d.d (1H, C^{2}H$_a$, ^2J 7.6, ^3J 6.7), 1.58 d.d (1H, C^{2}H$_b$, ^2J 7.6, ^3J 15.2), 1.76 d.d.d (1H, C^{1}H, ^3J 6.7, ^3J 7.6, ^3J 15.2), 3.68 d.d (1H, C $^{1'}$H, ^3J 7.6). ^{13}C NMR spectrum, δ$_C$, ppm: 25.26 (C^{2}H$_3$), 27.74 (C^{2}H$_2$), 37.99 (C^{1}H), 58.18 (C$^{1'}$H), 60.49 (C^{3}Cl$_2$). Mass spectrum, m/z, (I_{rel}, %): 170/172/ 174 (<1) [M]$^{+}$, 135/137/139 (10/9/3) $[M - $^.Cl]^+$, 156/158/160 (55/53/15), 143/145/147 (33/28/10), 134/ 136/138 (1/4/2), 111 (14), 109 (23), 101 (6), 78 (32), 76 (100), 65 (29), 63 (16).

2,2-DIMETHYL-3-CHLOROMETHYL-GEM-DICHLOROCYCLO-PROPANE (IB).

Yield 95%, colorless liquid, bp 58°C (3 mm Hg). ^{1}H NMR spectrum, δ, ppm (*J, Hz*): 1.29s (3H, C^{1}H$_3$), 1.42s (3H, C^{2}H$_3$), 1.66 d.d (1H, C^{3}H, ^3J 7.4, ^3J 8.2), 3.55 d.d (1H,

C^4H_a, ^2J 11.8, ^3J 8.2), 3.75d.d(1H, C^4H_b, ^2J 11.8, ^3J 7.4). ^{13}C NMR spectrum, δ_C, ppm: 16.89 ($C^{1'}H_3$), 24.65 ($C^{2'}H_3$), 30.63 (C^2), 39.31 (C^3H), 41.13 (C^4H_2), 70.28 (C^1Cl_2). Mass spectrum, m/z, (I_{rel}, %): 186 (1) $[M]^{+\cdot}$, 153/155 (20/14/2) $[M - {}^\cdot Cl]^+$, 137 (100), 109 (22), 101 (11), 79 (30), 65 (29), 53 (10), 39 (32), 76 (100), 65 (29).

2,3-BIS(CHLOROMETHYL)-GEM-DICHLOROCYCLOPROPANE (IC).

Yield 92%, colorless liquid, bp 100°C (3 mm Hg). ^1H NMR spectrum, δ, ppm (J, Hz): 2.25–2.28 m (2H, C^2H, C^3H), 3.70–3.72 m (4H, C^1H_2, C^4H_2). Mass spectrum, m/z, (I_{rel}, %): 205 (1) $[M]^+$, 170/172/174 (10/9/3) $[M - {}^\cdot Cl]^+$, 156/158/160 (55/53/15), 143/145/147 (33/28/10), 134/136/138 (37/23/4), 120/122/124 (34/22/5), 109/111/113 (100/63/10), 99/101 (34/13), 85/87/89 (32/21/15), 73/75 (24/22), 65 (18), 51 (21).

N-Alkylation of butyl- and diethylamine with *gem*-dichlorocyclopropanes Ia, Ib. A mixture of 15 mmol of amine and 5 mmol of *gem*-dichlorocyclopropane in 5 mL of DMSO was stirred during 4–20 h at 75–80°C. After cooling, the reaction mixture was washed with 20% aqueous NaOH, and the product was extracted with diethyl ether. The extract was washed with water and dried over K_2CO_3. After the solvent evaporation, the residue was distilled under reduced pressure and under nitrogen stream.

BUTYL[1-(2,2-DICHLOROCYCLOPROPYL)ETHYL]AMINE (IIIA).

Colorless liquid, bp 109°C (5 mm Hg). ^1H NMR spectrum, δ, ppm (J, Hz): 0.92 t (3H, $C^{4'}$ H_3, ^3J 7.2), 1.28–1.32 m (3H, $C^{2'}H_3$), 1.29–1.41 м (4H, $C^{3''}H_2$, C^3H_2), 1.43–1.55 m (3H, C^2 H_2, C^2H), 1.58s (1H, NH), 2.61– 2.68 m (2H, $C^{1''}H_2$), 2.87 d.d (1H, $C^\Gamma H$, ^3J 5.3, ^3J 12.9). Mass spectrum, m/z, (I_{rel}, %): 210/212 $[M]^{+\cdot}$ (<0.1), 166/168/170 $[M - {}^\cdot C_3H_7]^+$ (21/17/2), 137 (7), 101 (11), 86 (12), 69 (20), 65 (40), 44 (51), 41 (100), 40(10).

[1-(2,2-DICHLOROCYCLOPROPYL)ETHYL]DIETHYLAMINE (IIIB).

Colorless liquid, bp 97°C (5 mm Hg). ^1H NMR spectrum, δ, ppm (J, Hz): 1.04 t (6H, $C^{2'}$ H_3, $C^{2'}H_3$, ^3J 7.2), 1.21–1.25 m (2H, C^3H_2), 1.28–1.32 m (3H, $C^{2'}$ H_3), 2.52–2.65 m (5H, $C^{1'}$ H_2, C^1 H_2, C^1 H), 2.82 d.d (1H, $C^{1'}$ H, ^3J 5.1, ^3J 13.8). Mass spectrum, m/z, (I_{rel}, %): 207/209/211 $[M]^+$ (<1), 194/196/198 $[M - {}^\cdot CH_3]^+$ (10/8/2), 137 (3), 101 (13), 86 (100), 75 (7), 65 (38), 58 (42), 42 (52), 41(25).

BUTYL[(2,2-DICHLORO-3,3-DIMETHYLCYCLOPROPYL)-METHYL] AMINE (IVA).

Colorless liquid, bp 113°C (5 mm Hg). ^1H NMR spectrum, δ, ppm (J, Hz): 0.95 t (3H, $C^{4''}$ H_3, ^3J 7.2), 1.18s (3H, $C^{2''}H_3$), 1.31–1.53 m (9H, C^2 H_2, C^1 H_3, NH, C^3 H_2, C^1

H), 2.58–2.63 m (2H, $C^{1'}H_2$), 2.72 d (1H, $C^{1'}H$, ^3J 6.8). ^{13}C NMR spectrum, δ_C, ppm: 13.97 ($C^{4''}H_3$), 17.24 ($C^{2''}H_3$), 20.41 ($C^{3'}H_2$), 24.82 ($C^{1'}H_3$), 28.62 (C^3), 32.19 ($C^{2''}H_2$), 38.33 (C^1H), 46.17 ($C^{1'}H_2$), 49.35 ($C^{1''}H_2$), 70.99 (C^2Cl_2). Mass spectrum, m/z, (I_{rel}, %): 223/225/227 $[M]^{+\cdot}$ (<1), 190 (27), 180 (100), 166 (2), 151 (33), 114 (81), 86 (83), 79 (66), 57 (18), 44 (41), 41 (25), 30 (13).

[(2,2-DICHLORO-3,3-DIMETHYLCYCLOPROPYL)METHYL]-DIETHYLAMINE (IVB).

Colorless liquid, bp 103°C (5 mm Hg). ^1H NMR spectrum, δ, ppm (J, Hz): 1.02 t (6H, $C^{2'}H_3$, $C^{2'}H_3$, ^3J 7.1), 1.15s (3H, C^2H_3), 1.37s (3H, C^1H_3), 1.45 m (1H, C^1H), 2.51–2.54 m (6H, $C^{1'}H_2$, $C^{2'}H_2$, $C^{1''}H_2$). Mass spectrum, m/z, (I_{rel}, %): 223/225/227 $[M]^{+\cdot}$ (<0.1), 208 (8), 188 (10), 115 (9), 86 (100), 79 (21), 58 (32), 41 (8).

N-Alkylation of butyl- and benzylamine with *gem*-dichlorocyclopropane Ic. A mixture of 5 mmol of amine, 2.5 mmol of *gem*-dichlorocyclopropane Ic, 0.75 mmol of solid KOH, and 0.02 g of phase transfer catalyst (triethylbenzylammonium chloride) in 6.2 mL of DMSO was stirred during 4–20h at 75–80°C. After cooling down, the reaction mixture was washed with 20% aqueous NaOH, and the product was extracted with diethyl ether. The extract was washed with water and dried over K_2CO_3. After the solvent evaporation, the residue was distilled under reduced pressure and under nitrogen stream.

3-BUTYL-6,6-DICHLORO-3-AZABICYCLO[3.1.0]HEXANE (VA).

Colorless liquid, bp 108°C (5 mm Hg). ^1H NMR spectrum, δ, ppm (J, Hz): 0.95 t (3H, $C^{4''}H_3$, ^3J7.1), 1.25–1.41 m (4H, $C^{3'}H_2$, $C^{2'}H_2$), 2.29s (2H, C^2H, C^3H), 2.40–2.43 m (2H, $C^{2'}H_a$, $C^{1'}H_a$), 2.78–2.82 m (2H, $C^{1''}H_2$), 3.12–3.15 m (2H, $C^{2'}H_b$, $C^{1'}H_b$). Mass spectrum, m/z, (I_{rel}, %): 207/209/211 $[M]^{+\cdot}$ (<1), 192/194/196 $[M - {}^\bullet CH_3]^+$ (6/4/1), 114 (4), 86 (14), 79 (19), 57 (16), 44 (100), 41(81), 40(6).

3-BENZYL-6,6-DICHLORO-3-AZABICYCLO[3.1.0]HEXANE (VB).

Colorless liquid, bp 157°C (5 mm Hg). Physicochemical constants, NMR, and mass spectra parameters coincided with the literature data [4].

Competitive interaction of *n*-butylamine with monochloroalkyl-*gem*-dichlorocyclopropanes Ia and Ib. A mixture of 15 mmol of *n*-butylamine IIa and *gem*-dichlorocyclopropanes Ia and Ib (2.5 mmol each) in 5 mL of DMSO was stirred at 75–80°C. The reaction mixture was sampled every 30 min and analyzed by GLC. Relative reactivity of the amine was estimated from the rate of the final products accumulation; conversion of the starting compounds being below 25–35%.

N-BENZYL-(1,3-DIOXOLAN-4-YL)METHYLAMINE (VIIA).

A 50-ml reactor equipped with a mechanical stirrer, reflux condenser, and thermometer was charged with 0.08 mol (8.56 g) of benzylamine, 0.12 mol (6.72 g) of solid KOH, 0.05 mol (6.15 g) of 4-chloromethyl-1,3-dioxolane, 0.24 g of TEBAC, and 20 mL of DMSO. The reaction mixture was agitated and irradiated in a Sanyo EM-S1073W microwave oven (power 250 W) for 0.5 h. The temperature was 20–25°C at the beginning of a run and increased in the process to 35–40°C (outside the microwave irradiation zone).After the reaction was complete, the mixture was cooled to 20–25°C, washed with a twofold amount of water, and twice extracted with diethyl ether. The ether extract was shaken with a 10% solution of NaOH, washed with water to neutral reaction, dried with K_2CO_3, and ether was evaporated in a rotor evaporator. The residue was subjected to vacuum distillation in the atmosphere of nitrogen. Mass spectrum (electron impact, 70 eV), m/z (J_{rel}, %): 193 [M$^+$] (0), 192 (0.5), 162 (2), 148 (7), 120 (100), 91 (100), 77 (3), 65 (10). ^{1}H NMR spectrum (CDCl$_3$, δ, ppm, J, Hz): 2.75s (2N, C$^{1'}$N$_2$), 3.63 t (1N, C^{5}N$_b$, ^2J9.0, ^3J9.0), 3.84s (1N, C$^{1''}$N$_a$), 3.95 t (1N, C^{5}N$_b$, ^2J9.0, ^3J9.0), 4.20 kV (1H, C^{5}H, ^3J9.0), 4.86s (1H, C^{2}H$_a$), 5.01s (1H, C^{2}H$_b$), 7.20–7.40 m (5H, Ph–).

N-BENZYL-1,1-(2,2-DICHLOROCYCLOPROPYL)METHYL-AMINE (VIIB).

To a mixture of 0.1 mol (10.7 g) of benzylamine, 0.025 mol of 2-chloromethyl-1,1-dichlorocyclopropane was poured 30 mL of DMSO. The reaction mixture was agitated for 4 h at 70–75°C. On being cooled, it was washed with a 20% NaOH solution and extracted with ether; the extract was washed with water to neutral reaction and dried with K_2CO_3. After the solvent, ether, was evaporated, the residue was distilled in a vacuum in the atmosphere of nitrogen. Mass spectrum (electron impact, 70 eV), m/z (J_{rel}, %): 228/230/232 [M$^+$] (12/8/3), 194/196 (5/1), 165 (2), 151 (6), 147 (8), 146 (68), 133 (14), 132 (37), 120 (42), 118 (7), 104/106/108 (18/14/3), 91 (43), 90/92 (100/25), 89 (11), 77 (6), 65 (30). ^{1}H NMR spectrum (CDCl$_3$, δ, ppm, J, Hz): 1.17 t (1H, C^{2}H$_b$, ^2J7.4, ^3J7.4), 1.65 d.d (1H, C^{2}H$_a$, ^3J7.4, ^2J10.4), 1.85 d.d.d.d (1H, C^{3}H, ^3J7.4, 10.4, 8.3, 5.6), 2.75 d.d (1H, C$^{1'}$H$_a$, ^2J12.8, ^3J8.3), 2.9 d.d (1H, C$^{1'}$H$_b$, ^2J12.8, ^3J5.6), 3.85s (2H, C$^{1''}$H$_2$), 7.20– 7.40 m (5H, Ph–).

Interaction of secondary amines with chloro-sim-triazine (1:1). A mixture of 0.007 mol (1.29 g) of cyanuric chloride (VIa) and 0.007 mol of amine VIIa, VIIb [1.35 g of N-benzyl-(1,3-dioxolan-4-yl)methylamine or 1.61 g of N-benzyl-1–1-(2,2-dichlorocyclopropyl) methylamine] in 20 mL of dioxane was boiled under agitation for 3 h, then cooled to 20°C, a solution of 0.39 g (0.007 g-mol) of KOH in 5 mL of water was added, and the mixture was boiled under agitation for 1h. The mixture was cooled to 10–15°C, 100 mL of icy water was added, and the resulting mixture was washed with benzene (3 Ч 30 mL). The extract was dried over Na$_2$SO$_4$, concentrated to a volume of 5 mL, and chromatographed on a column with Al_2O_3 (H = 20 cm, d = 4.5 cm), with elution by a 20:1 mixture of benzene and methanol. This procedure was used to obtain compounds VIIIa and VIIIb.

N-BENZYL-4,6-DICHLORO-N-(1,3-DIOXOLAN-4-YLMETHYL)-1,3,5-TRIAZINE-2-AMINE (VIIIA).

Yield 73%, isolated by column chromatography (R_f = 0.65) (20:1 benzene–methanol as eluent). Mass spectrum (electron impact, 70 eV), m/z (J_{rel}, %): 340/342/343 [M^+] (0), 254/256/258 (24/14/3), 252/254/256 (10/7/4), 219/221 (3/1), 211 (5), 207 (4), 182 (7), 183 (3), 176/178/180 (6/3/0.5), 157 (11), 150 (11), 148 (7), 132 (4), 131 (42), 125 (13), 116 (8), 104 (57), 91 (100), 79 (35), 65 (50), 62/64 (42/20.5). ^{1}H NMR spectrum (CDCl$_3$, δ, ppm, J, Hz): 3.50 d.d (1H, C$^{1'}$H$_b$, ^2J 14.4, ^3J 8.1), 3.56 d.d (1H, C^{5}H$_b$, ^2J 8.4, ^3J 5.7), 3.80 d.d (1H, C$^{1'}$H$_a$, ^2J 14.4, ^3J 3.1), 3.98 d.d (1H, C^{5}H$_b$, ^2J 8.4, ^3J 6.2), 4.36 d.d.d.d (1H, C^{4}H, ^3J 8.1, 3.1, 5.7, 6.2), 4.85s (1H, C^{2}H$_a$), 5.06s (1H, C^{2}H$_b$), 4.79 d (1H, C$^{1''}$H$_a$, ^2J 15.1), 5.3 d (1H, C$^{1''}$H$_b$, ^2J 15.1), 7.20–7.40 m (5H, Ph–). ^{13}C NMR spectrum (CDCl$_3$, δ, ppm): 48.39 ($^{1'}$C); 51.54 ($^{1''}$C); 67.54 (^{5}C); 73.92 (^{4}C); 95.15 (^{2}C); 128.07 ($^{6'''}$C); 128.02 ($^{3'''}$C, $^{5'''}$C); 128.89 ($^{2'''}$C, $^{4'''}$C); 135.63($^{1'''}$C); 165.49 ($^{2''}$C); 170.32, 170.49 ($^{5''}$C, $^{4''}$C).

N-BENZYL-4,6-DICHLORO-N-[(2,2-DICHLOROCYCLO-PROPYL) METHYL]-1,3,5-TRIAZINE-2-AMINE (VIIIB).

Yield 80%, isolated by column chromatography (R_f = 0.78, 20:1 benzene–methanol as eluent). Mass spectrum, m/z (J_{rel}, %): 376/378/380/382 [M^+] (0), 268/270 (0.3/0.2), 242/244/246 (3.5/1.8/0.3), 207 (1), 192/194 (0.8/0.4), 132 (2), 104 (5), 92 (3), 91 (100), 87 (8.5), 65 (11). ^{1}H NMR spectrum (CDCl$_3$, δ, ppm, J, Hz): 1.2 t (1H, C^{2}H$_b$, ^2J 7.6, ^3J 7.6), 1.7 d.d (1H, C^{2}H$_a$, ^2J 7.6, ^3J 10.7), 1.9 d.d.d.d (1H, C^{3}H, ^3J 7.6, 10.7, 7.7, 5.8), 3.50 d.d (1H, C$^{1'}$H$_a$, ^2J 14.6, ^3J 7.7), 4.00 d.d (1H, C$^{1'}$H$_b$, ^2J 14.6, ^3J 5.8), 4.80 d (1H, C$^{1''}$H$_a$, ^2J 15.3), 5.30 d (1H, C$^{1''}$H$_b$, ^2J 15.3), 7.20–7.40 m (5H, Ph–). ^{13}C NMR spectrum (CDCl$_3$, δ, ppm, J, Hz): 25.71 (^{2}C), 28.52 (^{3}C), 47.10 ($^{1'}$C), 50.55 ($^{1''}$C), 59.80 (^{1}C), 128.31 ($^{4'''}$C), 127.76 ($^{3'''}$C, $^{5'''}$C), 128.97 ($^{2'''}$C, $^{4'''}$C), 135.52 ($^{1'''}$C), 165.42 ($^{2''}$C), 170.59 ($^{4''}$C, $^{6''}$C).

Interaction of secondary amines with chloro-sim-triazines (2:1). A mixture of 0.007 mol (1.29 g) of cyanuric chloride VIa and 0.014 g-mol of amine VIIa, VIIb [3.1 g of N-benzyl-(1,3-dioxolan-4-yl)methylamine or 3.22 g of N-benzyl-1–1-(2,2-dichlorocyclopropyl) methylamine] in 20 mL of 1,4-dioxane was boiled under agitation for 3 h, then it was cooled to 20°C, a solution of 0.78 g (0.014 g-mol) of KOH in 5 mL of water was added, and the mixture was boiled under agitation for 1h. Then the mixture was cooled to 10–15°C, 100 mL of icy water was added, and washed the resulting mixture with benzene (3 × 30 mL). The extract was dried over Na$_2$SO$_4$, concentrated to a volume of 5 mL, and chromatographed on a column with Al_2O_3 ($H = 20$ cm, $d = 4.5$ cm), with elution by a 40:1 benzene–ethanol mixture. This procedure was used to obtain compounds IXa and IXb.

N, N'-DIBENZYL-6-CHLORO-N, N'-BIS(1,3-DIOXOLAN-4-YLMETHYL)-1,3,5-TRIAZINE-2,4-DIAMINE (IXA).

Yield 76%, isolated by column chromatography ($R_f = 0.55$, 40:1 benzene–methanol mixture as eluent. Mass spectrum (electron impact, 70 eV), m/z (J_{rel}, %): 496/498 [M^+] (0), 266/268 (1.4/0.9), 218/220 (0.4/9.2), 207 (4), 190 (0.3), 176/178 (0,5/0.3), 131 (0.9), 104/106 (1.9/1), 91 (100), 92 (7), 87 (1.7), 65 (11), 62/64 (2/1.1). ^{1}H NMR spectrum (CDCl$_3$, δ, ppm, J, Hz): 3.20–4.10 m (8H, C$^{1"}$H$_2$, C$^{1"}$H$_2$, C^{5}H$_2$, C^{5}H$_2$); 4.35 m (2H, C^{4}H, C^{4}H), 4.85s (1H, C^{2}H$_a$, s, 1H, C^{2}H$_a$); 5.02s (1H, C^{2}H$_b$, s, 1H, C^{2}H$_b$); 4.79 d (2H, C$^{1"}$H$_a$, C$^{1"}$H$_a$, ^2J 15.6); 5.15, 5.20 d (1H, C$^{1"}$H$_b$, d, 1H, C$^{1"}$H$_b$, ^2J 13.5, ^2J 15.1); 7.20–7.40 m (10H, Ph–). ^{13}C NMR spectrum (CDCl$_3$, δ, ppm, J, Hz): 48.21, 48.79 (^{1}C, $^{1"}$C); 50.86, 51.32 ($^{1"}$C, $^{1"}$C); 67.32, 67.63 (^{5}C, ^{5}C); 74.91, 75.03 (^{4}C, ^{4}C); 94.85, 95.00 (^{2}C, ^{2}C); 128.09 ($^{4"}$C, $^{4"}$C); 127.59, 127.48 (^{3}C, ^{3}C, ^{5}C, ^{5}C); 127.70, 128.61 (^{2}C, ^{2}C, ^{6}C, ^{6}C), 137.44, 137.59 ($^{1"}$C, $^{1"}$C); 165.39, 165.61 ($2^"$C, $4^"$C); 167.71 ($6^"$C).

N, N'-DIBENZYL-6-CHLORO-N.N'-BIS[(2,2-DICHLORO-CYCLOPROPYL)METHYL]-1,3,5-TRIAZINE-2,4-DIAMINE (IXB).

Yield 80%, isolated by column chromatography ($R_f = 0.62$, benzene as eluent. Mass spectrum, m/z (J_{rel}, %): 569/571/573/575/577/579 [M^+] (0), 280/282/284 (2.6/1.7/0.3), 278/280/282 (3.7/2.7/1.7), 252/254/256 (2.2/1.6/0.3), 207 (0.5), 188/190/192 (3.7/2.8/0.7), 131 (3.7), 128/130 (4/2.1), 125/127 (2.5/1.7), 104 (6), 92 (10), 91 (100), 87 (7.3), 77 (3.4), 65 (12.5). ^{1}H NMR spectrum (CDCl$_3$, δ, ppm, J, Hz): 1.25 m (1H, C^{2}H$_a$, 1H, C^{2}H$_a$), 1.7 m (1H, C^{2}H$_b$, m, 1H, C^{2}H$_b$), 2.1 m (1H, C^{3}H, m, 1H, C^{3}H), 4 m (1H, C$^{1"}$H$_a$, 1H, C$^{1"}$H$_a$), 4.00 m (1H, C$^{1"}$H$_b$, 1H, C$^{1"}$H$_b$), 4.80d (1H, C$^{1"}$H$_a$, ^2J 15.4, d, 1H, C$^{1"}$H$_a$, ^2J 15.4), 5.30 d (1H, C$^{1"}$H$_b$, ^2J 15.4, d, 1H, C$^{1"}$H$_b$, ^2J 15.4), 7.20–7.40 m (5H, Ph"–, 5H, Ph"–). ^{13}C NMR spectrum (CDCl$_3$, δ, ppm, J, Hz): 25.31, 25.03 (^{2}C, ^{2}C); 29.07, 28.61 (^{3}C, ^{2}C); 46.20, 47.14 (^{1}C, $^{1"}$C); 49.72, 50.49 ($^{1"}$C, $^{1"}$C); 60.05, 60.13 (^{1}C, $^{1"}$C); 127.01, 127.42 ($^{4"}$C, $^{4"}$C); 127.49, 127.58 ($3^"$C, $3^"$C, $5^"$C, $5^"$C); 127.96, 128.33 ($2^"$C, $2^"$C, $6^"$C, $6^"$C), 136.97 ($1^"$C, $1^"$C); 165.09 ($2^"$C, $4^"$C); 170.16 ($6^{""}$C).

4.4   CONCLUSION

The similar yields to quantitative of secondary and tertiary amines, including *gem*-dichlorocyclopropane and cycloacetal fragments, are achieved at the simulation by microwave irradiation the reactions of appropriate alkyl chlorines with the primary and secondary amines.

KEYWORDS

- cyanuric chloride
- *gem*-dichlorocyclopropane and cycloacetal fragments
- microwave irradiation
- primary, secondary and tertiary amines

REFERENCES

1. Yukel'son, I.I., Technology of Basic Organic Synthesis, Moscow: "Khimiya" ("Chemistry," in Rus.) Publishing House, 1968, p. 712.
2. Bogomazova, A.A., Mikhailova, N.N., Zlotskii, S.S., Advances in Chemistry of *gem*-Dichloro-cyclopropanes, Saar-brücken: LAP LAMBERT Academic Publishing GmbH&Co. KG, 2011.
3. Zefirov, N.S., Kazimirchik, I.V., Lukin, K.L., Cycloaddition of Dichlorocarbene to Olefins, Moscow: "Nauka" ("Science," in Rus.) Publishing House, 1985.
4. Boswell, R.F., Bass, R.G., Journal of Organic Chemistry (in Rus.), 1975, V. 40(16), P2419.
5. Danilov, A.M., Additives and additions: Improvement of Ecological Characteristics of Petroleum Fuels, Moscow: "Khimiya" ("Chemistry," in Rus.) Publishing House, 1996.
6. Rat'kova, M.Yu., Danilov, A.M., Journal of Applied Chemistry (in Rus.), 1992, V. 65(11), p.2630.
7. Mur, V.I., Uspehi Khimii ("Successes of Chemistry," in Rus.), 1964, V. 33, # 2, p.182.

CHAPTER 5

PROGRESS IN PHOTOVOLTAIC TEXTILES: A COMPREHENSIVE REVIEW

M. KANAFCHIAN

University of Guilan, Rasht, Iran

CONTENTS

ABSTRACT

Today, energy is an important requirement for both industrial and daily life, as well as political, economical, and military issues between countries. While the energy demand is constantly increasing every day, existing energy resources are limited and slowly coming to an end. Due to all of these conditions, researchers are directed to develop new energy sources, which are abundant, inexpensive, and environmentally friendly. The solar cells, which directly convert sunlight into electrical energy, can meet these needs of mankind. This study reviews the efforts in incorporating of solar cells into textile materials.

5.1 INTRODUCTION

5.1.1 THE SIGNIFICANCE OF SOLAR ENERGY

There is increasing concern worldwide about the huge dependence on oil as a source of energy. Although coal resources are immense, their extraction and use tend to potentially damaging environmental problems. There are concerns too about nuclear fission as a source of energy, and controlled fusion has yet to be proven feasible. Consequently, there is growing interest in harnessing energy by other means that do not rely on the consumption of reserves, as with oil and coal, nor involve dangers (real and perceived) with utilizing nuclear fission. In recent years alternative renewable energies including that obtained by solar cells have attracted much attention due to exhaustion of other conventional energy resources especially fossil-based fuels. There are numerous examples of attempts to utilize solar energy for storage of chemical energy (mimicking natural photosynthesis) and for the storage of electrical energy in batteries. Nevertheless, greatest success has been achieved in the direct conversion of solar power into electrical energy, with the aid of photovoltaic devices (PV). Apart from using an endless source of energy, the application of photovoltaic devices offers a number of other advantages: the technology is clean and noiseless, most applicable and promising alternative energy using limitless sun light as raw material, maintenance costs are very small, and the technology is attractive for remote areas, which are difficult to supply by conventional means from a grid [1]. The only drawback is the initial price of commercially available solar panels. However, the decrease to 20% of the price in 1980 [2] already makes the solar energy a suitable solution for the remote areas with no electric network. This currently refers to almost 1.6 billion people, mainly in the developing countries. The total electric power-generation capacity (based on all available sources, including fossil fuels) installed worldwide was about 3500 GW in 2003 and it is expected to increase to almost 7000 GW in next 30 years [3]. Half of the new capacity will be installed in the developing countries, so the need for cheap and preferably renewable energy sources is obvious. Moreover, these countries typically have very good insolation conditions, which favors the solar energetics. According to the optimistic PV market scenario, almost 656 GWp of total solar power can be expected by 2030 if very large scale PV systems in the deserts are used [4].

5.1.2 HOW SOLAR CELL WORKS

Sunlight may be converted into electricity by heating water until it vaporizes, and then passing the steam through a conventional turbine set. This triple conversion of energy has losses at each step. PV cells are more efficient device to convert optical energy directly to electrical energy. To see how this works, we need a little semiconductor science. A typical PV cell is made up of special materials called semiconductors. These are typically made of silicon, which is doped to create a more powerful current. Electrons within the doped silicon have an uneven number making them unstable and looking to move around. When sunlight hits the cell, the semiconductor absorbs some of the energy of the rays. The energy then knocks electrons loose, allowing them to flow freely. PV cells also all have one or more electric field that acts to force electrons freed by light absorption to flow in a certain direction. This flow of electrons is a current, and by placing metal contacts on the top and bottom of the PV cell, we can draw that current off for external use to power a calculator. This current, together with the cell's voltage (which is a result of its built-in electric field or fields), defines the power (or wattage) that the solar cell can produce [5].

It is important to note that thermodynamics limits the efficiency for conversion of solar radiation by a simple photovoltaic cell to 32%, and today's crystalline cells can approach this; multiple junction cells have a thermodynamic limiting efficiency of 66% and triple junction cells have achieved 32%.. As mentioned above, in a semiconductor, valence electrons are freed into the solid, to travel to the contacts, when light below a wavelength specific to that solid is absorbed. This threshold wavelength must be selected to provide an optimum match to the spectrum of the light source; too short a wavelength will allow much of the radiation to pass through unabsorbed, too long a wavelength will extract only part of the energy of the shorter-wavelength photons. The typical solar spectrum requires a threshold absorption wavelength of 800 nm, in the near infra-red, to provide the maximum electrical conversion, but this still loses some 40% of the solar radiation. Further losses occur through incomplete absorption or reflection of even this higher energy radiation. The current that is delivered by a solar cell depends directly on the number of photons that are absorbed: the closer the match to the solar spectrum and the larger the cell area the better. As well as a current, electrical energy has a potential associated with it and this is determined by the internal construction of the cell, although it is ultimately limited by the threshold for electron release. In conventional silicon solar cells, there is an electrical field built into the cell by the addition of minute amounts of intentional impurities to the two cell halves; this produces what is known as a PN junction (Fig. 5.1). The PN junction is a built-in asymmetry in the solar cell, where an electric field ensures that the electron will travel in one direction, while the hole travels in the opposite direction. The junction separates the photon-generated electrons from the effective positive charges that pull them back and in so doing, generates a potential of 0.5 volts, approximately. This potential depends on the amounts of dopants, and on the actual semiconductor itself. Multiple junction cells have integrated layers stacked together during manufacture, producing greater voltages than single junctions. The power developed by a solar cell depends on the product of current and voltage. To operate any load will require a certain voltage and current, and this is tailored by adding cells in series and/or parallel, just as with conventional dry-cell batteries. Some power is lost before it reaches the intended load, by too high a resistance in the cells, their contacts and leads. Note that at least one of the electrical contacts must be semitransparent to allow light through, without interfering with the electrical current. Thus optimizing the design of a solar cell array involves optics, materials science, and electronics [6].

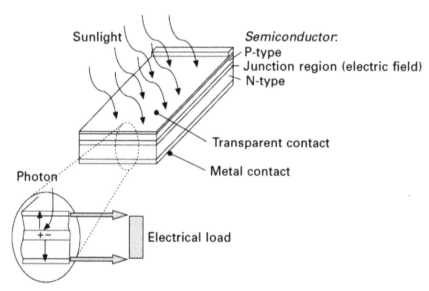

FIGURE 5.1   Schematic of a PV cell [6].

5.2   SOLAR CELL SUBSTANCES

5.2.1   SILICON

Silicon is the most known and the 2nd most widespread element in the Earth crust. Although it is one of the most studied and used elements, there exist many open questions and there even may be some interesting properties or new forms of silicon we do not know about yet. Thanks to the electronic industry and its rapid development in the second half of the twentieth century, we can make our lives easier by enjoying a rich variety of electronic devices based on silicon components. One of the critical properties that make silicon suitable as a solar cell material is that it is a semiconductor, possessing a band-gap. This band-gap is a range of energies that the electrons in the materials are not allowed to have. The electron can either have an energy placing it in its ground energy state in the valence band, or it can be in an excited state in the conduction band. The electron can transition from valence band to conduction band and back through excitation and recombination processes. The energy required for an excitation may come from a photon, being the smallest package of energy one can divide light into. The sunlight consists of photons with a wide range of energies. The energy of the photon corresponds to what we observe as the color of the light, where the blue light consists of photons with a higher energy, and the red light consists of photons with lower energy. The energy of the photon also corresponds to a wavelength of the light, where the blue light has a shorter wavelength, and the red light has a longer wavelength. When a photon hits the silicon, it may be absorbed by an electron in the silicon, providing enough energy for the electron to be excited from its ground energy state in the valence band to an excited state in the conduction band. Such an absorption process may only take place if the photon carries an energy corresponding to at least the band gap energy. The electron being excited will leave behind a hole in the valence band; an electron-hole pair is created. In a solar cell, the electron-hole pair moves by diffusion until it reaches the PN junction. As such, the electron may reach one of the electrical contacts, while the hole reaches the other contact, as a result of a combination of random diffusion and directional drift in an electric field. This is the principal mechanism for current generation in a solar cell. Only photons with high enough energy may be absorbed by the electrons. A photon with energy lower than the band-gap energy will not carry sufficient energy to lift the electron to the conduction band, and will as such not be absorbed in the semiconductor. Hence, its energy will not be converted into electricity. On the other hand, photons with high energy can create an electron-hole pair, lifting the electron high above the conduction band edge. However, all the excess energy that is put into the electron will be rapidly lost, as the electron will collide with other electrons or atoms, losing energy until it reaches the conduction band edge. This loss process is called thermalization. In a real solar cell, not all generated electron-hole pairs will contribute to current generation. There is always the chance that an electron finds a hole on its way to the contacts and relaxes back across the band gap, in a process called recombination. Recombination may happen slowly in the bulk of a high-quality silicon wafer, but it will always take place even in a perfect material. These unavoidable recombination mechanisms are termed intrinsic recombination mechanisms. In a more realistic material, recombination happens faster. Examples of recombination-active areas are crystal defects or impurities in the silicon, highly doped silicon, silicon crystal boundaries or wafer surfaces and metal-silicon interfaces, such as contacts. By combining intrinsic recombination mechanisms with spectrum losses, we reach a maximum efficiency of a solar cell, known as the Shockley-Queisser limit [7], which for silicon under a special irradiance is around 29% [8]. Currently, the record efficiency of a silicon solar cell is 25% [9], which is actually quite close to the theoretical maximum of 29% given by the Shockley-Queisser limit. Hence, for a further successful progress of the solar energetics we still need a lot of innovations leading, above all, to the competitive price of the solar cells and solar panels. Comparison of energy payback times shows that the thin film solar cells have potential for achieving low costs. The significant reduction of the raw material needs and faster production time make them an attractive alternative to the crystalline Si solar cells, although the efficiencies are still much lower. The theoretical limit for the efficiency of the single junction crystalline silicon solar cell given by the energy of its band gap is around 27% [10]. The disordered structure of the material in thin film solar cells leads to much lower efficiencies, however a significant improvements can be achieved by the application of a multi layered structure of stacked cells. So, the efficiency increase still remains to be an important task for the solar cell research.

5.2.2 CONDUCTIVE POLYMERS

Among the many literature of conducting materials, a tremendous amount of publications can be found in the field of conducting polymers [11]. In 1862, at the College of London Hospital, Letheby attained a partly conductive material by oxidizing aniline in a sulfuric acid solution. Better conductivity results were gained after the successful synthesis of polyacetylene [12]. Polyacetylene was first synthesized in 1958 as a highly crystalline high molecular weight conjugated polymer by Natta et al. after the polymerization of acetylene in hexane using an initiator. However, this received little attention due to the air sensitive, infusible and insoluble properties aside from the preparation method [13]. In the early 1970s, even though the inorganic explosive polymer polysulfurnitride was found to be very conductive at low temperatures, interest shifted to organic conductors. In 1974, by using a Ziegler-Natta catalyst, a silvery film of polyacetylene was prepared. In spite of its similar appearance to metals, this film is not conductive. In 1977, MacDiarmid, Shirakawa and Heeger modified the polyacetylene film through a partial oxidation treatment with oxidizing agents, such as halogens or AsF_5, making the film significantly conductive [14]. Since this breakthrough, many other conducting polymers have been synthesized and characterized, such as polyaniline (PANI), polythiophenes (PTh) and polypyrrole (PPy). In the beginning of the 1980s, as the interest in conductive materials significantly increased, the problems with early conducting polymers became clear. The lack of processability in comparison to plastic materials was a big disadvantage. In addition, insolubility, an infusible and brittle nature, and the lack of air stability of the conductive polymers prevented the integration of these conductive materials into new application areas. In the late 1980s, some of the problems, such as solubility and air durability, were overcome with modifications of the polymers [15]. In 2000, Heeger, MacDiarmid and Shirakawa were awarded a Nobel prize in chemistry in recognition of the significance of their research in the development of conducting polymers.

5.2.2.1 MOLECULAR STRUCTURE

Organic chemistry states that double bonds can be isolated, conjugated or cumulated [11]. Conjugated bonds require strict alternation of double and single bonds which is also seen in the structure of conducting polymers [15]. In the ground state, a carbon atom has an electron configuration of $1s^2 2s^2 2p^2$. It is expected for C to form only 2 bonds with the neighboring atoms since the 2s shell is completely occupied. However, it forms 4 bonds due to hybridization. In conjugated polymers, carbon makes sp^2 hybridization in which the 2s orbital pairs with two 2p orbitals to form sp^2 hybrid orbitals. At the end of hybridization, three sp^2 and one p orbitals are formed. Two of the three sp^2 orbitals on each carbon atom are bonded to another carbon atom next to it and the last sp^2 orbital is bonded either to hydrogen or any of the side groups. The bonding between these atoms is covalent bonding and in this case, it is called 'σ bond' which has a cylindrical symmetry around the internuclear axis. The unhybridized p orbitals of neighboring carbon atoms overlap and form a 'p bond' [16]. The p electrons are not strongly bound. As a result of the weak interaction between them, they can be readily delocalized, which in this situation, contribute electrical conductivity to the polymer [17]. According to the simple free electron molecular orbital model in the theories of Hünkel and Bloch, when the molecular chain length is long enough (showing metallic transport properties), p electrons are delocalized over the entire chain, which form a very small band gap. As a result, a conjugated polymer which has an alternation of double and single bonds can be conductive when conditions are right. In this case, the electrons delocalized over the conjugated chain are predicted to be evenly spaced out. Thus, all bonds are assumed to be equivalent [11]. However, under simplified conditions, the s and p bonds have different bonding lengths [15]. Thus, the conductivity of conjugated polymers cannot be explained with conventional theories used in semiconductor physics. In Fig. 5.2, the commonly studied conducting polymer structures are shown.

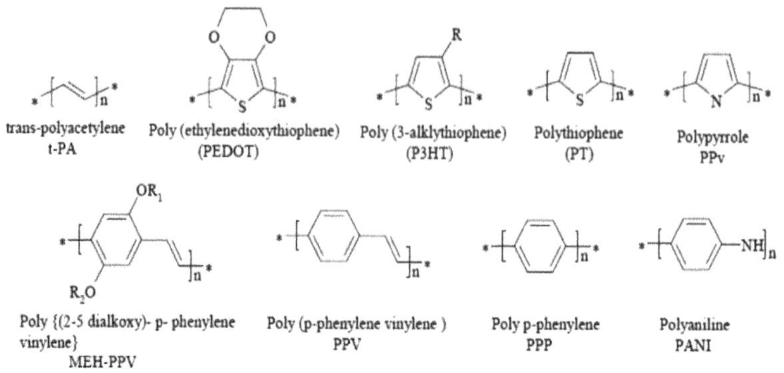

trans-polyacetylene   Poly (ethylenedioxythiophene)   Poly (3-alklythiophene)   Polythiophene   Polypyrrole
t-PA   (PEDOT)   (P3HT)   (PT)   PPv

Poly {(2-5 dialkoxy)- p- phenylene   Poly (p-phenylene vinylene)   Poly p-phenylene   Polyaniline
vinylene}   PPV   PPP   PANI
MEH-PPV

FIGURE 5.2   Molecular structure of some conducting polymers.

5.2.2.2   CONDUCTION MECHANISM

Materials can be generalized into three groups in terms of electrical conductivity: insulators, semiconductors and conductors. Conductive polymers are in the group of semiconductors. Electronic energy is important in determining the electric conductivity of materials. In metals, free electrons and electrons, which have similarly low binding energy, can move easily between atoms. The degenerated atomic orbitals overlap in all directions and form molecular orbitals in metals. In a solid state structure, the number of resulting molecular orbitals formed by neighboring atoms whose energy levels are equivalent is very high, up to 1022 for a 1 cm³ metal piece. This amount of molecular orbitals spaced together in a similar energy range leads to a formation of continuous band energies. In a metal atom, excluding an inert gas form, the valence orbitals are not fully filled, and give birth to molecular orbitals which are not fully filled either. On the other hand, above a certain energy level, which can be zero in the case of metals, all molecular orbitals appear to be empty. The energy states which are forbidden for electron occupation between the highest occupied (valence) and the lowest unoccupied (conduction) orbitals, is called the band gap [15].

Electrical conductivity depends highly on the band gap. If the gap is small, the electrons can be excited even with a small amount of thermal energy. In metals, there is either no gap or the bands are overlapped [18]. Materials with a band gap of 1–5 eV or below are considered semiconductors and generally, they have a conductivity range from 10–9 S/cm to 103 S/cm, which is lower than the metallic conductivity range of 106 S/cm. Most of the semiconductors are solid crystalline inorganic materials, but some conjugated polymers can also show semiconducting properties with appropriate doping. In 1955, Peierls described theoretically that for a one-dimensional metallic molecule with a partly filled band, the regular chain structure will never be stable. Peierls instability provides a better interpretation of the conduction mechanism in polymers so that even a small conformational distortion can change the band gap which directly affects the conductivity of the polymer [19]. In conductive polymers, there are no holes or free electron formation in the conduction bands, unlike crystalline semiconductors. The difference from inorganic semiconductors lies in the structure of the polymer backbone, which is deformed after the polymer is oxidized or reduced by the dopants. For example, after an electron is removed from the valence band of a conducting polymer, a cation is formed. The hole left in the valence band is not completely empty. In other words, it cannot delocalize completely as it is expected to do by the classical band theory. Instead, partial delocalization occurs which also causes structural deformation while spreading along the surrounding monomer units. The energy that was created by the electron is attempted to be balanced and this effort results in a new equilibrium condition [20]. A polaron has two defects: charge and defect. In this concept, conducting polymers can be classified into two main groups: degenerate ground state and nondegenerate ground state. The main difference between these polymers is apparent when they are subsequently charged. When conducting polymers with degenerate ground states are charged subsequently, a second polaron is formed. In this case, the polarons are independent and not bound to each other. As a result, they can be split into two phases of opposite orientation with equal energy. These are called solitons. Solitons could also be neutral, caused by isomerization. Although neutral solitons do not carry any charge, they can move along the chain and hop between localized states of adjacent polymer chains. For conducting polymers with nondegenerate ground states (such as PPy, PTh, polyphenylene, PANI, etc.) there is no soliton formation, but pairs of defects are created. These are called bipolarons. In this case, PPy can be given as an example for bipolaron formation due to its highly disordered structure. Bipolaron formation takes place when two polarons are formed on the same chain [20].

5.2.2.3   CONDUCTING POLYMER APPLICATIONS IN PV TECHNOLOGY

Conductive polymers have been expected to yield several applications due to their electrical properties and wide color variation despite their poor mechanical properties, they have the potential to replace metallic materials in many applications, which is attributable to their lightweight and semiconducting nature [21]. An application area for conducting polymers is definitely solar cell technology. It has been known for a long time that solar radiation has a great potential as an energy source which can be used in various ways [22]. Photovoltaic technology is based on conversion of sunlight into electricity. PV cells generate direct current electric power from semiconductors after they are illuminated by photons. Inorganic materials, silicon and other semiconductors are used in PV cells because by their nature, electrons in their structure can be excited from valence band into the conduction band when they are impinged upon by photons, and that generation leads the creation of electron-hole pairs. As a result, electrons move towards the P-type material and holes move towards the N-type material, leading to an electric current with an external circuit [23]. To lower the manufacturing cost, organic materials which can easily be processed are seen as an alternative way for PV cell production [24]. Organic solar cells differ from inorganic PV cells in their production technique, character of the materials used in the cell structure and the device design [25]. In a simple organic solar cell system, an organic semiconductor, which consists of donor and acceptor layers, is sealed between two metallic electrodes, indium tin oxide (ITO) and Al. MDMO-PPV: poly (2-methoxy-5-(3,'7'-dimethyloctyloxy)-1,4-phenyl-ene-vinylene), RRP3HT: regioregular poly(3-hexylthiophene), PCPDTBT: poly[2,6-(4,4-bis-(2-ethylhexyl)-4Hcyclopenta[2,1-b;3,4-b_]-di-thiophene)-alt-4,7(2,1,3benzothiadiazole) and PCBM: (6,6)-phenyl-C61-butyric acid methyl ester, which are some of the polymers used in the organic PV cell system. The mismatch between the electronic band structures of donor and acceptor is considered as the driving force of the electron transfer. Subsequent to sun illumination, an electron is excited from highest occupied molecular orbital (HOMO) to the lowest unoccupied molecular orbital (LUMO) of the donor. This photoinduced electron is first transferred from the excited state of the donor to the LUMO of the acceptor and then carried out to the Al electrode. Similarly, holes left in the HOMO of the donor are moved through the working electrode, ITO. Thus, with an external circuit, direct current is generated [26].

Among the wide range of conjugated polymers already developed, perhaps the diethoxy substituted thiophene poly(3,4-ethylenedioxythiophene) or PEDOT, also known under the trade name Baytron® or Clevios, is one of the most promising conducting polymer to date [27, 28]. It was synthesized by researchers at Bayer AG in Germany in 1988 with excellent conductivity (300 S/cm) [27, 29]. Since then, it has shown extraordinary scope for smart textiles, electronics, and optoelectronics applications [30]. PEDOT is highly conductive in its oxidized (doped) state, where the molecular backbone contains mobile carriers (holes). The chemical structures of 3,4-ethylenedioxythiophene (EDOT) monomer and PEDOT polymer are shown in Fig. 5.3. The presence of only two reactive hydrogen atoms at the 2 and 5 positions on the EDOT monomer ring gives PEDOT a very regular molecular structure, which makes it a stable polymer [31, 32].

PEDOT has a stiff conjugated aromatic backbone structure, which makes it insoluble and infusible in most organic and inorganic solvents [33–35]. Bayer resolved this problem by introducing polystyrene sulfonic acid (PSS), a water-soluble polyanion, during the polymerization of PEDOT as a charge balancing dopant [36]. This water-soluble PEDOT:PSS complex, as a commercial product known as BAYTRON PTM, has electrical and film-forming properties that somehow allow fairly stable, transparent conductive polymer coatings on a variety of substrates. However, PSS itself is a nonconducting material, which limits the conductivity of the

PEDOT:PSS complex to the ~1–10 S/cm range [37]. In order to use the conductive polymer PEDOT for electronics and optoelectronics applications, it should be applied or deposited as a thin film on the surface of different substrates. However, due to the lower solubility of PEDOT in most organic and inorganic solvents, the simplest coating techniques such as spray coating, spin casting, and solution casting are difficult to use with it. Deposition of PEDOT polymer directly on the surface of desired substrates could therefore be an efficient way of obtaining highly conductive, flexible, and thin uniform films.

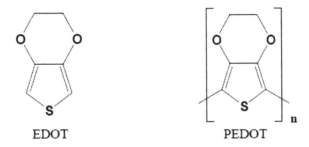

EDOT   PEDOT

FIGURE 5.3   Molecular structure of EDOT monomer and PEDOT polymer.

5.3   SMART TEXTILES

5.3.1   DEFINITION

Today, people want to make their lives more comfortable and talk a lot about intelligent devices. An intelligent device means a system that can evaluate the environment and respond to the results of the evaluation [38, 39]. These systems can give us information we need in any particular situation and help us to plan our everyday lives [40]. In order to make human life healthier, more comfortable, and safer, intelligent devices are being brought one stage closer to the individual by means of easy-to-use wearable interfaces [41–43]. In this regard, the multifunctional fabrics commonly known as smart textiles, electronic textiles (E-textiles), or textronics are good candidates for making our daily lives healthier, safer, and more comfortable [41]. The term smart textiles or textronics refers to interdisciplinary approaches in the process of producing and designing textile materials, which started about the year 2000 [44]. It is an emerging field of research that connects the textile industry with other disciplines such as information technology, microsystems, and materials science with elements of automation and metrology[44, 45].

Smart textiles are defined as "the synergic combination of electronic and conventionally used textile materials. They are able to sense the physical stimuli from the external environment and then perform some reactions against them" [46]. These external stimuli and the reactions of smart textiles can be in the form of electrical, thermal, chemical or magnetic signals. Smart textiles not only protect the human body from extreme environments, but can also monitor, do local computing, and communicate wirelessly [47]. E-textiles because of their multifunctional inter-activity in wearable devices that are adaptable, flexible, and conform to the human body are successful promoters of better quality of life and progress in the field of biomedicine, as well as in several other health-based disciplines such as rehabilitation, telemedicine, ergonomics, tele-assistance, and sports medicine [41, 48–51].

During the last decade, apart from biomedical applications, smart and interactive textiles have been widely used in technical clothing [52], flexible solar cell panels [14, 53], protective garments for electromagnetic shielding and static charge dissipation [54], heating elements [55], fabrics for dust and germ-free clothes [56, 57], pressure sensors [50, 58], chemical sensors [59], power sources [60], and wireless devices [61].

5.3.2 PV CELL IN TEXTILE

Conventional solar cells are mostly encased between glass substrates, giving the structure fragility, heaviness and rigidness that cause problems in their storage and transportations. Therefore, attention has turned to organic solar cells which are more flexible and lighter. The idea of manufacturing flexible solar cells is not novel. Several companies and research groups have been working on flexible solar cells that can generate enough electricity to run daily electronics. Konarka Technologies Inc. has recently introduced "power plastic" which is capable of running portable electronic devices and formed with a primary electrode, an organic polymer blend and a transparent electrode encased between the plastic substrate and the transparent packaging material. Another flexible solar cell design has been developed by Sky Station International Inc. In this research, a flexible sheet material is embedded with a solar cell for stratospheric vehicles. There are some other flexible solar cell devices designed for different applications, such as a solar tensile pavilion designed by Nicholas Goldsmith and a solar awning designed by Zebetakis et al. [62]. However, attaching conventional solar panels onto textiles or forming them within/on plastics does not meet expectations in terms of flexibility and wearability. To contribute flexibility and wearability to the system, textile materials can be directly used as structural elements that replace plastic based materials. Recently, Winther-Jensen et al., Krebs et al. and Biancardo et al. developed a strategy that involves thermal lamination of a thin layer of polyethylene (PE) onto textile followed by a plasma treatment which contributed adhesion to the surface and lastly, PEDOT formation on the top [63]. The main aspect of the textile based solar cells is the level of electrical conductivity on the surface, which enables them to reach an appropriate level of current collection efficiency.

As mentioned above, despite the many attractions of solar cells as vehicles for providing energy, the way in which they are constructed provides problems in application. Typically, solar cells are either encased between glass plates, which are rigid and heavy, or the cells are covered by glass. Glass plates are fragile, and so care has to be taken with their storage and transport. The rigid nature of solar cells requires their attachment to flat surfaces and, since they are used outdoors, they have to be protected from any atmospheric pollution and adverse weather. It is, therefore, not surprising that increasing attention has been turned to the construction of lighter, flexible cells, which can still withstand unfavorable environments, yet nevertheless maintain the durability required. Several examples have recently appeared of solar cells in plastic films. Examples are the 'power plastic' developed by Konarka Technologies Inc. [64] in which the cell is apparently coated or printed onto the plastic surface, and an integrated flexible solar cell developed at Sky Station International Inc. for high-altitude and stratospheric applications [65]. Iowa Thin Technologies Inc. have claimed the development of a roll-to-roll manufacturing method for integrating solar modules on plastic [66].

However, while the successful incorporation of solar cells into plastics represents an important step in the expansion of solar cell technology, still greater expansion of the technology would be achieved through their incorporation into textile fabrics, especially if the cells are fully integrated into the fabrics. Textiles are materials with a huge range of applications and markets, and they can be produced by a number of fabrication processes, all of which offer enormous versatility for tailoring fabric shape and properties. The different types of woven, knitted and nonwoven constructions that can nowadays be achieved seem almost infinite, and indeed technical uses are beginning to be found for crocheted and embroidered constructions [67, 69].

Some examples of the application of solar cells in textiles have now been reported. Architect, Nicholas Goldsmith, has designed a solar tensile pavilion which, while providing shade and shelter, can also capture sunlight, and transform it into electrical power [70]. The skin of the tent consists of amorphous silicon cells, encapsulated and laminated to contoured panels of the woven fabric. Zabetakis et al. have recently described the design of an awning, which provides protection from the sun and at the same time converts the incident sunlight into electricity [71]. A type of sail has been designed in which attached to the sailcloth are strips of flexible solar cells [72]. The incorporation of the solar cells has been designed to allow for expansion of the sailcloth when wind blows on it. Textiles offer further advantages too. Fabrics can be wound and rewound. Thus, Warema Renkhoff GmbH in Germany has developed a retractable woven canopy [73], in which energy is collected from the solar cells integrated into the canopy, and then stored. The stored energy is applied to the operation of the winding drive motor. In addition, textile fabrics can be readily installed into structures with complex geometries, a feature which opens up a number of potential applications.

5.3.3 PV TEXTILES BY PV FIBERS

5.3.3.1 CONDUCTIVE FIBERS

Generally, conductive fibers can be divided into two categories: Naturally and Treated conductive fibers.

I) Naturally conductive fibers
The fibers that can be produced purely from inherently conductive materials, such as metals, metal alloys, carbon sources, and conjugated polymers (ICPs), are associated with the class of naturally conductive fibers. In their pure form, these fibers have high conductivity values. these fibers can be divided into three categories:

a) Metallic fibers
Metallic fibers, as suggested by the name, are the first man-made fibers to be developed from metals or metal alloys. These fibers have very thin metal filaments with diameter ranging from 1 to 80 µm, which can be produced by the bundle-drawing or shaving process [74]. Metallic fibers have very high conductivity (106 S/cm) with a wide range of mechanical properties [75]. Despite the fact that they have extraordinary electro-mechanical properties, metallic fibers have limited textile applications because of their low flexibility, stiffness, high weight, high cost, low compatibility with other materials, and poor weaving properties [74, 76].

b) Carbon fibers
Carbon fibers, which were invented by Edison in 1879, are used as the most demanding materials in high-tech industrial applications such as structural composites in aerospace, transportation, and defense-related products. Carbon fibers are petroleum-based products and they can be produced from petroleum pitch and polyacrylonitrile (PAN) [77]. The heat treatment of PAN, also called graphitization, strongly influences the electrical and mechanical properties of carbon fibers. Carbon fibers have graphite structure, which means that they have conductivity values similar to those of metals, that is, 104–106 S/cm [78]. Carbon fiber composites are usually appropriate for structural applications when high strength, stiffness, lower weight, and extraordinary fatigue characteristics are required [79, 80]. For smart and interactive textile applications, carbon fibers cannot be easily integrated into knitted or weaved structures because of their high stiffness and brittleness. In addition, esthetic considerations and health-related issues are also strong reasons for the use of carbon fibers being limited in the clothing industry.

c) Conjugated polymer fibers
The conjugated polymers or ICPs are organic materials that conduct electricity. Due to their high conductivity, lower weight, and environmental stability, they have a very important place in the field of smart and interactive textiles. Several attempts

have been made to produce conductive fibers. Pomfret et al. [81] and Okuzaki et al. [82] produced polyaniline and PEDOT:PSS fibers, respectively, by using a one-step wet spinning process. However, it is very difficult to use these fibers for interactive applications. The major reasons are poor mechanical strength, brittleness, a lower production rate, and difficult processing. The production of pure PEDOT fibers with high conductivity values, from 150 to 250 S/cm, by a chemical polymerization method has also been reported, but due to their microscale size and brittle nature, useful applications could not be found [83].

II) Treated conductive fibers

The conductive fibers that can be produced by the combination of two or more materials, such as nonconductive and conductive materials, are known as treated conductive fibers or extrinsically conductive fibers. Special treatments involve the mixing, blending, or coating of inherently insulating materials such as polyethylene (PE), polypropylene (PP), polystyrene (PS), or textile fibers with highly conductive materials such as metals, carbon black, or ICPs. The conductive fibers obtained, also known as conductive polymer composites (CPCs), can have a combination of the electrical and mechanical properties of the treated materials. These fibers can be classified further as:

a) Conductive filled fibers

Conductive filled fibers are a class of conductive fibers that can be produced by adding conductive fillers such as metallic powder, carbon black, carbon nanotubes, or conjugated polymers to nonconductive polymers such as PP, PS, or PE [78, 84–86]. Usually, melt spinning and solution spinning techniques are used to produce filled conductive fibers. However, in order to get homogenous distribution of conductive particles or ICPs in polymers, the parent materials are well mixed before the spinning process. However, the conductive fibers produced by the solution spinning process have better electrical properties than those produced by the melt spinning process, but the need for large quantities of solvents, their separation, and possible health hazards has made this process obsolete [87]. The electrical conductivity values of melt-spun conductive fibers strongly depend on two parameters: volume loading of filler and filler shape. The fibrous conductive filler can give higher conductivity in melt-spun fibers than irregular and spherical particulates [88]. The melt spinning process is the most economical and least complex process for the production of filled conductive fibers, but the lower electrical and mechanical properties of conductive fibers limit their use in smart and interactive textile applications.

b) Conductive coated fibers

The conductive fibers that can be produced by coating insulating materials with highly conductive materials, such as metals, metal alloys, carbon black, carbon nanotubes, and ICPs, are known as conductive coated fibers [59, 89–91]. With coating processes, not only can the highest conductivities be achieved but also the mechanical properties of conductive fibers can be enhanced. Depending on the type of conductive materials, different coating techniques can be used to transform polymeric or textile fibers into electro-active fibers. To apply metallic coatings, sputtering, vacuum deposition, electroless plating, carbonizing, and filling or loading fibers are the most extensively used methods [74, 92]. High conductivities similar to those of metals (106 S/cm) can be achieved with these methods, but high cost, stiffness, brittleness, high weight, and lower levels of comfort limit their application in the textile field.

Carbon black, carbon nanotubes, and ICPs can be applied as a thin layer to a fiber surface by the solution casting, inkjet printing, in-situ polymerization, VPP, CVD, or PACVD methods [91, 93–96]. The thin layers of coating offer reasonably high conductivity and also preserve the flexibility and elasticity of substrate fibers [96]. However, the use of carbon black and carbon nanotubes for wearable applications is very limited because of health-hazard issues. On the other hand, due to their high conductivity, exceptional environmental stability, lower weight, and high compliance, ICPs are reasonable candidates for functionalization of different polymeric and textile materials. However, difficulty in processing, poor adhesion to substrate, and durability issues are some disadvantages of ICPs. By using suitable coating techniques and compatible substrates, these problems can be minimized [97].

5.3.3.2 PV FIBERS

A basic technique to manufacture PV textiles is based on the development of photovoltaic fibers using Si-based /organic semiconducting coating or incorporation of dye-sensitized cells (DSC). Availability of PV fiber offer more freedom in the selection of structure for various type of applications [98–102]. The development of photovoltaic fibers offers advantages to manufacture large area active surfaces and higher flexibility to weave or knit etc40. Although, the problem of manufacturing textile structure by using dye-sensitized cells (DSC-PV) fibers into textile structure is still alive and require a optimization with respect to textile manufacturing operations. In a typical research work the working electrode of DSCPV fiber is prepared by coating Ti wire with a porous layer of TiO_2. This working electrode is embedded in an electrolyte with titanium counter electrode. The composite structure is coated with a transparent cladding to ensure protection and structural integrity. The electrons from dye molecules are excited by photo energy and penetrated into the conduction band of TiO_2 and move to the counter electrode through external circuit and regenerate the electrolyte by happening of redox reaction. Ultimately the electrolyte regenerates the dye by means of reduction reaction.

The performance of DSC fiber is majorly depends on the grade of TiO_2 coating and its integration to Ti substrate. The integrity of Ti with TiO_2 will depend on the surface cleanliness and roughness of Ti, affinity between Ti and TiO_2 and other defects. The deposition of TiO$_2$ dye on Ti wire surface is performed by strategy. The integrity of coating on Ti substrate is tested by using peel test, tensile test, four point bending test and scratch test. The amount of discontinuities is measured by optional microscopy and SEM [103–109]. The photovoltaic potential of dye-sensitized solar cells (DSSC) of Poly(vinyl alcohol) (PVA) was improved by spun it into nanofibers by electrospinning technique using PVA solution containing silver nitrate (AgNO$_3$). The silver nanoparticles were generated in electrospun PVA nanofibers after irradiation with UV light of 310~380 nm wavelength. Electrospun PVA/Ag nanofibers have exhibited Isc, FF, Voc, and η showed the values of 11.9~12.5 mA/ cm^2, 0.55~10.59, 0.70~10.71 V, and 4.73~14.99%, respectively. When the silver was loaded upto 1% as dope additives in PVA solution, the resultant electrospun PVA/Ag nanofibers exhibited power conversion efficiency 4.99%, which is higher than that of DSSC using electrospun PVA nanofibers without Ag nanoparticles [110]. Ramier et al. concluded that the feasibility of producing textile structure from DSCPV fiber is quite good. The deposition of TiO$_2$ on flexible fiber is expected to be quite fruitful in order to maintain the structural integrity without comparing with PV performance [111]. Fiber based organic PV devices inroads their applications in electronics, lighting, sensing and thermoelectric harvesting. By successful patch up between commodity fiber and photovoltaic concept, a very useful and cost effective way of power harvesting is matured [112–114]. Coner et al. [115], have developed a photovoltaic fiber by deposition of small Molecular weight organic compound in the form of concentric layer on long fibers. They manufactured the optical photovoltaic (OPV) fiber by vacuum thermal evaporation (VTE) of concentric thin films upto 0.48 mm thickness on polyamide coated silica fiber. Different control devices are based on OPV cells containing identical layer structures deposited on polyimide substrates. The OPV based fiber cells were defined by the shape of the substrate and 1 mm long cathodes. All fiber surfaces were cleaned well prior to deposition. Lastly, they concluded that performance of OPV fiber cells from Indium tin oxide (ITO) is inferior in terms of changes in illumination angle, enabling OPV fiber containing devices to outperform its planar analog under favorable operating conditions. Light emitting devices are designed in such a way that becomes friendly to weave it. The light trapping on fiber surface can be improved by using external dielectric coating which is coupled with protective coating to enhance its service time. Successful PV fiber can be manufactured by opting appropriate material with more improved fabrication potential [116].

DSCs are low cost, applicable in wide range of application and simple to manufacture. These merits of DSCPV fiber make it a potential alternative to the conventional silicon and thin film PV devices55. DSC works on the principle of optoelectronically active cladding on an optical fiber. This group was manufactured two types of PV fiber using polymethylmethacrylate (PMMA) baltronic quality diameter 1.3 to 2.0 mm and photonium quality glass fiber with diameter 1.0 to 1.5 mm. Both virgin fiber were made electronically conductive by deposition of 130 nm thick layer of ZnO:Al by atomic layer deposition technique with the help of P400 equipment. The high surface area photoelectric film for DSC was prepared in two steps. In first step TiO_2 in the form of solution or paste having TiO_2 nanoparticle is deposited on electronically conductive surface. In the second step dry layer of TiO_2 is sintered at 450–500°C for 30 min to ensure proper adhesion to the fiber surface. PMMA fiber is suitable to survive upto 85°C. Hence mechanical compression is alternate technique to ensure the fixation.

Glass fiber is capable to withstand with sintering temperatures which inroads the possibilities of preparation of porous photoelectrodes on them. Commercially available TiO_2 paste was diluted to achieve appropriate viscosity with tarpin oil to make suitable for dip-coating. TiO_2 film was formed by dip coating and dried at room temperature upto 30 min before proper sintering between 475 to 500°C. Appropriately sintered fibers were then immersed in dye solution consisting of 0.32 mL of the cis-bis(isothiocyanate)bis(2,2-bipyridyl-4–4' dicarboxylato)-ruthenium(II) bis-tetrabutyl ammonium, Solaronix SA with trade name N719 dye in absolute ethanol for 48h. The dyeing of nonporous Polymethylmethacrylate (PMMA) fiber coated with nonporous TiO2 layer was performed in the same dye bath. After complete sanitization the excess dye was rinsed away with ethanol. A electrolyte solution was prepared with 0.5 M 4-tert butylpyridine and 0.5 M LiI, 0.05 M I_2 in methoxypropionitrile (MePRN) with 5 wt% polyvinylidene fluoride-hexafluoro-propylene(PVDE-HFP) added as gelatinizing agent as used by Wang et al. [117]. Finally gelatinized iodine electrolyte was added next with dip coating from hot solution. Lastly the carbon based counter electrode was coated by means of a gel prepared by exhaustive grinding of 1.4 g graphite powder and 0.49 grade carbon black simultaneously.

5.3.4   CONDUCTIVE LAYERS FOR PVS

Solar cells require two contacts in a sandwich configuration, with the active semiconductor between them. At least one of the contacts must be transparent to solar radiation, over the waveband which the semiconductor absorbs. Conventional crystalline silicon cells have a thick metal contact on the back surface, and a gridded metal contact on the front surface. They do not require complete coverage of the front surface because the top layer of silicon is sufficiently conducting ('heavily doped') to deliver the photo-current without significant resistance losses. Amorphous silicon cells, like most thin-film cells, have a more insulating top semiconductor layer and so require the whole surface to be contacted. This is achieved without blocking the incoming light, by a layer of transparent conducting oxide (TCO) such as ZnO or Indium tin oxide (ITO), often with other elements added to enhance the conductivity without reducing the transparency. In addition, a fine gridded contact is often superimposed to further improve the current collection efficiency.

It is also possible to build the cell with the light entering through the substrate: this is known as the superstrate configuration. Typically a glass sheet, coated with

TCO is then covered with thin-film semiconductor, which in turn is covered by an opaque metal layer. Textile substrates are unlikely to be sufficiently transparent that this construction may be used, although it is not necessary for the material to be transparent enough to show an image through it, only to be translucent. Hence the textile material is likely to be coated with a conducting layer that may be opaque. Alternatively, the textile may itself form part of the active cell structure, either enabling the conduction of photo-current generated within an imposed semiconductor, or in future materials, enabling photo-currents to be generated within the fabric (probably by semiconducting polymers). Organic solar cells or polymer solar cells still have a long process of development before they can be considered for this application, perhaps being a more intractable problem than the inverse application of light-emitting polymers. Unless the textile can be made from conducting fibers, an additional material is needed, that should not degrade the other desirable textile properties. It is possible, of course, to incorporate metal fibers within a fabric, but it is questionable whether these can make sufficiently good contact to the whole thin-film cell area. They may be helpful in addition to a thinner, continuous conductor. Although polycrystalline silicon can be sufficiently conducting to enable only partial coverage by the contacting layer, it generally requires much higher processing temperatures, before or after its deposition, than most textiles can withstand. We therefore require a continuous conducting layer to be placed over the whole textile surface before adding the semiconductor. The choice of materials (probably metals) is determined by electrical conductance (lower conductance requires thicker layers), compatibility with the substrate (chemically, physically, and during processing), and compatibility with the semiconductor (chemically and work function value). In addition, the mechanical behavior of the conductor in a cell that can flex or twist will depend on its composition, thickness and adhesion. There are several options to be considered, using both physical and chemical coating techniques, to obtain the correct composition and conformality with a nonplanar surface. These will deposit metals or TCOs at the limited temperatures allowed by the preferred textiles. Physical coating methods include sputtering from a solid target in an argon atmosphere and simple evaporation from a solid source (either thermally or with an electron beam), and both are widely used in the optics, electronics, and engineering industries. Chemical methods are based either on the decomposition of a volatile compound of the conductor by heat or electrical plasma, or on the deposition of a coating from a solution, usually driven by an electrical current. These methods are also in widespread use within industries. There are a few additional methods that are applicable for certain materials, such as dip- and spray- and spin-coating, that would be appropriate for the conducting polymers that are being developed now. These do not conduct well enough for photovoltaic cells, because they generate low voltage electrons that cannot overcome the resistance of a long path through a poor conductor. The compromise solution to retaining substrate flexibility with high electrical conductivity may be to use a conducting polymer layer on the textile, or incorporated within its structure, superimposed by a thin, more-conducting metallic layer. If this layer breaks during flexure then the minute gaps will be bridged by the underlying organic conductor, whose limited conductivity will not be a problem over such short paths. This also allows the semiconductor to be in contact with its preferred metal (say aluminum for silicon), avoiding the inclusion of resistive barriers that can exist at other semiconductor/conductor interfaces.

5.3.5   MANUFACTURING OF PV CELLS

ITO was used as a common transparent electrode in polymer-based solar cells due to its remarkable efficiency and ability of light transmission. However, it is quite expensive and generally too brittle to be used with flexible textile substrates. Therefore, highly conductive poly (3,4-ethylenedioxythiophene) doped with poly(styrene sulfonate) PEDOT:PSS, carbon nanotube or metal layers are used to substitute ITO electrode. This can be a promising way to develop PV textiles for smart application due to its low cost and easy application features for future photovoltaic textile applications. A group of scientists has demonstrated the fabrication of an organic photovoltaic device with improved power conversion efficiency by reducing lateral contribution of series resistance between subcells through active area partitioning by introducing a patterned structure of insulating partitioning walls inside the device. Thus, the method of the present invention can be effectively used in the fabrication and development of a next-generation large area organic thin layer photovoltaic cell device [118]. The manufacturing of organic photovoltaic cells can be possible at reasonable cost by two techniques:

5.3.5.1   ROLL-TO-ROLL COATING TECHNIQUE

A continuous roll-to-roll nanoimprint lithography (R2RNIL) technique can provide a solution for high-speed large-area nanoscale patterning with greatly improved throughput. In a typical process, four inch wide area was printed by continuous imprinting of nanogratings by using a newly developed apparatus capable of roll-to- roll imprinting (R2RNIL) on flexible web base. The 300 nm line width grating patterns are continuously transferred on flexible plastic substrate with greatly enhanced throughput by roll-to- roll coating technique. European Union has launched an European research project "HIFLEX" under the collaboration with Energy research Centre of the Netherland (ECN) to commercialize the roll to roll technique. Highly flexible Organic Photovoltaics (OPV) modules will allow the cost-effective production of large-area optical photovoltaic modules with commercially viable Roll-to-Roll compatible printing and coating techniques. Coatema, Germany with Renewable Technologies and Konarka Technologies has started a joint project to manufacture commercial coating machine. Coatema, Germany along with US Company Solar Integrated Technologies (SIT) has developed a process of hot-melt lamination of flexible photovoltaic films by continuous roll-to-roll technique [119]. Roll-to-roll (R2R) processing technology is still in neonatal stage. The novel innovative aspect of R2R technology is related to the roll to roll deposition of thin films on textile surfaces at very high speed to make photovoltaic process cost effective. This technique is able to produce direct pattern of the materials [120, 121].

5.3.5.2   THIN-FILM DEPOSITION TECHNIQUES

Various companies of the world have claimed the manufacturing of various photovoltaic thin films of amorphous silicon (a-Si), copper indium selenide (CIGS), cadmium telluride (CdTe) and dye-sensitized solar cell (DSSC) successfully. Thin film photovoltaics became cost effective after the invention of highly efficient deposition techniques. These deposition techniques offer more engineering flexibilities to increase cell efficiencies, reflectance and dielectric strength, as well as act as a barrier to ensure a long life of the thin film photovoltaics and create high vapor barrier to save the chemistry of these types of photovoltaics [122, 123]. A fiber shaped organic photovoltaic cell was produced by using concentric thin layer of small molecular organic compounds. Thin metal electrode are exhibited 0.5% efficiency of solar power conversion to electricity which is lower than 0.76% that of the planner control device of fiber shape organic PV cells. Results are encouraged to the researchers to explore the possibility of weaving these fibers into fabric form.

The thin film deposition of photovoltaic materials takes place by electron beam, resistance heating and sputtering techniques. These technologies differ from each other in terms of degree of sophistication and quality of film produced. A resistance-heated evaporation technology is relatively simple and inexpensive, but the material capacity is very small which restricts its use for commercial production line. Sputtering technique can be used to deposit on large areas and complex surfaces. Electron beam evaporation is the most versatile technique of vacuum evaporation and deposition of pure elements, including most metals, numerous alloys and compounds.

5.3.5.3   DYE-SENSITIZED PV

An exhaustive research on photovoltaic fibers based on dye-sensitized TiO_2-coated fibers has opened up various gateways for novel PV applications of textiles. The cohesion and adhesion of the TiO_2 layer are identified as crucial factors in maintaining PV efficiency after weaving operation. By proper control of tension on warp and weft fibers, high PV efficiency of woven fabrics is feasible. The deposition of thin porous films of ZnO on metalized textiles or textile-compatible metal wires by template assisted electro-deposition technique is possible. A sensitizer was adsorbed and the performance as photoelectrodes in dye-sensitized photovoltaic cells was investigated. The thermal instability of textiles restricts its use as photovoltaic material because process temperatures are needed to keep below 150°C. Therefore, the electrode position of semiconductor films from low-temperature aqueous solutions has become a most reliable technique to develop textile based photovoltaics. Among low-temperature solution based photovoltaic technologies; dye sensitized solar cell technology appears most feasible. If textile materials are behaved as active textiles, the maximum electrode distance in the range of 100 μm has to be considered. Loewenstein et al., and Lincot et al., have used Ag coated polyamide threads and fibers to deposit porous ZnO as semiconductor material. The crystalline ZnO films were prepared in a cathodic electrode position reaction induced by oxygen reduction in an aqueous electrolyte in presence of Zn2+ and eosin Y as structure-directing agent [124, 125].

Bedeloglu et al. [126], were used nontransparent nonconductive flexible polypropylene (PP) tapes as substrate without use of ITO layer. PP tapes were gently cleaned in methanol, isopropanol, and distilled water respectively and then dried in presence of nitrogen. 100 nm thick Ag layer was deposited by thermal evaporation technique. In next step, a thin layer of poly(3,4-ethylenedioxythiophene) doped: poly(styrene sulfonate) PEDOT: PSS mixture solution was dip coated on PP tapes. Subsequently, poly [2-methoxy-5-(3, 7- dimethyloctyloxy)-1–4-phenylene vinylene] and 1-(3-methoxycarbonyl)-propyl-1- phenyl(6,6)C61, MDMO: PPV: PCBM or poly(3-hexylthiophene) and 1-(3-methoxycarbonyl)- propyl-1-phenyl(6,6)C61, P3HT: PCBM blend were dip coated onto PP tapes. Finally, a thin layer of LiF (7 nm) and Al (10 nm) were deposited by thermal evaporation technique.

The enhanced conductivity will always useful to improve the photovoltaic potential of poly(3,4-ethylene dioxythiophene):poly(styrene sulfonate) (PEDOT:PSS). Photovoltaic scientific community found that the conductivity of poly(3,4-ethylene dioxythiophene): poly(styrene sulfonate) (PEDOT:PSS), film is enhanced by over 100-folds if a liquid or solid organic compound, such as methyl sulfoxide (DMSO), N, N-dimethylformamide (DMF), glycerol, or sorbitol, is added to the PEDOT:PSS aqueous solution. The conductivity enhancement is strongly dependent on the chemical structure of the organic compounds. The aqueous PEDOT: PSS can be easily converted into film form on various substrates by conventional solution processing techniques and these films have excellent thermal stability and high transparency in the visible range [127–130]. Some organic solvents such as ethylene glycol (EG), 2-nitroethanol, methyl sulfoxide or 1- methyl-2-pyrrolidinone are tried to enhance the conductivity of PEDOT: PSS. The PEDOT: PSS film which is soluble in water becomes insoluble after treatment with EG. Raman spectroscopy indicates that interchain interaction increases in EG treated PEDOT: PSS by conformational changes of the PEDOT chains, which change from a coil to linear or expanded-coil structure. The electron spin resonance (ESR) was also used to confirm the increased interchain interaction and conformation changes as a function of temperature. It was found that EG treatment of PEDOT: PSS lowers the energy barrier for charge among the PEDOT chains, lowers the polaron concentration in the PEDOT:PSS film by w 50%, and increases the electrochemical activity of the PEDOT: PSS film in NaCl aqueous solution by w100%. Atomic force microscopy (AFM) and contact angle measurements were used to confirm the change in surface morphology of the PEDOT:PSS film. The presence of organic compounds was helpful to increase the conductivity which was strongly dependent on the chemical structure of the organic compounds, and observed only with organic compound with two or more polar groups. Experimental data were enough to make a statement that the conductivity enhancement is due to the conformational change of the PEDOT chains and the driving force is the interaction between the dipoles of the organic compound and dipoles on the PEDOT chains [131].

5.3.6 CHARACTERIZATION OF PV TEXTILES

Characterization of various photovoltaic textiles is essential to prove its performance before send to the market. Various characterization techniques collectively ensure the perfect achievement of the targets to manufacture the desired product.

5.3.6.1 THICKNESS AND MORPHOLOGY OF PV TEXTILES

Scanning electron microscope is used to investigate the thickness and morphology of various donor, acceptor layers. Scanning electron microscopes from LEO Supra 35 and others can be used to measure the existence and thickness of various coated layers on various textile surfaces at nanometer level. Various layers on photovoltaic fibers become clearly visible with 50,000X magnification. The thickness of the layers can be seen from SEM photographs by bright interface line between the polymer anode and the photoactive layer.

5.3.6.2 CURRENT AND VOLTAGE

In order to characterize the photovoltaic fibers open circuit voltage, short circuit current density, current and voltage at the maximum power point under an illumination of 100 mW/cm^2 are carried out. In order to calculate the photovoltaic efficiency of photovoltaic textiles, current verses voltage study is essential. To achieve this target a computer controlled source meter equipped with a solar simulator under a range of illumination power is required with proper calibration. All photoelectrical characterizations are advised to conduct under nitrogen or argon atmosphere inside a glove box to maintain the preciseness of observations. The overall efficiency of the PV devices can be representing by following equation:

$$\eta = \frac{V_{oc} \times I_{sc} \times FF}{P_{in}} = \frac{P_{out}}{P_{in}} \quad (1)$$

where V_{oc} is the open circuit voltage (for $I=0$) typically measured in volt (V); I_{sc} is the short circuit current density (for $V=0$) in ampere /square meter (A/m²); P_{out} is the output electrical power of the device under illumination; P_{in} the incident solar radiation in (watt/square meter) W/m²; FF is the fill factor and can be explained by the following relationship:

$$FF = \frac{I_{mpp} \times V_{mpp}}{I_{sc} \times V_{oc}} \qquad (2)$$

where, V_{mpp} voltage at the maximum power point (MPP); I_{mpp} is the current at the maximum power point (MPP) where the product of the voltage and current is maximized.

To assure an objective measurement for precise comparison of various photovoltaic devices, characterization has to be performed under identical conditions. An European research group has used Keithley 236 source measure unit in dark simulated AM 1.5 global solar conditions at an intensity of 100 mW/cm². The solar simulator unit made by K.H. Steuernagel Lichttechnik GmbH was calibrated with the help of standard crystalline silicon diode. PV fibers were illuminated through the cathode side and I-V characteristics were measured. The semilogarithmic I-V curves demonstrate the current density versus voltage behavior of photovoltaic fibers under various conditions. It gives a comparative picture of voltage Vs current density as a function of various light intensities. Durisch et al., has developed a computer based testing instrument to measure the performance of solar cells under actual outdoor conditions. This testing system consist a sun-tracked specimen holder, digital multimeters, devices to apply different electronic loads and a computer based laser printer. Pyranometers, pyrheliometers and a reference cell is used to measure and record the insulation. This instrument is able to test wide dimensions of photovoltaic articles ranging from 3 mm 3 mm to 1 m 1.5 m. The major part of world's energy scientist community predicts that photovoltaic energy will play a decisive role in any sustainable energy future [132].

5.3.6.3 MECHANICAL CHARACTERIZATION

Textile substrates are subjected to different stresses under various situations. Hence usual tensile characterization is essential for photovoltaic textiles. For tensile testing of PV fibers, the constant rate of extension (CRE) based tensile testing machines are used at 1 mm per minute deformation rate using Linear Variable Differential Transformers (LVDT) displacement sensor. Fracture phenomenon is recorded by means of high resolution video camera integrated with tensile testers. To study about the adhesion and crack formation in coating on textile structures, generally 30 mm gauge length is used in case of photovoltaic fibers. Fiber strength measuring tensile tester, integrated with an appropriate optical microscope to record the images of specimen at an acquisition rate of about one frame per second is used to record the dynamic fracture of PV fibers. Different softwares are available to analyze the image data like PAXit, Clemex, and Digimizer etc.

5.3.6.4   *ABSORPTION SPECTRA OF SOLID FILMS*

Various spectrophotometers like Varian Carry 3G UV-Visible were used to observe the ultraviolet visible absorption spectra of photovoltaic films. The thin films are prepared to study the absorption spectrum of solid films. In a typical study, a thin film was prepared by spin coating of solution containing 10 mg of P3HT and 8 mg of PCBM and 4.5 mg of MDMO-PPV and 18 mg of PCBM (in case of 1:4)/ml with chlorobenzene as solvent. A typical absorption spectra of MDMO-PPV:PCBM and P3HT.

5.3.6.5   *X-RAY DIFFRACTION OF PHOTOVOLTAIC STRUCTURES*

Crystallization process is very common phenomenon that takes place during photovoltaic structure development. The content of crystalline and amorphous regions in photovoltaic structures influences the photoactivity of photovoltaic structures. X-ray diffraction technique is capable to characterize the amount of total crystallinity, crystal size and crystalline orientation in photovoltaic structures. Presently, thin film photovoltaics are highly efficient devices being developed in different crystallographic forms: epitaxial, microcrystalline, polycrystalline, or amorphous. Critical structural and microstructural parameters of these thin film photovoltaics are directly related to the photovoltaic performance. Various X-ray techniques like X-ray diffraction for phase identification, texture analysis, high-resolution x-ray diffraction, diffuse scattering, X-ray reflectivity are used to study the fine structure of photovoltaic devices [133].

5.3.6.6   *RAMAN SPECTROSCOPY*

The Raman Effect takes place when light rays incidents upon a molecule and interact with the electron cloud and the bonds of that molecule. Spontaneous Raman effect is a form of scattering when a photon excites the molecule from the ground state to a virtual energy state. When the molecule relaxes it emits a photon and it returns to a different rotational or vibrational state. Raman spectroscopy is majorly used to confirm the chemical bonds and symmetry of molecules. It provides a fingerprint to identify the molecules. The fingerprint region for organic molecules remains in the (wavenumber) range of 500–2000 cm^{-1}. Spontaneous Raman spectroscopy is used to characterize superconducting gap excitations and low frequency excitations of the solids. Raman scattering by anisotropic crystal offers information related to crystal orientation. The polarization of the Raman scattered light with respect to the crystal and the polarization of the laser light can be used to explore the degree of orientation of crystals [134]. The in situ morphological and optoelectronic changes in various photovoltaic materials can be observed by observing the changes in the Raman and photoluminescence (PL) feature with the help of a spectrometer. Various spectrums can be recorded at a definite integration time after avoiding any possibility of laser soaking of the sample67.

5.3.6 CHALLENGES TO BE MET

The active cell components, the semiconducting layers, therefore have a great part to play in determining the cell performance, not only from an efficiency point of view but also from the load matching perspective. The separate improvement of both output current and voltage demands effective optical absorption and efficient photogenerated charge separation. There may be unavoidable trade-offs in meeting both of these simultaneously. Textile substrates offer new ways of improving the optical absorption in thin layers by scattering light that passes through the layer back into the layer for a second chance of absorption. Rough surfaces in conventional silicon cells are sometimes a source of electrical problems but substrate texturing has always been seen as a possible tool for advanced thin-film cell design. Transferring some of the best features of amorphous silicon cell technology to a flexible textile substrate device [135] would provide more rapid evolution of the new cells than a separate development based on single crystal cell technology, given the processing temperature restraints of the preferred textiles. There are still plenty of challenges in ensuring efficient charge collection from a stack of thin layers on a nonplanar base. Finally the word textile still often suggests applications in clothing, but users would expect to have the same performance from a solar cloth as from a passive cloth, with respect to folding, wear, washing, abrasion, etc. Few of these properties have yet been tested with coatings of the type used in solar cells, nor have solar cells been assessed in such conditions, even if hermetically sealed with a thin polymer layer. Low cost production and easy processing of organic solar cells comparing to conventional silicon-based solar cells make them interesting and worth employing for personal use and large scale applications. Today, the smart textiles as the part of technical textiles using smart materials including PV materials, conductive polymers, shape memory materials, etc. are developed to mimic the nature in order to form novel materials with a variety of functionalities. The solar cell-based textiles have found its application in various novel field and promising development obtaining new features. These PV textiles have found its application in military applications, where the soldiers need electricity for the portable devices in very remote areas. The PV textile materials can be used to manufacture power wearable, mobile and stationary electronic devices to communicate, lighten, cool and heat, etc. by converting sun light into electrical energy. The PV materials can be integrated onto the textile structures especially on clothes, however, the best promising results from an efficient PV fiber has to be come which can constitute a variety of smart textile structures and related products [136]. Fossil fuels lead to the emission of CO_2 and other pollutants and consequently human health is under pressure due to adverse environmental conditions. In consequence of that renewable energy options have been explored widely in last decades [137, 138].

Unprecedented characteristics of PV cells attract maximum attention in comparison of other renewable energy options which has been proved by remarkable growth in global photovoltaic market4. Organic solar cells made of organic electronic materials based on liquid crystals, polymers, dyes, pigment etc. attracted maximum attention of scientific and industrial community due to low weight, graded transparency, low cost, low bending rigidity and environmental friendly processing potential [139, 140]. Various PV materials and devices similar to solar cells integrated with textile fabrics can harvest power by translating photon energy into electrical energy. The successful integration of solar cells into textiles has to take into account the type of fiber used and the method of textile fabrication. The selection of a type of fiber is strongly influenced by its ability to withstand prolonged irradiation by ultraviolet (UV) light. Fiber selection is also governed by the temperatures required to lay down the thin films comprising the solar cell, although it has been shown that nanocrystalline silicon thin films, an improved form of amorphous silicon, can be successfully deposited at temperatures as low as 200°C, and under the appropriate conditions even single crystal silicon may be grown, albeit epitaxially on silicon wafers [141]. These two factors of UV resistance and maximum temperature restrict the choice of commodity fibers. Commercial polyolefin fibers melt below 200°C. Cotton, wool, silk and acrylic fibers start to decompose below this temperature. Polyamide fibers are likely to be too susceptible to UV radiation. Polyethylene terephthalate (PET) fibers, however, are viable substrates. They melt at 260–270°C and exhibit good stability to UV light [142]. They are commercially attractive too because of their existing widespread use. Thus, fabrics composed of a large variety of PET grades are currently available. PET fibers possess good mechanical properties and are resistant to most forms of chemical attack. They are less resistant to alkalis, but a solar cell would not normally find use in an alkaline environment. Fabrics constructed from PET fibers should, therefore, be suitable as substrates for solar cells, while also possessing flexibility and conformability to any desired shape. Fabrics composed of glass fibers produced from E-glass formulations16 could also be used. One advantage could lie in their transparency, as with plate glass in conventional solar cells. Moreover, the price of E-glass is similar to that of PET [143]. E-glass fibers, however, suffer from poor resistance to acids and alkalis and are prone to flexural rupture. Other glasses are more stable to environments of extreme pH. S-glass is used for specialist sports equipment and aerospace components, but its price is approximately five times that of E-glass [144].

Many high-performance fibers, which readily withstand temperatures up to 300–400°C, could also be considered, although aramids would not be sufficiently stable to UV radiation. Examples of suitable high-performance fibers could be poly-benzimidazole (PBI) fibers, polyimide (PI) fibers and polyetheretherketone (PEEK) fibers. However, these fibers are expensive. There are clear commercial attractions, therefore, in adopting PET fibers. The type of textile fabric construction is also important to the performance of the solar cell. The type of construction affects the physical and mechanical properties of a fabric, and also its effectiveness as an electrical conductor. Where conduction in textile fabrics is required, woven fabrics are generally considered to be best in that they possess good dimensional stability and can be constructed to give desired flexibilities and conformations. Moreover, the yarn paths in woven structures are well ordered, which allows the design of complex woven fabric-based electrical circuits. Knitted structures, on the other hand, do not retain their shapes so well, and the rupture of a yarn may cause laddering. These problems are heightened if the shape of the fabric is continually changing, as in apparel usage. Nonwoven fabrics do not, as yet, generally possess the strength and dimensional stability of woven fabrics. More significantly, the construction of electrical circuits in them is limited, because their yarn paths are highly unoriented. Embroidery, however, may offer an opportunity for circuit design [145, 146].

5.4   CONCLUSION

In these times of uncertain energy sources and with the impacts that they are having on the atmosphere, clean, renewable solar energy is a good alternative. However, it is still more expensive than methods used currently by consumers, and inn order to be a viable option, it has to be attainable and affordable. With the use of different materials and thin film technologies, the use of photovoltaic solar energy for the average person is becoming more of a reality. The incorporation of photovoltaics into textiles has been explored new inroads for potential use in intelligent clothing in more smart ways. Incorporation of organic solar cells into textiles has been realized encouraging performances. The functionality of the photovoltaic textiles does not limited by mechanical stability of photovoltaics. Polymer-based solar cell materials and manufacturing techniques are suitable and applicable for flexible. The manufactured photovoltaic fibers may also be used to manufacture functional yarns by spinning and then fabric by weaving and knitting. Photovoltaic tents, curtains, tarpaulins and roofing are available to use the solar power to generate electricity in more green and clean fashion.

KEYWORDS

- **photovoltaic**
- **solar cell**
- **solar energy**
- **textile**

REFERENCES

1. Grätzel M., 'Photoelectrochemical cells,' *Nature*, 414, 2001, 338–344.
2. R.J. Hassett. *Future visions of U.S. photovoltaics industry*. In *Proceedings of the 4th IEA–PVPS international conference*, Osaka, Japan (2003). IEA-PVPS.
3. R. Sellers. *PV: Progress and Promise*. In *Proceedings of the 4th IEA–PVPS international conference*, Osaka, Japan (2003). IEA-PVPS.
4. K. Kurokawa. *PV Industries: Future Visions*. In *Proceedings of the 4th IEA–PVPS international conference*, Osaka, Japan (2003). IEA-PVPS.
5. Boutrois, J.P, Jolly, R., Petrescu, C. 1997, 'Process of Polypyrrole Deposit on Textile. Product Characteristics and Applications,' *Synthetic Metals*, vol. 85, 1405–1406.
6. Goetzberger A., Hebling C., Schock H.-W., 'Photovoltaic materials, history, status and outlook,' *Materials Science and Engineering: R: Reports*, 40, 2003, 1–46.
7. W. Shockley and H. J. Queisser, "Detailed Balance Limit of Efficiency of p-n Junction Solar Cells," *Journal of Applied Physics*, vol. 32, no. 3, 510–519, 1961.
8. M. J. Kerr, P. Campbell, and A. Cuevas, "Lifetime and efficiency limits of crystalline silicon solar cells," in *Proceedings of the 29th IEEE Photovoltaic Specialists Conference*, 2002, 438–441.
9. J. Zhao, A. Wang, and M. A. Green, "24·5% Efficiency silicon PERT cells on MCZ substrates and 24·7% efficiency PERL cells on FZ substrates," *Progress in Photovoltaics: Research and Applications*, vol. 7, no. 6, 471–474, Nov. 1999.
10. A.V. Shah, R. Platz and H. Keppner. *Thin-film silicon solar cells: A review and selected trends*. Sol. Energy Mater. Sol. Cells, 38, 501 (1995).
11. Roth, S., Carroll, D. 2004, *One-Dimensional Metals: Conjugated Polymers*, *Organic Crystals*, *Carbon Nanotubes*, 2nd edition.
12. Dinh, H.N. 1998, '*Electrochemical and structural studies of polyaniline film growth and degradation at different substrate surfaces*,' Ph.D. Thesis, University of Calgary.
13. Kuran, W. 2001, '*Principles of Coordination Polymerization, Coordination Polymerization of Alkynes*,' Wiley, 379.
14. Chiang, C.K, Fincher, C.R., Park, Y.W., Heeger, A.J., Shirakawa, H., Louis, E.J., Gau, S.C., MacDiarmid, A.G. 1977, 'Electrical Conductivity in Doped Polyacetylene,' *Physical Review Letters*, vol. 39, no. 17, 1098.
15. Neuendorf, A.J. 2003, '*High Pressure Synthesis of Conductive Polymers*,' Ph.D. Thesis, Faculty of Science, Griffith University.
16. Nardes, A.M. 2007, '*On the conductivity of PEDOT: PSS thin films*,' Ph.D. Thesis, Technische Universiteit Eindhoven, the Netherlands.
17. Kumar, D., Sharma, R.C. 1998, 'Advances in Conductive Polymers,' *European polymer journal*, vol. 34, 1053–1060.
18. Streetman, B., Banerjee, S. 2006, *Solid State Electronic Devices*, Prentice Hall.

19. Heeger, A.J. 2000, 'Semiconducting and Metallic Polymers: The Fourth Generation of Polymeric Materials,' *Nobel Lecture, The Nobel Prize in Chemistry 2000.*

20. George, P.M. 2005, *'Novel Polypyrrole Derivatives to Enhance Conductive Polymer-Tissue Interactions,* ' Ph.D. Thesis, Massachusetts Institute of Technology, USA.

21. Kim, B., Koncar, V., Devaux, E. 2004, 'Electrical Properties of Conductive Polymers: PET –Nanocomposites' Fibres,' *AUTEX Research Journal*, vol. 4, no. 1.

22. McDaniels, D.K. 1991, *The Sun.*, Krieger Pub Co.

23. Kasap, S.O. 2005, *Principles of Electronic Materials and Devices*, McGraw Hill Higher Education.

24. Spanggaard, H., Krebs, F.C. 2004, 'A brief history of the development of organic and polymeric photovoltaics,' *Sol. Energy Mater. Sol. Cells*, vol. 83, 125.

25. Hoppe, H. 2004, *'Nanomorphology – Efficiency Relationship in Organic Bulk Heterojunction Plastic Solar Cells,'* Ph.D. Thesis, Kepler University Linz.

26. Mayer, A.C., Scully, S.R., Hardin, B.E., Rowell, M.W., McGehee, M.D. 2007, 'Polymer Based Solar Cells,' *Materials Today*, vol. 10, no. 11.

27. Groenendaal, L., et al., *Poly(3,4-ethylenedioxythiophene) and Its Derivatives: Past, Present, and Future.* Advanced Materials, 2000. 12(7): p. 481–494.

28. Kirchmeyer, S. and K. Reuter, *Scientific importance, properties and growing applications of poly(3,4-ethylenedioxythiophene).* Journal of Materials Chemistry, 2005. 15(21): p. 2077–2088.

29. Jonas, F. and L. Schrader, *Conductive modifications of polymers with polypyrroles and polythiophenes.* Synthetic Metals, 1991. 41(3): p. 831–836.

30. Carpi, F. and D. Rossi, *Colours from electroactive polymers: electrochromic, electroluminescent and laser devices* based *on organic materials.* Optics Laser Technology, 2006. 38: p. 292–305.

31. Dietrich, M., et al., *Electrochemical and spectroscopic characterization of polyalkylenedioxythiophenes.* Journal of Electroanalytical Chemistry, 1994. 369(1–2): p. 87–92.

32. Cui, X. and D.C. Martin, *Electrochemical deposition and characterization of poly(3,4-ethylenedioxythiophene) on neural microelectrode arrays.* Sensors and Actuators B: Chemical, 2003. 89(1–2): p. 92–102.

33. Bhattacharya, A. and A. De, *Conducting polymers in solution – Progress toward processibility.* Journal of Macromolecular Science – Reviews in Macromolecular Chemistry and Physics, 1999. 39(1): p. 17–56.

34. Laforgue, A. and L. Robitaille, *Production of Conductive PEDOT Nanofibers by the Combination of Electrospinning and Vapor-Phase Polymerization.* Macromolecules, 2010. 43(9): p. 4194–4200.

35. Pei, Q., et al., *Electrochromic and highly stable poly(3,4-ethylenedioxythiophene) switches between opaque blue-black and transparent sky blue.* Polymer, 1994. 35(7): p. 1347–1351.

36. Asplund, M., H. Holst, and O. Inganäs, *Composite biomolecule/PEDOT materials for neural electrodes.* Biointerphases, 2008. 3(3): p. 83–93.

37. Talo, A., et al., *Polyaniline/epoxy coatings with good* anti*corrosion properties.* Synthetic Metals, 1997. 85(1–3): p. 1333–1334.

38. Norstebo, C.A., *Intelligent Textiles, Soft Products.* Journal of Future Materials, 2003 p. 1–14.

39. Das, S., *Mobility and Resource Management in Smart Home Environments*, in *Embedded and Ubiquitous Computing*, L. Yang, et al., Editors. 2004, Springer Berlin Heidelberg. p. 1109–1111.

40. Mattmann, C., F. Clemens, and G. Tröster, *Sensor for Measuring Strain in Textile.* Sensors, 2008. 8(6): p. 3719–3732.

41. Carpi, F. and D.D. Rossi, *Electroactive Polymer-Based Devices for e-Textiles in Biomedicine.* IEEE Transactions on Information Technology in Biomedicine, 2005. 9(3): p. 295–318.

42. De Rossi, D., A. Della Santa, and A. Mazzoldi, *Dressware: wearable hardware*. Materials Science and Engineering: C, 1999. 7(1): p. 31–35.

43. Billinghurst, M. and T. Starner, *Wearable devices: new ways to manage information*. Computer, 1999. 32(1): p. 57–64.

44. Zieba, J. and M. Frydrysiak, *Textronics- Electrical and Electronic Textiles. Sensors for Breathing Frequency Measurement* Fibers, Textiles in Eastern Europe, 2006. 14(5(59)): p. 43–48.

45. Marculescu, D., et al., *Electronic textiles: A platform for pervasive computing*. Proceedings of the IEEE, 2003. 91(12): p. 1995–2018.

46. Singh, M.K., *The state-of-art Smart Textiles*, 2004, http://www.ptj.com.pk/Web%20 2004/08–2004/Smart%20Textiles.html.

47. Shahhaidar, E., *Decreasing Power Consumption of a Medical E-textile*, in *World Academy of Science, Engineering and Technology* 2008.

48. Hung, K., Y.T. Zhang, and B. Tai. *Wearable medical devices for tele-home healthcare*. in *Engineering in Medicine and Biology Society, 2004. IEMBS '04. 26th Annual International Conference of the IEEE*. 2004.

49. Akita, J., et al., *Wearable electromyography measurement system using cable-free network system on conductive fabric*. Artificial Intelligence in Medicine, 2008. 42(2): p. 99–108.

50. Huang, C.-T., et al., *A wearable yarn*-based *piezo-resistive sensor*. Sensors and Actuators A: Physical, 2008. 141(2): p. 396–403.

51. McCann, J., R. Hurford, and A. Martin. *A design process for the development of innovative smart clothing that addresses end-user needs from technical, functional, esthetic and cultural view points*. in *Wearable Computers, 2005. Proceedings. Ninth IEEE International Symposium on*. 2005.

52. Mondal, S., *Phase change materials for smart textiles – An overview*. Applied Thermal Engineering, 2008. 28(11–12): p. 1536–1550.

53. Fan, X., et al., *Wire-Shaped Flexible Dye-sensitized Solar Cells*. Advanced Materials, 2008. 20(3): p. 592–595.

54. Yajima, T., M.K. Yamada, and M.S. Tanaka, *Protection effects of a silver fiber textile against electromagnetic interference in patients with pacemakers*. Journal of Artifical Organics, 2002. 5: p. 175–178.

55. Yang, X., et al., *Vapor phase polymerization of 3,4-ethylenedioxythiophene on flexible substrate and its application on heat generation*. Polymers for Advanced Technologies, 2011. 22(6): p. 1049.

56. Xue, P., et al., *Electrically conductive yarns* based *on PVA/carbon nanotubes*. Composite Structures, 2007. 78(2): p. 271–277.

57. Dastjerdi, R. and M. Montazer, *A review on the application of inorganic nanostructured materials in the modification of textiles: Focus on antimicrobial properties*. Colloids and Surfaces B: Biointerfaces, 2010. 79(1): p. 5–18.

58. Rothmaier, M., M.P. Luong, and F. Clemens, *Textile Pressure Sensor Made of Flexible Plastic Optical Fibers*. Sensors, 2008. 8: p. 4318–4329.

59. Kincal, D., et al., *Conductivity switching in polypyrrole-coated textile fabrics as gas sensors*. Synthetic Metals, 1998. 92: p. 53–56.

60. Bessette, R.R., et al., *Development and characterization of a novel carbon fiber based cathode for semifuel cell applications*. Journal of Power Sources, 2001. 96(1): p.40- 244.

61. Meoli, D. and T. May-Plumlee, *Interactive Electronic Textile Development: A review of Technologies*. Journal of Textile and Apparel, Technology and Management, 2002. 2(2).

62. Mattila, H.R. 2006, *Intelligent Textiles and Clothing*, CRC Press.

63. Krebs, F.C., Biancardo, M., Winther-Jensen, B., Spanggrad, H., Alstrup, J. 2005, 'Strategies for incorporation of polymer photovoltaics into garments and textiles,' *Solar Energy Materials, Solar Cells*, vol. 90, 1058–1067.

64. Konarka Technologies, Inc., *Konarka Builds Power Plastic*, http://www.konarka.com/ technology/photovoltaics.php.

65. Lee Y.-C., Chen S. M.-S., Lin Y.-L., Mason B.G., Novakovskaia E.A., Connell, V.R., *Integrated Flexible Solar Cell Material and Method of Production* (Sky Station International Inc.), US Patent 6,224,016, May 2001.

66. Iowa Thin Film Technologies, Inc., *Revolutionary Package of Proprietary Solar/ Semiconductor Technologies*, http://www.iowathinfilm.com/technology/index.html.

67. Mowbray J.L., 'Wider choice for narrow fabrics,' *Knitting International*, 2002, 109, 36–38.

68. Ellis J.G., 'Embroidery for engineering and surgery' in *Proceedings of the Textile Institute World Conference*, Manchester, 2000.

69. Karamuk E., Mayer J., Düring M., Wagner B., Bischoff B., Ferrario R., Billia M., Seidl R., Panizzon R., Wintermantel E., 'Embroidery technology for medical textiles' in *Medical Textiles*, editor Anand S., Woodhead Publishing Limited, Cambridge, 2001.

70. Goldsmith N., 'Solar tensile pavilion, ' http://ndm.si.edu/EXHIBITIONS/sun/2/ obj_tent. mtm.

71. Zabetakis A., Stamelaki A., Teloniati T., 'Solar textiles: Production and distribution of electricity coming from the solar radiation. Applications.' in *Proceedings of Conference on Fibrous Assemblies at the Design and Engineering Interface*, INTEDEC 2003, Edinburgh, 2003.

72. Muller, H.-F., 'Sailcloth arrangement for sails of water-going vessels, ' US Patent Office, 6237521, May 2001.

73. Martin T., 'Sun shade canopy has solar energy cells within woven textile material that is wound onto a roller' (Warema Renkhoff GmbH), German Patent Office, DE10134314, January 2003.

74. Meoli, D. and T. May-Plumlee, *Interactive Electronic Textile Development: A Review of Technologies.* Journal of Textile and Apparel, Technology and Management, 2002. 2(2).

75. Tibtech. *conductive yarns and fabrics for energy transfer and heating devices in SMART textiles and composites.* http://www.tibtech.com/metal_fiber_composition.php.

76. Skrifvars, M. and A. Soroudi, *Melt spinning of carbon nanotube modified polypropylene for electrically conducting nanocomposite* fibers. Solid State Phenomena, 2008. 151: p. 43–47.

77. Hegde, R.R., A. Dahiya, and M.G. Kamath. *Carbon Fibers.* 2004 http://web.utk.edu/~mse/ Textiles/CARBON%20FIBERS.htm.

78. Dalmas, F., et al., *Carbon nanotube-filled polymer composites. Numerical simulation of electrical conductivity in three-dimensional entangled fibrous networks.* Acta Materialia, 2006. 54(11): p. 2923–2931.

79. Hunt, M.A., et al., *Patterned Functional Carbon Fibers from Polyethylene.* Advanced Materials, 2012. 24(18): p. 2386–2389.

80. Chung, D.D.L., ed. *Carbon Fiber Composites.* 1994, Elsevier. 227.

81. Pomfret, S.J., et al., *Electrical and mechanical properties of polyaniline fibers produced by a one-step wet spinning process.* Polymer, 2000. 41(6): p. 2265–2269.

82. Okuzaki, H., Y. Harashina, and H. Yan, *Highly conductive PEDOT/PSS microfibers fabricated by wet-spinning and dip-treatment in ethylene glycol.* European Polymer Journal, 2009. 45(1): p. 256–261.

83. Woonphil, B., et al., *Synthesis of highly conductive poly(3,4-ethylenedioxythiophene) fiber by simple chemical polymerization.* Synthetic Metals, 2009. 159(13): p. 1244- 1246.

84. Saleem, A., L. Frormann, and A. Iqbal, *Mechanical, Thermal and Electrical Resisitivity Properties of Thermoplastic Composites Filled with Carbon Fibers and Carbon Particles.* Journal of Polymer Research, 2007. 14(2): p. 121–127.

85. Bigg, D.M., *Mechanical and conductive properties of metal fiber-filled polymer composites.* Composites, 1979. 10(2): p. 95–100.

86. Show, Y. and H. Itabashi, *Electrically conductive material made from CNT and PTFE.* Diamond and Related Materials, 2008. 17(4–5): p. 602–605.

87. Shirakawa, H., et al., *Synthesis of electrically conducting organic polymers: Halogen derivatives of polyacetylene, (CH).* Journal of the Chemical Society, Chemical Communications, 1977: p. 578–580.

88. Bigg, D.M., *Conductive polymeric compositions.* Polymer Engineering, Science, 1977. 17(12): p. 842–847.

89. Zabetakis, D., M. Dinderman, and P. Schoen, *Metal-Coated Cellulose Fibers for Use in Composites Applic*able to *Microwave Technology.* Advanced Materials, 2005. 17(6): p. 734–738.

90. Xue, P. and X.M. Tao, *Morphological and Electromechanical Studies of Fibers Coated with Electrically Conductive Polymer.* Journal of Applied Polymer Science, 2005. 98: p. 1844–1854.

91. Negru, D., C.-T. Buda, and D. Avram, *Electrical Conductivity of Woven Fabrics Coated with Carbon Black Particles.* Fibers, Textiles in Eastern Europe, 2012. 20(1(90)): p. 53–56.

92. Sen, A.k., *Coated Textiles Principles and Applications* J. Damewood, Editor 2001, Technomic Publishing Company, Inc.

93. Hohnholz, D., et al., *Uniform thin films of poly 3,4-ethylenedioxythiophene (PEDOT) prepared by in-situ deposition.* Chemical Communications, 2001(23): p. 2444–2445.

94. Rozek, Z., et al., *Potential applications of nanofiber textile covered by carbon coatings.* Journal of Achievements in Materials and Manufacturing Engineering, 2008. 27(1): p. 35–38.

95. Tao, X.M., *Smart Fibers, Fabrics and Clothing.* The Textile Institute, England, 2001.

96. Kaynak, A., et al., *Characterization of conductive polypyrrole coated wool yarns.* Fibers and Polymers, 2002. 3(1): p. 24–30.

97. Jönsson, S.K.M., et al., *The effects of solvents on the morphology and sheet resistance in poly(3,4-ethylenedioxythiophene)-polys*tyrene*sulfonic acid (PEDOT-PSS) films.* Synthetic Metals, 2003. 139(1): p. 1–10.

98. Rajahn M, Rakhlin M, Schubert M B "Amorphous and heterogeneous silicone based films" MRS Proc. 664, 2001.

99. Schubert MB, Werner J H, Mater. Today 9(42), 2006.

100. Drew C, Wang X Y, Senecal K, J of Macro. Mol. Sci Pure A 39, 2002, 1085.

101. Baps B, Eder K M, Konjuncu M, Key Eng. Mater. 206–213, 2002, 937.

102. Gratzel M, Prog. Photovolt: Res. Appl. 8, 2000, 171.

103. Verdenelli M, Parole S, Chassagneux F, Lettof J M, Vincent H and Scharff J P, J of Eur. Ceram. Soc. 23, 2003, 1207.

104. Xie C, Tong W, Acta Mater, 53, 2005, 477.

105. Muller D, Fromm E Thin Mater. Solid Films 270, 1995, 411.

106. Hu M S and Evans A G, Acta Mater, 37, 1998, 917.

107. Yang Q D, Thouless M D, Ward S M, J of Mech Phys Solids, 47, 1999, 1337.

108. Agrawal D C, Raj R, Acta Mater, 37, 1989, 1265.

109. Rochal G, Leterrier Y, Fayet P, Manson J Ae, Thin Solid Films 437,2003, 204.

110. Park S H, Choi H J, Lee S B, Lee S M L, Cho S E, Kim K H, Kim Y K, Kim M R and Lee J K "Fabrications and photovoltaic properties of dye-sensitized solar cells with electrospun poly(vinyl alcohol) nanofibers containing Ag nanoparticles" Macromolecular Research 19(2), 2011, 142–146.

111. Ramiera J., Plummera C.J.G., Leterriera Y., Mansona J.-A.E., Eckertb B., and Gaudianab R. "Mechanical integrity of dye-sensitized photovoltaic fibers" Renewable Energy 33 (2008), 314–319.

112. Hamedi M, Forchheimer R and Inganas O Nat Mat. 6, 2007, 357.

113. Bayindir M, Sorin F, Abouraddy A F, Viens J, Hart S D, Joannopoulus J D, Fink Y Nature 431, 2004, 826.

114. Yadav A, Schtein M, Pipe K P J of Power Sources 175, 2008, 909.

115. Coner O, Pipe K P Shtein M Fibre based organic PV devices Appl. Phys. Letters 92, 2008, 193306.

116. Ghas A P, Gerenser L J, Jarman C M, Pornailik J E, Appl. Phys. Lett. 86, 2005, 223503.

117. Wang P, Zakeeruddin S M, Gratzel M and Fluorine J Chem. 125, 2004, 1241.

118. Yu J W, Chin B D, Kim J K and Kang N S Patent IPC8 Class: AH01L3100FI, USPC Class: 136259, 2007.

119. http://www.coatema.de/ger/downloads/veroeffentlichungen/news/0701_Textile%20Month_GB.pdf.

120. Krebs F C., Fyenbo J and J́rgensen Mikkel "Product integration of compact roll-to-roll processed polymer solar cell modules: methods and manufacture using flexographic printing, slot-die coating and rotary screen printing" J. Mater. Chem., 20, 2010, 8994–9001.

121. Krebs F C., Polymer solar cell modules prepared using roll-to-roll methods: Knife-overedge coating, slot-die coating and screen printing Solar Energy Materials and Solar Cells 93, (4), 2009, 465–475.

122. Luo P, Zhu Cand Jiang G Preparation of CuInSe2 thin films by pulsed laser deposition the Cu–In alloy precursor and vacuum selenization, Solid State Communications, 146 (1–2), 2008, 57–60.

123. Solar energy: the state of art Ed: Gordon J Pub: James and James Willium Road London, 2001.

124. Loewenstein T., Hastall A., Mingebach M., Zimmermann Y., Neudeck A. and Schlettwein D., Phys. Chem. Chem. Phys., 2008, 10,1844.

125. Lincot D. and Peulon S., J. Electrochem. Soc., 1998, 145, 864.

126. Bedeloglua A, Demirb A, Bozkurta Y, Sariciftci N S, Synthetic Metals 159 (2009) 2043–2048.

127. Pettersson LAA, Ghosh S, Inganas O. Org Electron 2002; 3: 143.

128. Kim JY, Jung JH, Lee DE, Joo J. Synth Met 2002; 126: 311.

129. Kim WH, Ma̋kinen AJ, Nikolov N, Shashidhar R, Kim H, Kafafi ZH. Appl Phys Lett 80, (2002), 3844.

130. Jonsson SKM, Birgerson J, Crispin X, Greczynski G, Osikowicz W, van der Gon AWD, SalaneckWR, Fahlman M. Synth. Met 1, 2003; 139.

131. Ouyang J, Xu Q, Chu C W, Yang Y, Lib G, Shinar Joseph S On the mechanism of conductivity enhancement in poly(3,4-ethylenedioxythiophene): poly(styrene sulfonate) film through solvent treatment" Polymer 45, 2004, 8443–8450.

132. Durisch W, Urban J and Smestad G "Characterization of solar cells and modules under actual operating conditions" WERS 1996, 359–366.

133. Wang W, Xia G, Zheng J, Feng L and Hao R "Study of polycrystalline ZnTe(ZnTe:Cu) thin films for photovoltaic cells." Journal of Materials Science: Materials in Electronics 18(4), 2007, 427–431.

134. Khanna, R.K. "Raman-spectroscopy of oligomeric SiO species isolated in solid methane." Journal of Chemical Physics 74 (4), (1981) 2108.

135. Koch, C., Ito M., Schubert M., 'Low temperature deposition of amorphous silicon solar cells,' Solar Energy Materials and Solar Cells, 68, 2001, 227–236.

136. Aernouts, T. 19th European Photovoltaics Conference, June 7–11, Paris, France, 2004.

137. Lund P D Renewable energy 34, 2009, 53.

138. Yaksel I Renewable energy 4, 2008, 802.

139. Gunes S, Beugebauer H and Saricftci N S Chem Rev. 107, 2007, 1324.

140. Coakley KM, ,cGehee M D, Chem Mater. 16,2004,4533.

141. Ji J.-Y., Shen T.-C., 'Low-temperature silicon epitaxy on hydrogen-terminated Si surfaces,' *Physical Review B* 70, 2004, 115–309.

142. Moncrieff R.W., *Man-made Fibres*, Newnes-Butterworths, 6th edn, London and Boston, 1975.

143. Hearle J.W.S., 'Introduction' in *High-performance Fibres*,' editor Hearle J.W.S., Woodhead Publishing Limited, Cambridge, 2001, 1–22.

144. Jones F.R., 'Glass fiber' in *High-performance Fibres*, editor Hearle J.W.S., Woodhead Publishing Limited, Cambridge, 2001, 191–238.

145. Abdelfattah M.S., 'Formation of textile structures for giant-area applications' in *Electronics on Unconventional Substrates – Electrotextiles and Giant-Area Flexible Circuits*, MRS Symposium roceedings Volume 736, Materials Processing Society, Warrendale, Pennsylvania, USA, 2003, 25–36.

146. Bonderover E., Wagner S., Suo Z., 'Amorphous silicon thin transistors on kapton fibers,' in *Electronics on Unconventional Substrates – Electrotextiles and Giant-Area Flexible Circuits*, MRS Symposium Proceedings Volume 736, Materials Processing Society, Warrendale, Pennsylvania, USA, 2003, 109–114.

CHAPTER 6

MODERN APPLICATIONS OF NANOENGINEERED MATERIALS IN TEXTILE INDUSTRIES

SHIMA MAGHSOODLOU and AREZOO AFZALI

University of Guilan, Rasht, Iran

CONTENTS

ABSTRACT

The field of nanotechnology can be broadly or narrowly applied to any industry, and the textiles market is no exception. In essence, there should be less emphasis on the catch phrase "nanotechnology" and more emphasis on the diverse technologies to create nanofibers, modify the surface of fibers and fabrics, and incorporate small particles into or onto fibers and nanofibers. With the commercial success of nanofibers production and many surface modification technologies, and the growing academic interest in all areas, no doubt, be not only many new technologies that can be used, but also many new commercial products to enhance people's lives and keep them safe.

6.1 INTRODUCTION

The fast development and changes in life style has attracted peoples towards a more comfort and luxurious life. People are moving towards small, safer, cheaper and fast working products which not only reduces the work load but also help them to carry out their works at a much greater pace with minimum efforts. There have been development of gadgets that are much smaller in size like microchip, nano-capsules, carbon tubes, memory cards, pen drives, etc., which reduces the problems of transport, and storage and are also much faster and reliable by which we can carry out more of our work in less time. In the formation and development of such products nanotechnology plays a very important and vital role [1, 2].

The term nanotechnology (sometimes shortened to "nanotech") comes from nanometer – a unit of measure of 1 billionth of a meter of length. The concept of Nanotechnology was given by Nobel Laureate Physicist Richard Feynman, in 1959. Nanotechnology is defined as the understanding, manipulation, and control of matter at the length scale on nanometer, such that the physical, chemical, and biological properties of materials (individual atoms, molecules and bulk matter) can be engineered, synthesized or altered to develop the next generations of improved materials, devices, structures, and systems [3].

Generally, nanotechnology deals with structures that are sized between 1 to 100 nm in at least one dimension and involves developing materials or devices possessing dimension within size. Nanotechnology creates structure that have excellent properties by controlling atoms and molecules, functional materials, devices and systems on the nanometer scale by involving precise placement of individual atoms [4]. Although nanotechnology is a relatively recent development in scientific research, the development of its central concepts happened over a longer period of time. The emergence of nanotechnology in the 1980s was caused by the convergence of experimental advances. The early 2000s also saw the beginnings of commercial applications of nanotechnology, although these were limited to bulk applications of nanomaterials, such as, the silver nano-platform for using silver nanoparticles as an antibacterial agent, nanoparticles-based transparent sunscreens, and carbon nano-tubes for stain-resistant textiles [5, 6].

Throughout history, the textiles sectors have been used worldwide in a wide range of consumer applications. Natural fibers, such as, cotton, silk, and wool, along with synthetic fibers, such as, polyester and nylon, continue to be the most widely used fibers for apparel manufacturing. Synthetic fibers are mostly suitable for do-mestic and industrial applications, such as, carpets, tents, tires, ropes, belts, cleaning cloths, and medical products. Natural and synthetic fibers generally have different characteristics, which make them ideally suitable mainly for apparel. Depending on the end-use application, some of those characteristics may be good, while the others may not be as good to contribute to the desired performance of the end product [7, 8].

As stated previously, nanotechnology brings the possibility of combining the merits of natural and synthetic fibers, such that advanced fabrics that complement the desirable attributes of each constituent fiber can be produced. In the last de-cade, the advent of nanotechnology has spurred significant developments and in-novations in this field of textile technology. By using nanotechnology, there have been developments of several fabric treatments to achieve certain enhanced fabric attributes, such as, superior durability, softness, tear strength, abrasion resistance, durable-press and wrinkle-resistance. The (nano)-treated core component of a core-wrap bi-component fabric provides high strength, permanent antistatic behavior, and durability, while the traditionally treated wrap component of the fabric provides desirable softness, comfort, and esthetic characteristics [9, 10].

6.2   THE TEXTILE INDUSTRY BACKGROUND

The origin of the textile industry is lost in the past. Fine cotton fabrics have been found in India, dating from some 6–7000 years ago, and fine and delicate linen fabrics have been found from two to three thousand years ago, at the height of the Egyptian civilizations. More recent archaeological excavations, among some of Eu-rope's oldest Stone Age sites, have found imprints of textile structures, dating back some 25,000 years, but in the humid conditions obtaining in these more northerly areas, all traces of the actual textiles have long disappeared, unlike those from the dry areas of India and Egypt [11].

Until more recent times, the spinning of the yarns and the weaving of the fabrics were generally undertaken by small groups of people, working together – often as a family group. However, during the Roman occupation of England, the Romans established a factory' at Winchester, for the production, on a larger scale, of warm Woolen blankets, to help reduce the impact of the British weather on the soldiers from southern Europe [12].

In the family context, it generally fell to the female side to undertake the spin-ning, while the weaving was the domain of the men. Spinning was originally done using the distaff to hold the unspun fibers, which were then teased out using the fingers and twisted into the final yarn on the spindle. In the 1530s, in Brunswick, a 'spinning wheel' was invented, with the wheel driven by a foot pedal, giving better control and uniformity to the yarns produced. Often, great skill was developed, as shown by the records of a woman in Norwich, who spun one pound of combed wool into a single yarn measuring 168,000 yards, and from the same weight of cotton, spun a yarn of 203,000 yards. In today's measures this is equivalent to a cotton count of 240, or approximately 25 decitex. Cotton count is the number of hanks of 840 yards (768 meters) giving a total weight of 1 lb (453.6 g). A Tex is a measurement of the linear density of a yarn or cord, being the weight in grams of a 1,000 m length; a decitex is the weight in grams of a 10,000 m length [13, 14].

By the eighteenth century, small cooperatives were being formed for the production of textiles, but it was really only with the mechanization of spinning and weaving during the Industrial Revolution, that mass production started.

Up to this time, both spinning and weaving were essentially hand operations. Handlooms were operated by one person, passing the weft (the transverse threads) by hand, and performing all the other stages of weaving manually. In 1733, John Kay invented the 'flying shuttle, ' which enabled a much faster method for inserting the weft into the fabric at the loom and greatly increased the productivity of the weavers [15].

Until the advent of the flying shuttle, the limiting factor in the production chain for fabrics was the output of the individual weaver, but this now changed and with the more rapid use of the yarns, their production became the limiting factor in the total process. In 1764, this was partly resolved by the invention, by James Hargreaves, of the 'Spinning Jenny,' which was developed further by Sir Richard Arkwright, with his water spinning frame, in 1769, and then in 1779, by Samuel Crompton, with his 'spinning mule' [16].

Alongside these developments in spinning, similar changes were taking place in the weaving field, with the invention of the power loom by Edmund Cartwright, in 1785.

With this increase in mechanization of the whole industry, it was logical to bring the production together, rather than keeping it widely spread throughout the homes of the producers. Accordingly, factories were established [17]. The first of such was in Doncaster in 1787, with many power looms powered by one large steam engine. Unfortunately, this was not a financial success, and the mill only operated for about 3 or 4 years [18].

Meanwhile, other mills were being established, in Glasgow, Dumbarton and Manchester. A large mill was erected at Knott Mill, Manchester, although this burnt down after only about 18 months. The first really successful mill was opened in Glasgow in 1801 [19].

However, this industrialization was not to everyone's liking; many individuals were losing their livelihoods to the mass production starting to come from the increasing number of mills. This led to a backlash from the general public, resulting in the Luddite Riots in 1811–1812, when bands of masked people under the leadership of 'King Ludd' attacked the new factories, smashing all the machinery therein. It was only after very harsh suppression, resulting in the hanging or deportation of convicted Luddites in 1813, that this destruction was virtually stopped. However, there were still some outbreaks of similar actions in 1816, during the depression following the end of the British war with France, and this intermittent action only finally stopped when general prosperity increased again in the 1820s [20].

Following this, the textile industry expanded considerably, particularly in the areas where the raw materials were readily available. For example, the Woolen mills in East Anglia, where there was good grazing for the sheep, and in West Yorkshire and Eastern Lancashire, where either coal was available for powering the new steam engines, or where fast flowing streams existed to provide the energy source for water-powered mills. The main Woolen textile production developed in Yorkshire, as it was easier and cheaper to transport the raw wool there, than to carry the large quantities of coal required to power the mills to the wool growing areas [21]. In Lancashire, with the ports of Liverpool and Manchester close by for the importation of cotton from America, the cotton industry grew and flourished. However, in the 1860s, due to the American Civil War, the supply of cotton from America dried up and caused great hardship among the cotton towns of south and east Lancashire [22].

On account of this, and with the great strides being made in chemistry, research was begun to find ways of making artificial yarns and fibers. The first successful artificial yarn was the Chardonnet 'artificial silk, ' a cellulosic fiber regenerated from spun nitrocellulose. Further developments lead to the cuprammonium process and then to the viscose process for the production of another cellulosic, rayon [23]. This latter viscose was fully commercialized by Courtaulds in 1904, although it was not widely used in rubber reinforcement until the 1920s, with the development of the balloon type [24].

Research continued into fiber-forming polymers, but the next new fully synthetic yarn was not discovered until the 1930s, when Wallace Hume Carothers, working for DuPont, discovered and developed nylon. This was first commercialized in 1938 and was widely developed during the 1940s to become one of the major yarn types used. Continuing research led to the discovery of polyester in 1941, and over the ensuing decades, polyolefin fibers (although because of their low melting/softening temperatures, these are not used as reinforcing fibers in rubbers) and aramids [25].

As the chemical industry greatly increased the types of yarns available for textile applications, so the machinery used in the industry was being developed. Whereas the basic principles of spinning and weaving have not significantly changed over the millennia, the speed and efficiency of the equipment used for this has been vastly been improved. In weaving, the major changes have been related to the method of weft insertion; the conventional shuttle has been replaced by rapiers, air and water jets, giving far higher speeds of weft insertion [26].

Other methods of fabric formation have similarly been developed, such as, the high speed knitting machines and methods for producing fabric webs [27].

The use of nanotechnology in the textile industry made it multifunctional which can produce fibers with variable functions and applications such as UV protection, antiodor, antimicrobial etc. In many cases smaller amounts of the additive are required, for the saving on resources. The success of Nanotechnology and its potential applications in textiles lies in various fields where new methods are combined with multifunctional textile systems, durable etc. without affecting the inherent properties of the textiles including softness, flexibility, washability, etc. Keeping the above factors in consideration, the present review highlighted the use of nanotechnology in textile industry and textile engineering, the types and methods of preparation of different nano composites used in textiles [28, 29].

6.3   TEXTILE TYPES

Textiles can be made from many materials. They are classified on the basis of their component fibers into, animal (wool, silk), plant (cotton, flax, jute), mineral (asbestos, glass fiber), and synthetic (nylon, polyester, acrylic). They are also classified as to their structure or weave, according to the manner in which warp and weft cross each other in the loom. Textiles are made in various strengths and degrees of durability, from the finest gossamer to the sturdiest canvas. The relative thickness of fibers in cloth is measured in deniers. Microfiber refers to fibers made of strands thinner than one denier [30, 31].

6.3.1   *ANIMAL* TEXTILES

The main animal fiber used for textiles is wool. Animal textiles are commonly made from hair or fur of animals. Silk is another animal fiber produces one of the most luxurious fabrics. Sheep supply most of the wool, but members of the camel family and some goats also furnish wool. Wool, commonly used for warm clothing, refers to the hair of the domestic goat or sheep and it is coated with oil known as lanolin, which is water-proof and dirt-proof making a comfortable fabric for dresses, suits, and sweaters. The term woolen refers to raw wool, while *worsted* refers to the yarn spun from raw wool. Cashmere, the hair of the Indian Cashmere goat, and mohair, the hair of the North African Angora goat, are types of wool known for their softness. Other animal textiles made from hair or fur is alpaca wool, vicuña wool, llama wool, and camel hair. They are generally used in the production of coats, jackets, ponchos, blankets, and other warm coverings. Angora refers to the long, thick, soft hair of the Angora rabbit [32, 33].

Silk is an animal textile made from the fibers of the cocoon of the Chinese silk-worm. It is spun into a smooth, shiny fabric prized for its sleek texture. Silk comes from cocoons spun by silkworms. Workers unwind the cocoons to obtain long, natural filaments. Fabrics made from silk fibers have great luster and softness and can be dyed brilliant colors. Silk is especially popular for saree, scarfs and neckties [34].

6.3.2 *PLANT* TEXTILES

Plants provide more textile fibers than do animals or minerals. Cotton fibers produce soft, absorbent fabrics that are widely used for clothing, sheets, and towels. Fibers of the flax plant are made into linen. The strength and beauty of linen have made it a popular fabric for fine tablecloths, napkins, and handkerchiefs [35].

Grass, rush, hemp, and sisal are all used in making rope. In the first two cases, the entire plant is used for this purpose, while in the latter two; only fibers from the plant are used. Coir (coconut fiber) is used in making twine, floor mats, door mats, brushes, mattresses, floor tiles, and saking. Straw and bamboo are both used to make hats. Straw, a dried form of grass, is also used for stuffing, as is kapok. Fibers from pulpwood trees, cotton, rice, hemp, and nettle are used in making paper. Cotton, flax, jute, and modal are all used in clothing. Piña (pineapple fiber) and ramie are also fibers used in clothing, generally with a blend of other fabrics such as cotton. Seaweed is sometimes used in the production of textiles. A water soluble fiber known as *alginate* is produced and used as a holding fiber. When the cloth is finished, the alginate is dissolved, leaving an open area [36, 37].

6.3.3 *MINERAL* TEXTILES

Asbestos and basalt fiber are used for vinyl tiles, sheeting, and adhesives, "transite" panels and siding, acoustical ceilings, stage curtains, and fire blankets. Glass fiber is used in the production of spacesuits, ironing board and mattress covers, ropes and cables, reinforcement fiber for motorized vehicles, insect netting, flame-retardant and protective fabric, soundproof, fireproof, and insulating fibers. Metal fiber, metal foil, and metal wire have a variety of uses, including the production of "cloth-of-gold" and jewelry [38, 39].

6.3.4 *SYNTHETIC* TEXTILES

Most manufactured fibers are made from wood pulp, cotton linters, or petrochemicals. Petrochemicals are chemicals made from crude oil and natural gas. The chief fibers manufactured from petrochemicals include nylon, polyester, acrylic, and olefin. Nylon has exceptional strength, wears well, and is easy to launder. It is popular for hosiery and other clothing and for carpeting and upholstery. Such products as conveyor belts and fire hoses are also made of nylon. All synthetic textiles are used primarily in the production of clothing [37, 40].

- Polyester fiber is used in all types of clothing, either alone or blended with fibers such as cotton.
- Acrylic is a fiber used to imitate wools, including cashmere, and is often used in place of them.
- Nylon is a fiber used to imitate silk and is tight-fitting; it is widely used in the production of pantyhose.
- Lycra, spandex, and tactel are fibers that stretch easily and are also tight-fitting. They are used to make active wear, bras, and swimsuits.
- Olefin (Polypropylene or Herculon) fiber is a thermal fiber used in active wear, linings, and warm clothing.
- Lurex is a metallic fiber used in clothing embellishment.
- Ingeo is a fiber blended with other fibers such as cotton and used in clothing. It is prized for its ability to wick away perspiration.

6.4 TEXTILE PRODUCTION

6.4.1 *PRODUCTION* METHODS

Most textiles are produced by twisting fibers into yarns and then knitting or weaving the yarns into a fabric. This method of making cloth has been used for thousands of years. But throughout most of that time, workers did the twisting, knitting, or weaving largely by hand. With today's modern machinery, textile mills can manufacture as much fabric in a few seconds as it once took workers weeks to produce by hand [41]. The production of textiles are done by different methods and some of the common production methods are listed as follows [42]:

(i) *Weaving* (by machine as well as by hand) (ii) *Knitting* (iii) *Crochet* (iv) *Felt* (fibers are matted together to produce a cloth (v) *Braiding* (vi) *Knotting. Weaving* is a textile production method that involves interlacing a set of vertical threads (called the warp) with a set of horizontal threads (called the weft). This is done on a machine known as a loom, of which there are a number of types. Some weaving is still done by hand, but a mechanized process is used most often. Tapestry, sometimes classed as embroidery, is a modified form of plain cloth weaving [43].

Knitting and *crocheting* involve interlacing loops of yarn, which are formed either on a knitting needle or crochet hook, together in a line. The two processes differ in that the knitting needle has several active loops at one time waiting to interlock with another loop, while crocheting never has more than one active loop on the needle [44].

Other specially prepared fabrics not woven are *felt* and *bark* (or tapa) cloth, which are beaten or matted together, and a few in which a single thread is looped or plaited, as in crochet and netting work and various laces. *Braiding* or *plait-*

ing involves twisting threads together into cloth. Knotting involves tying threads together and is used in making macramé [45].

Most textiles are now produced in factories, with highly specialized power looms, but many of the finest velvets, brocades, and table linens are still made by hand. Lace is made by interlocking threads together independently, using a backing and any of the methods described above, to create a fine fabric with open holes in the work. Lace can be made by hand or machine. The weaving of carpet and rugs is a special branch of the textile industry.

Carpets, rugs, velvet, velour, and velveteen are made by interlacing a secondary yarn through woven cloth, creating a tufted layer known as a nap or pile [46].

6.4.2 PRODUCTION OF COTTON CLOTHES

In the 1700s, English textile manufacturers developed machines that made it possible to spin thread and weave cloth into large quantities. Today, the United States, Russia, China and India are major producers of cotton. When cotton arrives at a textile mill, several blenders feed cotton into *cleaning machines*, which mix the cotton, break it into smaller pieces and remove trash. The cotton is sucked through a pipe into *picking machines*. Beaters in these machines strike the cotton repeatedly to knock out dirt and separate lumps of cotton into smaller pieces. Cotton then goes to the *carding machine*, where the fibers are separated. Trash and short fibers are removed. Some cotton goes through a *comber* that removes more short fibers and makes a stronger, more lustrous yarn. This is followed by spinning processes which do three jobs: *draft* the cotton, or reduce it to smaller structures, straighten and parallel the fibers and lastly, put twist into the yarn. The yarns are then made into cloth by weaving, knitting or other processes [47, 48].

Some of the properties of cotton are discussed as follows; (i) soft and comfortable, (ii) wrinkles easily, (iii) absorbs perspiration quickly, (iv) good color retention and good to print on, and (v) strong and durable [49].

6.4.3 PRODUCTION OF WOOL

The processing of wool involves four major steps. First comes shearing, followed by sorting and grading, making yarn and lastly, making fabric. This is followed by grading and sorting, where workers remove any stained, damaged or inferior wool from each fleece and sort the rest of the wool according to the quality of the fibers. Wool fibers are judged not only on the basis of their strength but also by their *fineness* (diameter), length, *crimp* (waviness) and color [50].

The wool is then scoured with detergents to remove the *yolk* and such impurities as sand and dust. After the wool dries, it is *carded*. The carding process involves passing the wool through rollers that have thin wire teeth. The teeth untangle the fibers and arrange them into a flat sheet called a *web*. The web is then formed into narrow ropes known as *silvers*. After carding, the processes used in making yarn vary slightly, depending on the length of the fibers. Carding length fibers are used to make woolen yarn. Combing length fibers and French combing length fibers are made into *worsted yarn* [51].

Woolen yarn, which feels soft, has a fuzzy surface and it is heavier than worsted. While worsted wool is lighter and highly twisted, it is also smoother, and is not as bulky, thus making it easier to carry or transport about. Making worsted wool requires a greater number of processes, during which the fibers are arranged parallel to each other. The smoother the hard-surface worsted yarns, the smoother the wool it produces, meaning, less fuzziness. Fine worsted wool can be used in the making of athletics attire, because it is not as hot as polyester, and the weave of the fabric allows wool to absorb perspiration, allowing the body to "breathe." Wool manufacturers knit or weave yarn into a variety of fabrics. Wool may also be dyed at various stages of the manufacturing process and undergo finishing processes to give them the desired look and feel [52, 53].

The finishing of fabrics made of woolen yarn begins with *fulling*. This process involves wetting the fabric thoroughly with water and then passing it through the rollers. Fulling makes the fibers interlock and mat together. It shrinks the material and gives it additional strength and thickness. Worsteds go through a process called *crabbing* in which the fabric passes through boiling water and then cold water. This procedure strengthens the fabric.

The exclusive features of cotton fabric are (i) hard wearing and absorbs moisture, (ii) does not burn over a flame but smolders instead, (iii) lightweight and versatile, (iv) does not wrinkle easily, and (v) resistant to dirt and wear and tear [54].

6.4.4   PRODUCTION OF SILK

Silkworms are cultivated and fed with mulberry leaves. Some of these eggs are hatched by artificial means such as an incubator, and in the olden times, the people carried it close to their bodies so that it would remain warm. Silkworms that feed on smaller, domestic tree leaves produce the finer silk, while the coarser silk is produced by silkworms that have fed on oak leaves. From the time they hatch to the time they start to spin cocoons, they are very carefully tended to. Noise is believed to affect the process, thus the cultivators try not to startle the silkworms. Their cocoons are spun from the tops of loose straw. It will be completed in two to three days' time. The cultivators then gather the cocoons and the chrysales are killed by heating and drying the cocoons. In the olden days, they were packed with leaves and salt in a jar, and then buried in the ground, or else other insects might bite holes in it. Modern machines and modern methods can be used to produce silk but the old-fashioned hand-reels and looms can also produce equally beautiful silk [55, 56].

The properties of silk includes; (i) versatile and very comfortable, (ii) absorbs moisture, (iii) cool to wear in the summer yet warm to wear in winter, (v) easily dyed, (v) strongest natural fiber and is lustrous, and (vi) poor resistance to sunlight exposure [57].

6.4.5   *PRODUCTION* OF NYLON MATERIALS

Nylon is made by forcing molten nylon through very small holes in a device called 'spinneret'. The streams of nylon harden into filament once they come in contact with air. They are then wound onto bobbins. These fibers are drawn (stretched) after they cool. Drawing involves unwinding the yarn or filaments and then winding it around another spool. Drawing makes the molecules in each filament fall into parallel lines. This gives the nylon fiber strength and elasticity. After the whole drawing process, the yarn may be twisted a few turns per yard or meters as it is wound onto spools. Further treatment to it can give it a different texture or bulk [58].

Properties of the nylon: (i) it is strong and elastic, (ii) it is easy to launder, (iii) it dries quickly, (iv) it retains its shape, and (v) it is resilient and responsive to heat setting [59].

4.6   PRODUCTION OF POLYESTER

Polyesters are made from chemical substances found mainly in petroleum. Polyesters are manufactured in three basic forms – *fibers*, *films* and *plastics*. Polyester fibers are used to make fabrics. Poly ethylene terephthalate or simply PET is the most common polyester used for fiber purposes. This is the polymer used for making soft drink bottles. Recycling of PET by remelting and extruding it as fiber may saves much raw materials as well as energy. PET is made by ethylene glycol with either terephthalic acid or its methyl ester in the presence of an antimony catalyst. In order to achieve high molecular weights needed to form useful fibers, the reaction has to be carried out at high temperature and in a vacuum [60].

6.5   BASIC NANOTECHNOLOGY

Two main approaches are used in nanotechnology that is, the bottom up approach and the top-down approach. In case of the "bottom-up" approach, the different type of materials and the instruments are made up from different types of molecular components which combine themselves by chemical ways basing on the mechanism of molecular recognition. In case of the "top-down" approaches, various nano-objects are made from various types of components without atomic-level control. Materials reduced to the nanoscale can show different properties compared to what they exhibit on a macro scale, enabling unique applications. The basic premise is that properties can dramatically change when a substance's size is reduced to the nanometer range. For instance, ceramics which are normally brittle can be deformable when their size is reduced, opaque substances become transparent (copper); stable materials turn combustible (aluminum); insoluble materials become soluble (gold) [61, 62].

Nanoparticles can be prepared from a variety of materials such as proteins, polysaccharides and synthetic polymers. The selection of materials in mainly depended on factors like size of the nanoparticles required, inherent properties such as aqueous, solubility and stability, surface characteristics i.e. charge and permeability, degree of biodegradability, biocompatibility and toxicity, release of the desired product, antigenicity of the final product etc. Polymeric nanoparticles have been prepared most frequently be three methods: (1) dispersion of the performed polymers; (2) polymerization of the monomers; and (3) ionic gelation of the hydrophobic or hydrophilic polymers. However, techniques like supercritical fluid technology and particle repulsion in non wetting templates (PRINT) have been also used in modern days [63–65].

6.6 NANOTECHNOLOGY IN TEXTILE INDUSTRY AND TEXTILE ENGINEERING

Of the many applications of nanotechnology, textile industry has been currently added as one of the most benefited sector. Application of nanotechnology in textile industry has tremendously increased the durability of fabrics, increase its comfortness, hygienic properties and have also reduces its production cost. Nanotechnology also offers many advantages as compare to the conventional process in term of economy, energy saving, ecofriendliness, control release of substances, packaging, separating and storing materials on a microscopic scale for later use and release under control condition [66]. The unique and new properties of nanotechnology have attracted scientists and researchers to the textile industry and hence the use of nanotechnology in the textile industry has increased rapidly. This may be due to the reason that textile technology is one of the best areas for development of nanotechnology. The textile fabrics provide best suitable substrates where a large surface area is present for a given weight or a given volume of fabric. The synergy between nanotechnology and textile industry uses this property of large interfacial area and a drastic change in energetic is experienced by various macromolecules or super molecules in the vicinity of a fiber when changing from wet state to a dry state [67].

The application of nanoparticles to textile materials have been the objective of several studies aimed at producing finished fabrics with different functional performances. Nanoparticles can provide high durability for treated fabrics as they posse large surface area and high surface energy that ensure better affinity for fabrics and led to an increase in durability of the desired textile function. The particle size also plays a primary role in determining their adhesion to the fibers. It is reasonable to except that the largest particle agglomerates will be easily removed from the fiber surface, while the smallest particle will penetrate deeper and adhere strongly into the fabric matrix [68, 69]. Thus, decreasing the size of particles to nano-scale dimensional, fundamentally changes the properties of the material and indeed the entire substance.

A whole variety of novel nanotech textiles are already on the market at this moment. Areas where nanotech enhanced textiles are already seeing some applications include sporting industry, skincare, space technology and clothing as well as materials technology for better protection in extreme environments. The use of nanotechnology allows textiles to become multifunctional and produce fabrics with special functions, including antibacterial, UV-protection, easy-clean, water- and stain repellent and antiodor [68].

6.7. TECHNOLOGICAL USES

Textile materials are materials for the daily use. Besides, textiles are also play a vital role in fashion shows. Technically, they are applied in variety of our life savers including safety belt, and the airbags in the cars, bullet-proof vests protect against weapons, used as Implant material in medical applications. Recently, PU foams that can be combined with TPU films are used as excellent material for functional medical wound dressing which are not only help wounds to heal but also allow the wound to breathe by permeate the water-vapor [70, 71].

6.7.1 *ADVANCEMENTS* IN TEXTILE PRODUCTION

Technological advances during the past decade have opened many new doors for the Textile and Apparel industries, especially in the area of rapid prototyping and related activities. During the past decade, the textile and apparel complex has been scrambling to adjust to a rapidly changing business environment. Textiles and yarns have been around for thousands of years but in the last 50 years, progress in the technology has been most remarkable. The application of textiles and yarns have move beyond clothing and fabrics and they are increasingly used in high value-added applications such as composites, filtration media, gas separation, sensors and biomedical engineering. With the emergence of nanotechnology, the users of textiles and yarns are switching their attention to the production of nanometer diameter fibers [72, 73].

6.7.2 *BODY SCANNING*

The development of 3 dimensional body scanning technologies may have significant potential for use in the apparel industry. First, this technology has the potential of obtaining an unlimited number of linear and nonlinear measurements of human bodies (in addition to other objects) in a matter of seconds. Because an image of the body is captured during the scanning process, the location and description of the measurements can be altered as needed in mere seconds, as well. Second, the measurements obtained using this technology has the potential of being more precise and reproducible than measurements obtained through the physical measurement process. Third, with the availability of an infinite number of linear and nonlinear measurements the possibility exists for garments to be created to mold to the 3 dimensional shapes of unique human bodies. Finally, the scanning technology allows measurements to be obtained in a digital format that could integrate automatically into apparel CAD systems without the human intervention that takes additional time and can introduce error [74, 75].

6.7.3   APPAREL CAD

Adoption of CAD/CAM technology over the past few decades has increased the speed and accuracy of developing new products, reducing the manpower required to complete the development process. Unfortunately, this technology has also encouraged manufacturers to simplify the design of garments, allowing a more efficient use of materials and making mass production much easier. These systems initially only made an effort to adapt traditional manual methods instead of encouraging innovation in design or fit adaptations. Current developments in the area of information technology help build on the traditional CAD/CAM functions and offer a new way of looking at and using the systems for design and product development [76, 77].

6.7.4   NANOFIBER FABRICATION

As with all new technologies, polymeric nanofibers have brought with it a new beginning to the understanding of polymeric fibers. One apparent advantage of nanofibers is the huge increase in the surface area to volume ratio. Given the huge potential of nanofibers, the key is to use a technique that is able to easily fabricate nanofibers out of most, if not all the different type of polymers. A number of processing techniques such as drawing [78], template synthesis [79], phase separation [80], self-assembly [81], electrospinning [82], etc. have been used to prepare polymer nanofibers in recent years. The drawing is a process similar to dry spinning in fiber industry, which can make one-by-one very long single nanofibers. The template synthesis, as the name suggests, uses a nanoporous membrane as a template to make nanofibers of solid (a fibril) or hollow (a tubule) shape. The phase separation consists of dissolution, gelation, and extraction using a different solvent, freezing, and drying resulting in nanoscale porous foam. The process takes relatively long period of time to transfer the solid polymer into the nano-porous foam. The self-assembly is a process in which individual, preexisting components organize themselves into desired patterns and functions.

6.7.5 ELECTROSPINNING

Electrospinning is a process where continuous fibers with diameters in the submicron range are produced through the action of an applied electric field imposed on a polymer solution. Textiles made from these fibers have high surface area and small pore size, making them ideal materials for use in protective clothing. There are currently several applications of electrospinning that are being investigated, including: The fabrication of transparent composites reinforced with nanofibers. The effect of processing conditions on the morphology of polymers has been investigated. The scaling up of current production techniques is one of the main areas of research in which we have ongoing interest is the Adhesives, Permselective membranes, Antifouling coatings, Active protective barriers against chemical and biological threats are few examples of nanofibers produced through electrospinning [83].

6.7.5.1 USING ELECTROSPUN NANOFIBERS: BACKGROUND AND TERMINOLOGY

The three inherent properties of nanofibrous materials that make them very attractive for numerous applications are their high specific surface area (surface area/unit mass), high aspect ratio (length/diameter) and their biomimicking potential. These properties lead to the potential application of electrospun fibers in such diverse fields as high-performance filters, absorbent textiles, fiber-reinforced composites, biomedical textiles for wound dressings, tissue scaffolding and drug-release materials, nano- and microelectronic devices, electromagnetic shielding, photovoltaic devices and high-performance electrodes, as well as a range of nanofiber-based sensors [84, 85].

In many of these applications the alignment, or controlled orientation, of the electrospun fibers is of great importance and large-scale commercialization of products will become viable only when sufficient control over fiber orientation can be obtained at high production rates. In the past few years research groups around the world have been focusing their attention on obtaining electrospun fibers in the form of yarns of continuous single nanofibers or uniaxial fiber bundles. Succeeding in this will allow the processing of nanofibers by traditional textile processing methods such as weaving, knitting and embroidery. This, in turn, not only will allow the significant commercialization of several of the applications cited above, but will also open the door to many other exciting new applications [86, 87].

Incorporating nanofibers into traditional textiles creates several opportunities. In the first instance, the replacement of only a small percentage of the fibers or yarns in a traditional textile fabric with yarns of similar diameter, but now made up of several thousands of nanofibers, can significantly increase the toughness and specific surface area of the fabric without increasing its overall mass [88]. Alternatively, the complete fabric can even be made from nanofiber yarns. This has important implications in protective clothing applications, where lightweight, breathable fabrics with protection against extreme temperatures, ballistics, and chemical or biological agents are often required. On an esthetic level, nanofiber textiles also exhibit extremely soft handling characteristics and have been proposed for use in the production of artificial leather and artificial cashmere [85]. In biomedical applications the similarity between certain electrospun polymeric nanofibers and the naturally occurring nanofibrous structures of connective tissues such as collagen and elastin gives rise to the opportunity of creating artificial biomimicking wound dressings and tissue engineering scaffolds. Several studies on nonwoven nanofiber webs of biocompatible polymers have already shown potential in this area. Simple three-dimensional constructs for vascular prostheses have also been manufactured by electrospinning onto preformed templates. Although these initial studies show that enhanced cell adhesion, cell proliferation and scaffold vascularization can be obtained on porous, nonwoven nanofiber webs, the simplicity of the constructs and the fragile nature of nonwoven webs still limit their applicability to small areas [89, 90].

Creating complex three-dimensional scaffold structures with fibers aligned in a controlled fashion along the directions of the forces that are usually present in dynamic tissue environments, as for instance in muscles and tendons, will lead to significant improvements in the performance of tissue engineering scaffolds. With continuous nanofiber yarns it will become possible to create such aligned fiber structures on a large scale, simply by weaving. In addition, the age-old techniques of knitting and embroidery can then be applied to create very complicated, three-dimensional scaffolds, with precisely controlled porosity, and yarns placed exactly along the lines of dynamic force [91].

Several other fields will also benefit from the availability of continuous yarns from electrospun fibers. Owing to the high fiber-aspect ratios and increased fiber–matrix adhesion caused by the high specific surface areas, aligned nanofiber yarns can lead to stronger and tougher, lightweight, fiber reinforced composite materials. The incorporation of nanofiber-based sensors into textiles can lead to new opportunities in the fields of smart and electronic textiles. Aligned nanofiber yarns of piezoelectric polymers and other microactuator materials may lead to better performance in advanced robotics applications [92].

Since the revival of electrospinning in the early 1990s, several research groups have worked on controlling the orientation of electrospun fibers. Those who have worked in the field of electrospinning over the past decade have come from various disciplinary backgrounds, including physics, chemistry and polymer science, chemical and mechanical engineering, and also from the traditional textiles field. The result of this has been that literature on the topic of electrospinning, and especially yarns from electrospun fibers, is plagued with terminology from different disciplines, which often leads to misunderstanding and even self-contradictory statements. So, for instance, in paper on the electrospinning of individual fibers of

a novel polymer some authors might use the term yarn when they are actually referring to an individual fiber, or authors might refer to spinning a filament when they
are actually spinning a yarn [93].

- Fiber – a single piece of a solid material, which is flexible and fine, and has a
  high aspect ratio (length/diameter ratio).
- Filament – a single fiber of indefinite length.
- Tow – an untwisted assembly of a large number of filaments; tows are cut up
  to produce staple fibers.
- Sliver – an assembly of fibers in continuous form without twist. The assembly
  of staple fibers, after carding but before twisting, is also known as a sliver.
- Yarn – a generic term for a continuous strand of textile fibers or filaments in a
  form suitable for knitting, weaving or otherwise intertwining to form a fabric.
- Staple fiber – short-length fibers, as distinct from continuous filaments, which
  are twisted together (spun) to form a coherent yarn. Most natural fibers are
  staple fibers, the main exception being silk which is a filament yarn. Most
  artificial staple fibers are produced in this form by slicing up a tow of continuous filaments.
- Staple fiber yarn – a yarn consisting of twisted together (spun) staple fibers.
- Filament yarn – a yarn normally consisting of a bundle of continuous filaments. The term also includes monofilaments.
- Core-spun yarn – a yarn consisting of an inner core yarn surrounded by staple
  fibers. A core-spun yarn combines the strength and/or elongation of the core
  thread and the characteristics of the staple fibers that form the surface.
- Denier – a measure of linear density: the weight in grams of 9000 meters of
  yarn.
- Tex – another measure of linear density: the weight in grams of 1000 meters
  of yarn.

6.7.5.2   CONTROLLING FIBER ORIENTATION

As stated in the previous section, achieving control over the orientation of electrospun fibers is an important step towards many of their potential applications.
However, if one considers the fact that fiber formation occurs at very high rates
(several hundreds of meters of fiber per second) and that the fiber formation process
coincides with a very complicated three-dimensional whipping of the polymer jet
(caused by electrostatic bending instability), it becomes clear that controlling the
orientation of fibers formed by electrospinning is no simple task [94].

Various mechanical and electrostatic approaches have been taken in efforts to
control fiber alignment.

- Spinning onto a rapidly rotating surface

Several research groups have been routinely using this technique to obtain reasonably aligned fibers. The rapid rotation of a drum or disk and coinciding high

linear velocity of the collector surface allows fast take-up of the electrospun fibers as they are formed. The 'point-to-plane' configuration of the electric field does, however, lead to fiber orientations that deviate from the preferred orientation.

A special instance of the rotating drum set-up involves spinning onto a rapidly rotating sharp-edged wheel, which uses an additional electrostatic effect, since the sharp edge of the wheel creates a stronger converging electrostatic field, or a 'point-to-point' configuration, which has a focusing effect on the collected fibers (Fig. 6.1). This in turn leads to better alignment of the fibers [95, 96].

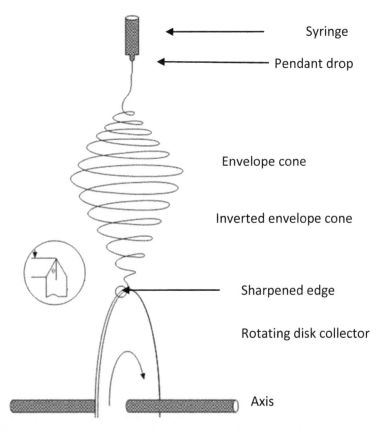

Syringe

Pendant drop

Envelope cone

Inverted envelope cone

Sharpened edge

Rotating disk collector

Axis

FIGURE 6.1   Converging electrostatic field on sharp-edged wheel electrode.

The gap alignment effect-uniaxially aligned arrays of electrospun fibers can be obtained through the gap alignment effect, which occurs when charged electrospun fibers are deposited onto a collector that consists of two electrically conductive substrates, separated by an insulating gap. This electrostatic effect (Fig. 6.2) has been observed by various groups. Recently this was investigated in more detail. Briefly, the lowest energy configuration for an array of highly charged fibers between two conductive substrates, separated by an insulating gap, is obtained when fibers align parallel to each other [97].

6.7.5.3 *PRODUCING* **NONCONTINUOUS OR SHORT YARNS**

Both spinning onto a rapidly rotating collector and the gap alignment effect have been used to obtain short yarns for experimental purposes [98].

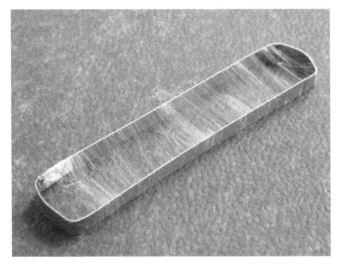

FIGURE 6.2   Aligned fibers obtained through the gap alignment effect.

Ring of aligned
nanofibers

FIGURE 6.3   Aligned fiber tows on rotating disk collector.

- Rotating collector method

Fibers can be spun onto a rapidly rotating disk, where the shearing force of the rotating disk led to aligned fibrous assemblies with good orientation.

These oriented fibers could then be collected and manually twisted into a yarn [99].

- Twisted yarns obtained from tows spun on a rotating disk collector

The stress–strain behavior of the yarns can be examined and the modulus, ultimate strength and elongation at the ultimate strength can be measured as a function of twist angle.

- Gap alignment method

short yarns can be made by quickly passing a wooden frame through the electrospinning jet several times (for up to an hour), in a process also known as 'combing,' resulting in a tow of reasonably aligned fibers, which were then 'gently twisted' to form a yarn [100].

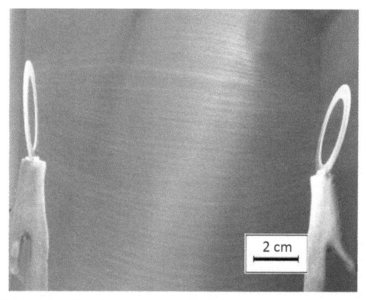

FIGURE 6.4   Fibers aligned between two parallel ring collectors.

6.7.5.4   PRODUCING CONTINUOUS YARNS

A common misconception in recent electrospinning literature is that the first literature on electrospinning dates back to 1934 when Formhals patented a method for manufacturing yarns from electrospun fibers. Some of the first publications on electrospinning date back as far as 1902, when first Cooley and then Morton patented processes for dispersing fluids. In both patents, the authors describe processes for producing very fine artificial fibers by delivering a solution of a fiber-forming material, such as, pyroxylin, a nitrated form of cellulose, dissolved in alcohol or ether, into a strong electric field [101, 102].

The reason for this oversight is unclear, but it could possibly be blamed on differences in terminology since the term 'electrospinning' has only become popular with the revival of the process in the mid-1990s [100]. Another more puzzling oversight, which has recently led several authors to bemoan the lack of processes for making continuous yarns from electrospun fibers, is that Formhals actually registered a series of seven patents over a period of ten years between 1934 and 1944 and that all these patents describe processes and/or improvements to processes for the manufacture of continuous yarns from electrospun fibers. Since the youngest of these patents is more than 60 years old, one could speculate that these processes did not really work, which would explain the absence of commercially available electrospun fiber yarns. An alternative explanation could be that, since Formhals lived in Mainz, Germany, and since the last patent application was filed in 1940, the disruption of World War II and the ensuing years simply led to the processes being forgotten. Closer inspection of the patents, aided by more recent knowledge of the electrospinning process, also leads us to believe that at least some of the described processes are viable and that they deserve further consideration. The patents of Formhals show a gradual evolution of his yarn production process over time and in many instances he applied the same fundamental practical aspects of electrospinning that have reoccurred in the recent literature. These included obtaining aligned fibers by spinning onto conductive strips or rods that were separated in space from each other by an insulating material (gap alignment effect), increasing production rates by using multiple spinnerets, regulating the electric field between the source and the collector by adding additional electrostatic elements, using corona discharge to discharge the electrospun fibers, and posttreating the electrospun fibers by submerging them in a liquid bath [103].

• Rotating dual-collector yarn

Formhals's original patent relates to the manufacture of slivers of cellulose acetate fibers by electrospinning from a cogwheel source onto various collector setups. In these collector set-ups fibers are first spun onto a rotating collector and then removed in a continuous fashion, onto a second take-up roller. The first of these collector set-ups consisted of a solid conductive wheel or ring with string attached to the edge. In this set-up the wheel was rotated for a short period while fibers were spun onto its edge. The process was then stopped, the string was loosened and then drawn over rollers and/or through twist-imparting rings to a second take-up roller, and the spinning process restarted. The newly spun sliver or semitwisted yarn was then drawn off continuously onto the second take-up roller. Another collector set-up consisted of a looped metal belt with fixtures to push or blow the fiber sliver off the belt before the fiber sliver was collected on a second take-up roller. The concept of using multiple spinnerets for increasing production rates was also introduced in this patent.

Formhals later identified several problems related to his first design and hence in subsequent patents he made various additions and/or alterations to the original design, which were intended to eliminate these problems. These problems and their solutions included the following [92, 104]:

• Problem: Fibers flying to-and-fro between the source and the collector

Solution: In his second patent, Formhals claimed that one cause of the fibers flying to-and-fro between the jet and collector was that the collector was at too high a voltage of opposite polarity and that resulting corona discharge from the collector reversed the charge of the fibers while they were passing between the jet and the collector. This in turn caused them to change direction and fly back to the source. He proposed to eliminate this problem by adding a voltage regulator on the collector-side of the circuit in order to down-regulate the voltage of the collector.

• Problem: Fibers not drying sufficiently between source and collector.

Solution: Formhals later designed various additions to his spinning system for regulating the shape and intensity of the electrical field in the vicinity of the spinning source.30 This was done in order to direct the formed fibers along a longer, predetermined and constant path towards the collecting electrode and was achieved by placing, in close proximity to the fiber stream, conductive strips, wires, plates and screens, which were connected to the same potential as the fiber source. These additions allowed a more thorough drying of fibers before they were deposited on the collector.

An additional problem, which is not specifically discussed in Formhals's patents, but which can be foreseen when examining his first collector design,25 is that fiber alignment would be less than ideal when spinning onto a solid wheel or belt. It appears, however, that Formhals did encounter this problem and that he overcame it by using the gap alignment effect in the design of subsequent collectors. The design consisted of a picket-fence-like belt, with individual, pointed electrodes, separated from each other by an air gap.

• Multi-collector yarn

In this patented process from the Korea Research Institute of Chemical Technology, continuous slivers or twisted yarns of different polymers, but especially of polyamide–polyimide copolymers, are claimed to be obtained by electrospinning first onto one stationary or rotating plate or conductive mesh collector, where the charges on the fibers are neutralized, and then continuously collecting the fibers from the first onto a second rotating collector.

A diagram depicting the process is given in Figure 6.5. The underlying principle of this process closely resembles the rotating dual-collector yarn process patented by Formhals in 1934 [105].

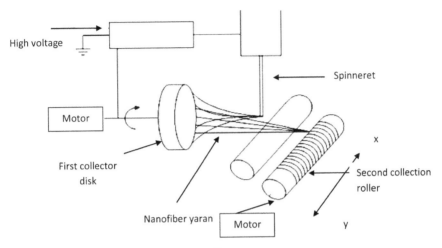

FIGURE 6.5   Multi-collector yarn process diagram.

• Core-spun yarn

In his 1940 patent, Formhals described a method for making composite yarns by electrospinning onto existing cotton, wool or other preformed yarn. It was also proposed that a sliver of fibers, such as, wool, could be coated with the electrospun fibers before twisting the product into an intimately blended yarn [103].

• Staple fiber yarn

Formhals developed a method for controlling the length of the electrospun fibers with the main objective being to manufacture fibers with a controlled and comparatively short length. This goal was achieved by modulating the electric field using a spark gap. In this modulation, it was preferable to periodically switch the field strength to at least 35% and preferably 20% of its original voltage in order to interrupt the electrospinning process for a short period and thereby create a sliver of short fibers, which could then be spun into a staple fiber yarn [103].

• Continuous filament yarn

Instead of spinning directly onto the counter electrode, Formhals altered the process so the polarity of the charge on the fibers was changed before reaching the counter electrode. This was achieved by using high voltage of opposite polarity on a sharp-edged or thin wire counter electrode. The high voltage led to corona discharge, which initially reduced and eventually inverted the charge on the fiber while it was traveling from the source to the collector [106].

This caused the fibers to turn away from the collector electrode and they could then be intercepted at a point below the counter electrode and rolled up as a continuous filament yarn on a take-up roller. In a final improvement on the system, the entire spinning apparatus was encased in a box with earthed conductive siding. This avoided build-up of charges in the panels, which could lead to disturbance of the electric field inside the spinning chamber and disruption of the spinning process. In addition, variable voltage power on the source and collector electrodes allowed the tuning and moving of the position of the neutral zone in which the yarn formation process takes place, which in turn allowed better control over the continuity of the spinning process [107, 108].

• Self-assembled yarn

The self-assembled yarn process was developed by Ko et al. at Drexel University. When a solution of pure polymer, or a polymer-containing polymer blend, was electrospun onto a solid conductive collector under appropriate conditions, the fibers did not deposit on the collector in the form of a flat nonwoven web as is usually observed. Instead, initial fibers deposited on a relatively small area of the collector and then subsequent fibers started accumulating on top of them and then on top of each other, forming a self-assembled yarn structure that rapidly grew upwards from the collector towards the spinneret. The formation of a self-assembled yarn is illustrated in Fig. 6.6. The self-assembling yarn, suspended in the space between the Spinneret and the collector, continued to grow in this fashion until it reached a critical point somewhere in the vicinity of the spinneret. At this critical point, a branched tree-like fiber structure formed and newly formed fibers deposited on the branches of the tree. The yarn could then be collected by slowly taking up the fibers collected on one of the tree branch structures, or by slowly moving the target electrode away from the spinneret. Post-processing of the yarn, including twisting, could be done in a second step [109].

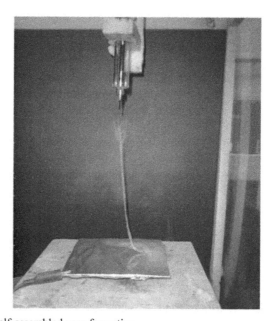

FIGURE 6.6 Self-assembled yarn formation.

It was proposed that the charge on the electrospun fibers, which is induced through the high voltage in the spinneret, is dissipated through the evaporation of the solvent during the electrospinning process, so that the fibers are essentially neutral when they reach the collector electrode. This could explain why the initial fibers deposit on such a small area on the collector. If the fibers on the collector are charged, they repel incoming fibers leading to an expanding random web. Neutral fibers would not have the same repelling effect on incoming fibers and so the fibers would collect on a smaller area. Neutral fibers, deposited on top of each other, and therefore closer to the spinneret than the target electrode surface, also form an attractive target for incoming fibers. This would explain why subsequent fibers selectively deposit on the tip of the self-assembling yarn [85].

• Conical collector yarn

A method for the production of hollow fibers by the electrospinning process was reported by Kim et al. The conventional electrospinning device was modified to include a conical collector and an air-suction orifice to generate hollow and void-containing, uniaxially aligned electrospun fibers. Use of the conical collector allowed for the collection of aligned yarns with diameters of approximately 157 nm [110].

• Spin-bath collector yarn

In this recently published method, developed by our group at Stellenbosch University, continuous uniaxial fiber bundle yarns are obtained by electrospinning onto the surface of a liquid reservoir counter electrode. The web of electrospun fibers, which forms on the surface of the spin-bath, is drawn at low linear velocity (ca. 0.05 m/s) over the liquid surface and onto a take-up roller. A diagrammatic representation of the electrospinning set-up is given in Fig. 6.7. All the yarns obtained using this method exhibit very high degrees of fiber alignment and bent fiber loops are observed in all the yarns [47].

The process of yarn formation is illustrated in Fig. 6.8. It can be described in three phases. In the first phase, a flat web of randomly looped fibers forms on the surface of the liquid. In the second phase, when the fibers are drawn over or through the liquid, the web is elongated and alignment of the fibers takes place in the drawing direction. The third phase consists of drawing the web off the liquid and into air. The surface tension of the remaining liquid on the web pulls the fibers together into a three-dimensional, round yarn structure.

The average yarn obtained in a single-spinneret electrospinning set-up contains approximately 3720 fibers per cross-section and approximately 180 m of yarn can be spun per hour. The yarns obtained are very fine, with calculated linear densities in the order of 10.1 denier. Although higher linear densities can be obtained by reducing the yarn take-up rate, this is accompanied by a decrease in fiber alignment within the yarn. Currently investigations are focused on various options to overcome these challenges by, for instance, combining aligned yarns from multiple spinnerets into single yarns.

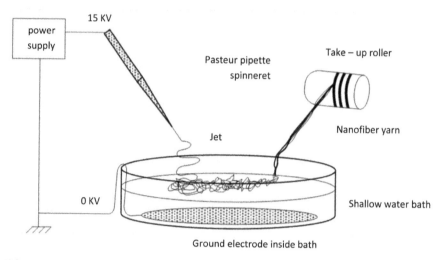

FIGURE 6.7   Yarn-spinning set-up with grounded spin-bath collector electrode.

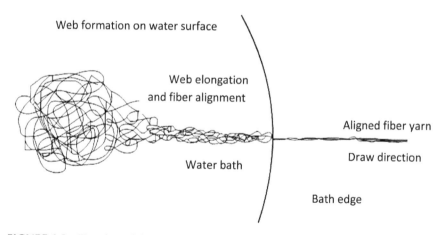

FIGURE 6.8   Top view of the yarn formation process.

- Twisted nonwoven web yarn

This method, patented by Raisio Chemicals Korea Inc. involves electrospinning nanofibers through multiple nozzles to obtain a nonwoven nanofiber web, either directly in a ribbon form or in a larger form, which is then cut into ribbons, and subsequently passing the nanofiber web ribbons through an air twister to obtain a twisted nanofiber yarn. A diagram depicting the process is given in Fig. 6.9 [92].

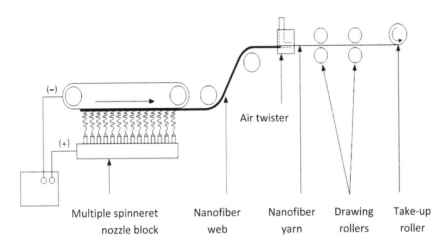

Multiple spinneret   Nanofiber   Nanofiber   Drawing   Take-up
nozzle block   web   yarn   rollers   roller

FIGURE 6.9   Spinning process used to prepare a twisted nonwoven web yarn.

- Grooved belt collector yarn

In a recent patent by Kim and Park, a ribbon-shaped nanofiber web is prepared by electrospinning onto a collector consisting of an endless belt- type nonconductive plate with grooves formed at regular intervals along a lengthwise direction and a conductive plate inserted into the grooves of the nonconductive plate. The nanofiber webs are electrospun onto the conductive plates in the grooves and later separated from the collector, focused, drawn and wound into a yarn [92].

- Vortex bath collector yarn

In this patent pending process developed at the National University of Singapore, a basin with a hole at the bottom is used to allow water to flow out in such a manner that a vortex is created on the water surface.

Electrospinning is carried out over the top of the basin so that electrospun fibers are continuously deposited on the surface of the water. Owing to the presence of the vortex, the deposited fibers are drawn into a bundle as they flow through the water vortex. Generally, a higher feed rate or multiple spinnerets are required to deliver sufficient fibers on the surface of the water so that the resultant fiber yarn has sufficient strength to withstand the drawing and winding process. Figure 6.10 shows the set-up used for the yarn drawing process.

Yarn drawing speeds as high as 80 m/min have been achieved and yarns made of poly(vinylidene fluoride) (PVDF) and polycaprolactone have been fabricated using this process [103].

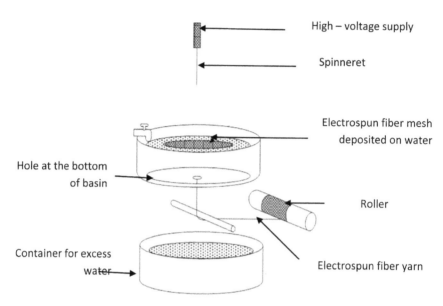

FIGURE 6.10   Vortex bath collector yarn process.

6.3.5.12   GAP-SEPARATED ROTATING ROD YARN

This method developed by Doiphode and Reneker at the University of Akron uses the gap alignment effect in a very similar way to the work published by Dalton et al. discussed before. The process is described with reference to Figure 6.11. Fibers are electrospun between a 2 mm metal rod on the right and a hollow 25 mm metal rod with a hollow hemisphere attached to its end on the left. Both the geometries are grounded and placed at a distance of a few centimeters. Fibers are collected across the gap between these two collector surfaces and are given a twist by rotating the hemispherical collector. Yarn collected in this manner on the tip of the metal rod can be translated away from the rotating collector, thereby drawing the yarn and producing yarn continuously. Yarns with lengths up to 30 cm were produced by this method and the creators of the process believe that optimizing the winding mechanism can lead to production of continuous yarns [111].

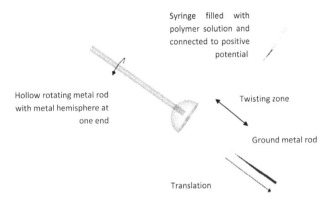

FIGURE 6.11   Schematic diagram for gap-separated rotating rod yarn set-up.

• Conjugate electrospinning yarn

Methods for making continuous nanofiber yarns based on the principle of conjugate electrospinning were recently published and patented by Xinsong Li et al. at South-east University in Nanjing as well as Luming Li and co-workers at Tsinghua University in Beijing. In conjugate electrospinning, two spinnerets or two groups of spinnerets are placed in an opposing configuration and connected to high voltage of positive and negative polarity respectively. The process is presented diagrammatically in Fig. 6.12. Oppositely charged fiber jets are ejected from the spinnerets and Coulombic attraction leads the oppositely charged fibers to collide with each other. The collision of the fibers leads to rapid neutralization of the charges on the fibers and rapid decrease in their flying speeds. In the processes described by both groups, the neutralized fibers are then collected onto take-up rollers to form yarns. Each continuous yarn contains a large quantity of nanofibers, which are well aligned along the longitudinal axis of the yarn. Conjugate electrospinning works for a variety of polymers, composites and ceramics [29].

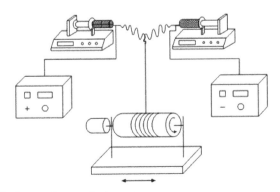

FIGURE 6.12   Conjugate electrospinning set-up.

6.8 NANOPARTICLES IN TEXTILES FINISHING

Fabric treated with nanoparticles of TiO_2 and MgO replaces fabrics with active carbon, previously used as chemical and biological protective materials. The photocatalytic activity of TiO_2 and MgO nanoparticles can break harmful and toxic chemicals and biological agents. These nanoparticles can be preengineered to adhere to textile substrates via spray coating or electrostatic methods. Textiles with nanoparticles finishing are used to convert fabrics into sensor-based materials which has numerous applications. If nanocrystalline epiezoceramic particles are incorporated into fabrics, the finished fabric can convert exerted mechanical forces into electrical signals enabling the monitoring of bodily functions such as heart rhythm and pulse if they are worn next to skin [112–114].

6.9 FABRIC FINISHING BY USING NANOTECHNOLOGY

Finishing of textile fabrics is made of natural and synthetic fibers to achieve desirable surface texture, color and other special esthetic and functional properties, has been a primary focus in textile manufacturing industries. In the last decade, the advent of nanotechnology has spurred significant developments and innovations in this field of textile technology [115]. Fabric finishing has taken new routes and demonstrated a great potential for significant improvements by applications of nanotechnology. The developments in the areas of surface engineering and fabric finishing have been highlighted in several researches. There are many ways in which the surface properties of a fabric can be manipulated and enhanced, by implementing appropriate surface finishing, coating, and/or altering techniques, using nanotechnology. A few representative applications of fabric finishing using nanotechnology are schematically displayed in the Fig. 6.13.

Nanotechnology provides plenty of efficient tools and techniques to produce desirable fabric attributes, mainly by engineering modifications of the fabric surface. For example, the prevention of fluid wetting towards the development of water or stain-resistant fabrics has always been of great concerning textile manufacturing.

The basic principles and theoretical background of "fluid-fabric" surface interaction are well described in recent manuscript. It has been demonstrated that by altering the micro and nano-scale surface features on a fabric surface, a more robust control of wetting behavior can be attained. The alteration in the fabric's surface properties enables to exhibit the "Lotus-Effect," which demonstrates the natural hydrophobic behavior of a leaf surface. This sort of surface engineering, which is capable of replicating hydrophobic behavior, can be used in developing special chemical finishes for producing water and/or stain- resistant fabrics [116, 117]. In recent years, several attempts have been made by researchers and industries to use similar concepts of surface-engineered modifications through nanotechnology to develop high performance textile and smart textile. The concept of surface engineering and nano-textile develops hydrophobic fabric surfaces that are capable of repelling liquids and resisting stains, while complementing the other desirable fabric attributes, such as, breathability, softness, and comfort [118].

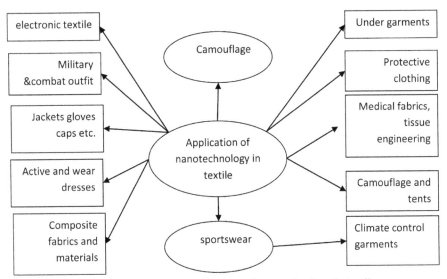

FIGURE 6.13   Some representative applications of nanotechnology in textiles.

6.10   TEXTILE MODIFICATION

Considering special advantages and high potentialities of the application of nano-structured materials in textile industry, especially for producing high performance textiles, here we reviewed the application of nano-structured materials for antibacterial modification of textile and polymeric materials. The modification of textile fibers is carried out by commonly used chemical or electro-chemical application methods. Many of the classical textile finishing techniques (e.g., hydrophobization, easy-care finishing) that are already used since decades are among these methods. Modification of textiles via producing polymeric nano-composites and also surface modification of textiles with metallic and inorganic nanostructured materials are developed due to their unique properties. Considering the fact that fiber and film processing are the most difficult procedures of molding polymeric materials, bulk modification of continuous multifilament yarns is an extremely sensitive process. However, achieving optimum process conditions will present an economical technique [119, 120].

Different methods have been used for surface modifications of textiles by using poly carboxylic acids as spacers for attaching TiO_2 nano-particles to the fabrics [121] and argon plasma grafting nano-particles on wool surface [122]. Plasma pretreatment has been used for the generation of active groups on the surface to be combined with TiO_2 nanoparticles [123]. The radical groups on the surface have also been generated using irradiation of the textile surfaces with UV light to bond the nano-particles [124]. Deposition of nanoparticles from their metallic salt solution on the surface pretreated with RF-plasma and vacuum-UV [125].

Nanotechnology holds great potential in the textile and clothing industry offering enhanced performance of textile manufacturing machines and processes so as to overcome the limitations of conventional methods. Nanofibers have good properties such as high surface area, a small fiber diameter, good filtration properties and high permeability. Nanofibers can be obtained via electro-spinning application or bicomponent extrusion (islands in the sea technique) [126]. One of the interesting areas for the application of nanotechnology in the textile industry are coating and finishing processes of textiles which is done by the techniques like sol-gel [127] and plasma [128]. Nanotech enhanced textiles include sporting industry, skincare, space technology and clothing as well as material technology exhibiting better healthcare systems, protective clothing and integrated electronics. By using nanotechnology, textiles with self-cleaning surfaces have attracted much attention which is created by the lotus effect. In brief, nanoscaled structures similar to those of a lotus leaf create a surface that causes water and oil to be repelled, forming droplets, which will simply roll of the surface, taking any with them [62].

With the advent of nanoscience and technology, a new area has developed in the area of textile finishing called "Nanofinishing." Growing awareness of health and hygiene has increased the demand for bioactive or antimicrobial and UV-protecting textiles. Coating the active surfaces cause UV blocking, antimicrobial, flame retardant, water repellant and self-cleaning properties. The UV-blocking property of a fabric is enhanced when a dye, pigment, delustrant, or ultraviolet absorber finish is present that absorbs ultraviolet radiation and blocks its transmission through a fabric to the skin. Metal oxides like ZnO as UV-blocker are more stable when compared to organic UV-blocking agents. For antibacterial finishing, ZnO nanoparticles scores over nano-silver in cost-effectiveness, whiteness, and UV-blocking property [129, 130].

6.11 TYPES OF NANOMATERIALS

6.11.1 *NANOCOMPOSITE* FIBERS

A composite is a material that combines one or more separate components. Composites are designed to exhibit the best properties of each component. A large variety of systems combining one, two and three dimensional materials with amorphous materials mixed at the nanometer scale [131]. Nanostructure composite fibers are intensively used in automotive, aerospace and military applications. Nanocomposite fibers are produced by dispersing nanosize fillers into a fiber matrix. Due to their large surface area and high aspect ratio, nanofillers interact with polymer chain movement and thus reduce the chain mobility of the system. Being evenly distributed in polymer matrices, nanoparticles can carry load and increase the toughness and abrasion resistance. Most of the nanocomposite fibers use fillers such as nanosilicates, metal oxide nanoparticles, GNF as well as single-wall and multiwall CNT [132, 133]. Some novel CNT reinforced polymer composite materials have been developed, which can be used for developing multifunctional textiles having superior strength, toughness, lightweight, and high electrical conductivity [72].

6.11.2   *CARBON* NANOFIBERS AND CARBON NANOPARTICLES

Carbon nanofibers and carbon black nanoparticles are among the most commonly used nanosize filling materials. Nanofibers can be defined as fibers with a diameter of less than 1 mm or 1000 nm and are characterized as having a high surface area to volume ratio and a small pore size in fabric form [66]. Carbon nanofibers can effectively increase the tensile strength of composite fibers due to its high aspect ratio, while carbon black nanoparticles can improve their abrasion resistance and toughness. Several fiber-forming polymers used as matrices have been investigated including polyester, nylon and polyethylene with the weight of the filler from 5 to 20% [134].

There are numerous applications in which nanofibers could be suited. The high surface area to volume ratio and small pore size allows viruses and spore-forming bacterium such as anthrax to be trapped. Filtration devices and wound dressings are just some of the applications in which nanofibers could be used. Researchers are investigating textile materials made from nanofibers which can act as a filter for pathogens (bacteria, viruses), toxic gasses, or poisonous or harmful substances in the air. Medical staff, fire fighters, the emergency services or military personnel could all benefit from protective garments made from nanofibers materials [135].

6.11.3   *CLAY* NANOPARTICLES

Clay nanoparticles are resistant to heat, chemicals and electricity, and have the ability to block UV light. Incorporating clay nanoparticles into a textile can result in a fabric with improved tensile strength, tensile modulus, flexural strength and flexural modulus. Nanocomposite fibers which use clay nanoparticles can be engineered to be flame, UV light resistant and anticorrosive. Although there have been a number of flame retardant finishes available since the 1970's, the emission of toxic gasses when set ablaze make them somewhat hazardous. Clay nanoparticles have been incorporated into nylon to impart flame retardant characteristics to the textile without the emission of toxic gas. The addition of clay nanoparticles has made polypropylene dyeable. Metal oxide nanoparticles of TiO_2, Al_2O_3, ZnO and MgO exhibit photocatalytic ability, electrical conductivity, UV absorption and photo-oxidizing capacity against chemical and biological species. The main research efforts involving the use of nanoparticles of metal oxides have been focused on antimicrobial, self-decontaminating and UV blocking applications for both military protection gears and civilian health products [66, 72]. Nylon fibers filled with ZnO nanoparticles can provide UV shielding function and reduce static electricity on nylon fibers. A composite fiber with nanoparticle of TiO_2 or MgO can provide self-sterilizing function [134].

6.11.4   CARBON NANOTUBES

Carbon Nanotube is a tubular form of carbon with diameter as small as nanometer (nm). A carbon nanotube is configurationally equivalent to a two dimensional graphene sheet rolled into a tube. They can be metallic or semiconducting, depending on chirality. CNT are one of the most promising materials due to their high strength and high electrical conductivity. CNT consists of tiny shell(s) of graphite rolled up into a cylinder(s) [69, 136]. CNT exhibit 100 times the tensile strength of steel at one-sixth weight, thermal conductivity better than all but the purest diamond, and an electrical conductivity similar to copper, but with the ability to carry much higher currents. The potential applications of CNTs include conductive and high-strength composite fibers, energy storage and energy conversion devices, sensors, and field emission displays. Possible applications include screen displays, sensors, aircraft structures, explosion-proof blankets and electromagnetic shielding. The composite fibers have potential applications in safety harnesses, explosion-proof blankets, and electromagnetic shielding applications. Continuing research activities on CNT fibers involve study of different fiber polymer matrices such as polymethylmethacrylate (PMMA) and polyacrylonitrile (PNA) as well as CNT dispersion and orientation in polymers [137].

6.11.5   NANOCELLULAR FOAM STRUCTURE

Polymeric materials with nanosize porosity exhibit lightweight, good thermal insulation, as well as high cracking resistance at high temperature without sacrifices in mechanical strength. By choosing the pretreatment condition to the fiber, the transverse mechanical properties of the composite can be also enhanced through the molecular diffusion across the interface between the fiber and the matrix. The nanocomposites clearly surpass the mechanical properties of most comparable cellulosic materials, their greatest advantage being the fact that they are fully bio-based and biodegradable, but also of relatively high strength. A potential application of cellular structure is to encapsulate functional compounds such as pesticides and drugs inside of the nanosize cells. One of the approaches to fabricate nanocellular fibers is to make use of a thermodynamic instability during supercritical carbon dioxide extrusion and reduce the size of the cellular fibers that can be used as high-performance composite fibers as well as for sporting and aerospace materials [72, 136].

6.12   PROPERTIES OF NANO-TEXTILE FIBERS

6.12.1   *WATER* REPELLENCE

The water-repellent property of fabric created by nano-whiskers, which are hydro-carbons and 1/1000 of the size of a typical cotton fiber, when added to the fabric create a peach fuzz effect without lowering the strength of cotton. The spaces between the whiskers on the fabric are smaller than the typical drop of water, but still larger than water molecules; water thus, remains on the top of the whiskers and above the surface of the fabric. However, liquid can still pass through the fabric, if pressure is applied to it [117, 118].

Nanosphere impregnation involving a three-dimensional surface structure with gel forming additives which repel water and prevent dirt particles from attaching themselves are also used. Once water droplets fall onto them, water droplets bead up and, if the surface slopes slightly, will roll off. As a result, the surfaces stay dry even during a heavy shower. Furthermore, the droplets pick up small particles of dirt as they roll, and so the leaves of the lotus plant keep clean even during light rain. By altering the micro and nano-scale surface features on a fabric surface, a more robust control of wetting behavior can be attained. It has been demonstrated that by combining the nanoparticles of hydroxylapatite, TiO_2, ZnO and Fe_7O_3 with other organic and inorganic substances, the audio frequency plasma of fluorocarbon chemical was applied to deposit a nanoparticulate hydrophobic film onto a cotton fabric surface to improve its water repellent property. This sort of surface engineering, which is capable of replicating hydrophobic behavior, can be used in developing special chemical finishes for producing water-and/or stain- resistant fabrics while complementing the other desirable fabric attributes, such as, breathability, softness and comfort. The surfaces of the textile fabrics can be appreciably modified to achieve considerably greater abrasion resistance, UV resistance, electromagnetic and infrared protection properties [138, 139].

6.12.2   *UV* PROTECTION

Inorganic UV blockers are more preferable to organic UV blockers as they are non-toxic and chemically stable under exposure to both high temperatures and UV [140, 141]. Inorganic UV blockers are usually certain semiconductor oxides such as TiO_2, ZnO, SiO_2 and Al_2O_3. Among these semiconductor oxides, TiO_2 and ZnO are commonly used. It was determined that nano-sized titanium dioxide and zinc oxide are more efficient at absorbing and scattering UV radiation than the conventional size, and are thus better to provide protection against UV rays. This is due to the fact that nano-particles have a larger surface area per unit mass and volume than the conventional materials, leading to the increase of the effectiveness of blocking UV radiation [68, 140]. Various researchers have worked on the application of UV blocking treatment to fabric using nanotechnology. UV blocking treatment for cotton fabrics are developed using the sol-gel method. A thin layer of titanium dioxide is formed on the surface of the treated cotton fabric which provides excellent UV protection; the effect can be maintained after 50 home launderings [141]. Apart from titanium dioxide, zinc oxide nano rods of 10 to 50 nm in length are also applied to cotton fabric to provide UV protection. According to the studies on the UV blocking effect, the fabric treated with zinc oxide nanorods were found to have demonstrated an excellent UV protective factor (UPF) rating. This effect can be further enhanced by using a different procedure for the application of nanoparticles on the fabric surface. When the process of padding is used for applying the nanoparticles on to the fabric, the nanoparticles get applied not only on the surface alone but also penetrates into the interstices of the yarns and the fabric, that is, some portion of the nanoparticles get penetrate into the fabric structure. Such Nanoparticles which do not stay on the surface may not be very effective in shielding the UV rays. It is worthwhile that only the right (face) side of the fabric gets exposed to the rays and therefore, this surface alone needs to be covered with the nanoparticles for better UV protection. Spraying (using compressed air and spray gun) the fabric surface with the nanoparticles can be an alternate method of applying the nanoparticles [118].

6.12.3 ANTIMICROBIAL

Although many antimicrobial agents are already in used for textile, the major classes of antimicrobial for textile include organo-silicones, organo-metallics, phenols and quaternary ammonium salts. The bi- phenolic compounds exhibit a broad spectrum of antimicrobial activity. For imparting antibacterial properties, nano-sized silver, titanium dioxide, zinc oxide, triclosan and chitosan are used [118]. Nano-silver particles have an extremely large relative surface area, thus increasing their contact with bacteria or fungi and vastly improving their bactericidal and fungicidal effectiveness. Nano-silver is very reactive with protein and shows antimicrobial properties at concentrations as low as 0.0003 to 0.0005%. When contacting bacteria and fungi, it will adversely affect cellular metabolism and inhibits cell growth. It also suppresses respiration, the basal metabolism of the electron transfer system, and the transport of the substrate into the microbial cell membrane. Furthermore, it inhibits the multiplication and growth of those bacteria and fungi which cause infection, odor, itchiness and sores [140]. Some synthetic antimicrobial nano particles which are used in textiles are as follows. Triclosan, a chlorinated bi- phenol, is a synthetic, nonionic and broad spectrum antimicrobial agent possessing mostly antibacterial alone with some antifungal and antiviral properties. Chitosan, a natural biopolymer, is effectively used as antibacterial, antifungal, antiviral, nonallergic and biocompatible. ZnO nanoparticles have been widely used for their antibacterial and UV-blocking properties.

6.12.4 ANTISTATIC

An antistatic agent is a compound used for treatment of materials or their surfaces in order to reduce or eliminate buildup of static electricity generally caused by the triboelectric effect. The molecules of an antistatic agent often have both hydrophilic and hydrophobic areas, similar to those of a surfactant; the hydrophobic side interacts with the surface of the material, while the hydrophilic side interacts with the air moisture and binds the water molecules [133]. As synthetic fibers provide poor antistatic properties, research work concerning the improvement of the antistatic properties of textiles by using nanotechnology has been at large. It was determined that nano-sized particles like titanium dioxide, zinc oxide whiskers, nanoantimony-doped tin oxide and silanenanosol could impart antistatic properties to synthetic fibers. Such material helps to effectively dissipate the static charge which is accumulated on the fabric [118]. On the other hand, silanenanosol improves antistatic properties, as the silica gel particles on fiber absorb water and moisture in the air by amino and hydroxyl groups and bound water. Electrically conductive nano-particles are durably anchored in the fibrils of the membrane of Teflon, creating an electrically conductive network that prevents the formation of isolated chargeable areas and voltage peaks commonly found in conventional antistatic materials. This method can overcome the limitation of conventional methods, which is that the antistatic agent is easily washed off after a few laundry cycles [62].

6.12.5 WRINKLE RESISTANCE

To impart wrinkle resistance to fabric, resin is commonly used in conventional methods. However, there are limitations to applying resin, including a decrease in the tensile strength of fiber, abrasion resistance, water absorbency and dye-ability, as well as breathability. To overcome the limitations of using resin, some researchers employed nano-titanium dioxide and nano-silica to improve the wrinkle resistance of cotton and silk respectively [118]. Nano-titanium dioxide employed with carboxylic acid as a catalyst under UV irradiation to catalyzes the cross-linking reaction between the cellulose molecule and the acid. On the other hand, nano-silica when applied with maleic anhydride as a catalyst could successfully improve the wrinkle resistance of silk [116]. More over the wrinkle recovery of the fabrics can also be improved to a great extent by imparting techniques like padding and exhaustion beside the use of nano-materials to the fabrics. Studies also have suggest that treatment of fabrics with microwaves are more wrinkle resistant as comparable to oven curing, because it generates higher frequency and volumetric heating which minimizes the damage from over drying.

6.13 SMART TEXTILES

All applications textile materials have a vast number of clear advantages:
- They are omnipresent, everybody is familiar with them
- They are easy to use and to maintain
- Clothes have a large contact with the body
- They make us look nice
- They are extremely versatile in terms of raw materials used, arrangement of the fibers, finishing treatments, shaping etc.
- They can be made to fit typical applications where textile structures are to be preferred are:
  - Long term or permanent contact without skin irritation,
  - Home applications,
  - Applications for children: in a discrete and careless way,
  - Applications for the elderly: discretion, comfort and esthetics are important.

The multifunctional textiles such as fashion and environmental protection, ballistic and chemical protection, flame protection are all passive systems. The smart textiles are a new generation of fibers, yarns, fabrics and garments that are able to sense stimuli and changes in their environments, such as, mechanical, thermal, chemical, electrical, magnetic and optical changes, and then respond to these changes in predetermined ways. They are multifunctional textile systems that can be classified into three categories of passive smart textiles, active smart textiles and very smart textiles [142].

The functionalities of smart textiles can be classified in 5 groups: sensoring, data processing, actuation, communication, energy.

At this moment, most of the progress has been achieved in the area of sensoring. Many type of parameters can be measured:
- Temperature
- Biopotentials: cardiogram, myographs, encephalographs
- Acoustic: heart, lungs, digestion, joints
- Ultrasound: blood flow
- Biological, chemical
- Motion: respiration, motion
- Pressure: blood
- Radiation: IR, spectroscopy
- Odor, sweat
- Mechanical skin parameters
- Electric (skin) parameters

Some of these parameters are well known, like cardiogram and temperature. Nevertheless, permanent monitoring also opens up new perspectives for these traditional parameters too. Indeed today evaluation is usually based on standards for global population groups. Permanent monitoring supported by self learning devices will allow the set up of personal profiles for each individual, so that conditions deviating from normal can be traced the soonest possible. Also diagnosis can be a lot more accurate.

Apart from the actual measuring devices data processing is a key feature in this respect. These types of data are new. They are numerous with multiple complex interrelationships and time dependent. New self learning techniques will be required.

Actuation is another aspect. Identification of problems only makes sense when followed by an adequate reaction. This reaction can consist of reporting or calling for help, but also drug supply and physical treatment. A huge challenge in this respect is the development of high performance muscle like materials.

Smart textiles is a new aspect in textile that is a multidiscipline field of research in many sciences and technologies such as textile, physics, chemistry, medicine, electronics, polymers, biotechnology, telecommunications, information technology, microelectronics, wearable computers, nanotechnology and microelectromechanical machines. Shape memory materials, conductive materials, phase change materials, chromic materials, photonic fibers, mechanical responsive materials, intelligent coating/membranes, micro and nanomaterials and piezoelectric materials are applied in smart textiles [143].

The objective of smart textile is to absorb a series of active components essentially without changing its characteristics of flexibility and comfort. In order to make a smart textile, firstly, conventional components such as sensors, devices and wires are being reshaped in order to fit in the textile, ultimately the research activities trend to manufacture active elements made of fibers, yarns and fabrics structures. Smart textiles are ideal vehicle for carrying active elements that permanently monitor our body and the environment, providing adequate reaction should something happen [144].

The smart textiles have some of the capabilities such as biological and chemical sensing and responding, power and data transmission from wearable computers and polymeric batteries, transmitting and receiving RF signals and automatic voice warning systems as to 'dangers ahead' that may be appropriate in military applications. Other than military applications of smart textiles, mountain climbers, sportsmen, businessmen, healthcare and medical personnel, police, and firemen will be benefitted from the smart textiles technologies.

A smart textile can be active in many other fields. Smart textiles as a carrier of sensor systems can measure heart rate, temperature, respiration, gesture and many other body parameters that can provide useful information on the health status of a person. The smart textiles can support the rehabilitation process and react adequately on hazardous conditions that may have been detected. The reaction can consist of warning, prevention or active protection. After an event has happened, the smart textile is able to analyze the situation and to provide first aid [145].

Wearable electronics and photonics, adaptive and responsive structures, biomimetics, bioprocessing, tissue engineering and chemical/drug releasing are some of the research areas in integrated processes and products of smart textiles. There are some areas that the research activities have reached the industrial application. Optical fibers, shape memory polymers, conductive polymers, textile fabrics and composites integrated with optical fiber sensors have been used to monitor the health of major bridges and buildings. The first generation of wearable motherboards has been developed, which has sensors integrated inside garments and is capable of detecting injury and health information of the wearer and transmitting such information remotely to a hospital. Shape memory polymers have been applied to textiles in fiber, film and foam forms, resulting in a range of high performance fabrics and garments, especially sea-going garments. Fiber sensors, which are capable of measuring temperature, strain/stress, gas, biological species and smell, are typical smart fibers that can be directly applied to textiles. Conductive polymer-based actuators have achieved very high levels of energy density. Clothing with its own senses and brain, like shoes and snow coats which are integrated with Global positioning system and mobile phone technology, can tell the position of the wearer and give him/her directions. Biological tissues and organs, like ears and noses, can be grown from textile scaffolds made from biodegradable fibers [146, 147].

6.13.1 PHASE CHANGE MATERIALS

Phase change materials are thermal storage materials that are used to regulate temperature fluctuations. The thermal energy transfer occurs when a material changes from a solid to a liquid or from a liquid to a solid. This is called a change in state, or phase.

Incorporating microcapsules of PCM into textile structures improves the thermal performance of the textiles. Phase change materials store energy when they change from solid to liquid and dissipate it when they change back from liquid to solid. It would be most ideal, if the excess heat a person produces could be stored intermediately somewhere in the clothing system and then, according to the requirement, activated again when it starts to get chilly.

The most widespread PCMs in textiles are paraffin-waxes with various phase change temperatures (melting and crystallization) depending on their carbon numbers. The characteristics of some of these PCMs are summarized in Table 6.1. These phase change materials are enclosed in microcapsules, which are 1–30 μm in diameter. Hydrated inorganic salts have also been used in clothes for cooling applications. PCM elements containing Glauber's salt (sodium sulfate) have been packed in the pockets of cooling vests.

PCM can be applied to fibers in a wet-spinning process, incorporated into foam or embedded into a binder and applied to fabric topically, or contained in a cell structure made of a textile reinforced synthetic material [148, 149].

TABLE 6.1   Phase Change Materials

Phase change material	Melting temperature °C	Crystallization temperature °C	Heat storage capacity in J/g
Eicosane	36.1	30.6	246
Nonadecane	32.1	26.4	222
Octadecane	28.2	25.4	244
Heptadecane	22.5	21.5	213
Hexadecane	18.5	16.2	237

In manufacturing the fiber, the selected PCM microcapsules are added to the liquid polymer or polymer solution, and the fiber is then expanded according to the conventional methods such as dry or wet spinning of polymer solutions and extrusion of polymer melts.

Fabrics can be formed from the fibers containing PCM by conventional weaving, knitting or nonwoven methods, and these fabrics can be applied to numerous clothing applications.

In this method, the PCMs are permanently locked within the fibers, the fiber is processed with no need for variations in yarn spinning, fabric knitting or dyeing and properties of fabrics (drape, softness, tenacity, etc.) are not altered in comparison with fabrics made from conventional fibers. The microcapsules incorporated into the fibers in this method have an upper loading limit of 5–10% because the physical properties of the fibers begin to suffer above that limit, and the finest fiber is available. Due to the small content of microcapsules within the fibers, their thermal capacity is rather modest, about 8–12 J/g.

Usually PCM microcapsules are coated on the textile surface. Microcapsules are embedded in a coating compound such as acrylic, polyurethane and rubber latex, and applied to a fabric or foam. In lamination of foam containing PCMs onto a fabric, the selected PCMs microcapsules can be mixed into a polyurethane foam matrix, from which moisture is removed, and then the foam is laminated on a fabric. Typical concentrations of PCMs range from 20% to 60% by weight. Microcapsules should be added to the liquid polymer or elastomer prior to hardening. After foaming (fabricated from polyurethane) microcapsules will be embedded within the base material matrix. The application of the foam pad is particularly recommended because a greater amount of microcapsules can be introduced into the smart textile. In spite of this, different PCMs can be used, giving a broader range of regulation temperatures. Additionally, microcapsules may be anisotropically distributed in the layer of foam. The foam pad with PCMs may be used as a lining in a variety of clothing such as gloves, shoes, hats and outerwear. Before incorporation into clothing or footwear the foam pad is usually attached to the fabric, knitted or woven, by any conventional means such as glue, fusion or lamination [150, 151].

The PCM microcapsules are also applied to a fibrous substrate using a binder (e.g., acrylic resin). All common coating processes such as knife over roll, knife over air, screen-printing, gravure printing, dip coating may be adapted to apply the PCM microcapsules dispersed throughout a polymer binder to fabric. The conventional pad–mangle systems are also suitable for applying PCM microcapsules to fabrics. The formulation containing PCMs can be applied to the fabric by the direct nozzle spray technique.

There are many thermal benefits of treating textile structures with PCM microcapsules such as cooling, insulation and thermo regulating effect. Without phase change materials the thermal insulation capacity of clothing depends on the thickness and the density of the fabric (passive insulation). The application of PCM to a garment provides an active thermal insulation effect acting in addition to the passive thermal insulation effect of the garment system. The active thermal insulation of the PCM controls the heat flux through the garment layers and adjusts the heat flux to the thermal circumstances. The active thermal insulation effect of the PCM results in a substantial improvement of the garment's thermo-physiological wearing comfort [142]. Intensity and duration of the PCM's active thermal insulation effect depend mainly on the heat-storage capacity of the PCM microcapsules and their applied quantity. In order to ensure a suitable and durable effect of the PCM, it is necessary to apply proper PCM in sufficient quantity into the appropriate fibrous substrates of proper design.

The PCM quantity applied to the active wear garment should be matched with the level of activity and the duration of the garment use. Furthermore, the garment construction needs to be designed in a way which assists the desired thermo-regulating effect. Thinner textiles with higher densities readily support the cooling process. In contrast, the use of thicker and less dense textile structures leads to a delayed and therefore more efficient heat release of the PCM. Further requirements on the textile substrate in a garment application include sufficient breathability, high flexibility, and mechanical stability [152].

In order to determine a sufficient PCM quantity, the heat generated by the human body has to be taken into account carrying out strenuous activities under which the active wear garments are worn. The heat generated by the body needs to be entirely released through the garment layers into the environment. The necessary PCM quantity is determined according to the amount of heat which should be absorbed by the PCM to keep the heat balance equalized. It is mostly not necessary to put PCM in all parts of the garment. Applying PCM microcapsules to the areas that provide problems from a thermal standpoint and thermoregulating the heat flux through these areas is often enough. It is also advisable to use different PCM microcapsules in different quantities in distinct garment locations [151].

6.13.1.1 APPLICATIONS OF TEXTILES CONTAINING PCMS

Fabrics containing PCMs have been used in a variety of applications including apparel, home textiles and technical textiles (Table 6.2).

Phase change materials are used both in winter and summer clothing. PCM is used not only in high-quality outerwear and footwear, but also in the underwear, socks, gloves, helmets and bedding of world-wide brand leaders. Seat covers in cars and chairs in offices can consist of phase change materials.

Currently, phase change materials are being used in a variety of outdoor apparel items such as smart jackets, vests, men's and women's hats and rainwear, outdoor active-wear jackets and jacket lining, golf shoes, trekking shoes, ski and snowboard gloves, boots, earmuffs and protective garments. In protective garments, the absorption of body heat surplus, insulation effect caused by heat emission of the PCM into the fibrous structure and thermo-regulating effect, which maintains the microclimate temperature nearly constant are the specified functions of PCM contained smart textile [153].

The addition of PCMs to fabric-backed foam significantly increases the weight, thickness, stiffness, flammability, insulation value, and evaporative resistance value of the material. It is more effective to have one layer of PCM on the outside of a tight-fitting, two layer ensemble than to have it as the inside layer. This may be because the PCMs closest to the body did not change phase [154].

TABLE 6.2 Application of PMs in Textiles

Casual clothing	Underwear, Jackets, sport garments
Professional clothing	Fire fighters protective clothing, Bullet proof fabrics, Space suits, Sailor suits
Medical uses	Surgical gauze, Bandage, Nappies, Bed lining, Gloves, Gowns, Caps, Blankets
Shoe linings	Ski boots, Golf shoes
Building materials	IIn proofing concrete
Life style apparel	Elegant fleece vests
Other uses	Automotive interiors, Battery warmers

PCM protective garments should improve the comfort of workers as they go through these environmental step changes (e.g., warm to cold to warm, etc.). For these applications, the PCM transition temperature should be set so that the PCMs are in the liquid phase when worn in the warm environment and in the solid phase in the cold environment. The effect of phase change materials in clothing on the physiological and subjective thermal responses of people would probably be maximized if the wearer was repeatedly going through temperature transients (i.e., going back and forth between a warm and cold environment) or intermittently touching hot or cold objects with PCM gloves [154].

One example of practical application of PCM smart textile is cooling vest. This is a comfort garment developed to prevent elevated body temperatures in people who work in hot environments or use extreme physical exertion. The cooling effect is obtained from the vest's 21 PCM elements containing Glauber's salt which start absorbing heat at a particular temperature (28°C). Heat absorption from the body or from an external source continues until the elements have melted. After use the cooling vest has to be charged at room temperature (24°C) or lower. When all the PCMs are solidified the cooling vest is ready for further use [155].

A new generation of military fabrics feature PCMs which are able to absorb, store and release excess body heat when the body needs it resulting in less sweating and freezing, while the microclimate of the skin is influenced in a positive way and efficiency and performance are enhanced [156].

In the medical textiles field, a blanket with PCM can be useful for gently and controllably reheating hypothermia patients. Also, using PCMs in bed covers regulates the micro climate of the patient [157].

In domestic textiles, blinds and curtains with PCMs can be used for reduction of the heat flux through windows. In the summer month's large amounts of heat penetrate the buildings through windows during the day. At night in the winter months the windows are the main source of thermal loss. Results of the test carried out on curtains containing PCM have indicated a 30% reduction of the heat flux in comparison to curtains without PCM [158].

6.13.2 *SHAPE* MEMORY MATERIALS

Shape memory materials are able to 'remember' a shape, and return to it when stimulated, for example, with temperature, magnetic field, electric field, pH-value and UV light. An example of natural shape memory textile material is cotton, which expands when exposed to humidity and shrinks back when dried. Such behavior has not been used for esthetic effects because the changes, though physical, are in general not noticeable to the naked eye. The most common types of such SMMs materials are shape memory alloys and polymers, but ceramics and gels have also been developed. When sensing this material specific stimulus, SMMs can exhibit dramatic deformations in a stress free recovery. On the other hand, if the SMM is prevented from recovering this initial strain, a recovery stress (tensile stress) is induced, and the SMM actuator can perform work. This situation where SMM deforms under load is called restrained recovery [159].

Because of the wide variety of different activation stimuli and the ability to exhibit actuation or some other predetermined response, SMMs can be used to control or tune many technical parameters in smart material systems in response to environmental changes –such as shape, position, strain, stiffness, natural frequency, damping, friction and water vapor penetration. Both the fundamental theories and engineering aspects of SMMs have been investigated extensively and a rather wide variety of different SMMs are presently commercial materials. Commercialized shape memory products have been based mainly on metallic SMAs, either taking advantage of the shape change due the shape memory effect or the super-elasticity of the material, the two main phenomena of SMAs. SMPs and shape memory gels are developed at a quick rate, and within the last few years also some products based on magnetic shape memory alloys have been commercialized. SMC materials, which can be activated not only by temperature but also by elastic energy, electric or magnetic field, are mainly at the research stage [160].

6.13.2.1   APPLICATIONS OF TEXTILES CONTAINING SMMS

There are many potential applications of shape memory polymers in industrial components like automotive parts, building and construction products, intelligent packing, implantable medical devices, sensors and actuators, etc. SMPs are used in toys, handgrips of spoons, toothbrushes, razors and kitchen knives, also as an automatic choking device in small-size engines. One of the most well-known examples of SMP is a clothing application, a membrane called Diaplex. The membrane is based on polyurethane based shape memory polymers developed by Mitsubishi Heavy Industries [161].

Polyurethane is an example of shape memory polymers which is based on the formation of a physical cross-linked network as a result of entanglements of the high molecular weight linear chains, and on the transition from the glassy state to the rubber-elastic state. Shape memory polyurethane is a class of polyurethane that is different from conventional polyurethane in that these have a segmented structure and a wide range of Tg. The long polymer chains entangle each other and a three-dimensional network is formed. The polymer network keeps the original shape even above Tg in the absence of stress. Under stress, the shape is deformed and the deformed shape is fixed when cooled below Tg. Above the glass transition temperature polymers show rubber-like behavior [162].

The material softens abruptly above the glass transition temperature Tg. If the chains are stretched quickly in this state and the material is rapidly cooled down again below the glass transition temperature the polynorbornene chains can neither slip over each other rapidly enough nor become disentangled. It is possible to freeze the induced elastic stress within the material by rapid cooling. The shape can be changed at will. In the glassy state the strain is frozen and the deformed shape is fixed. The decrease in the mobility of polymer chains in the glassy state maintains the transient shape in polynorbornene. The recovery of the material's original shape can be observed by heating again to a temperature above Tg. This occurs because of the thermally induced shape-memory effect. The disadvantage of this polymer is the difficulty of processing because of its high molecular weight [162, 163]. Some of the shape memory polymers are suitable for textiles applications are shown in Table 6.3.

TABLE 6.3   Some of the Shape Memory Polymers For Textiles Applications

Polymer	Physical interactions	Form
	Original shape	Transient shape
Polynorbomene	Chain entanglement	Glassy state
Polyurethane	Microcrystal	Glassy state
Polyethylene/nylon-6 graft co-polymer	Crosslinking	Microcrystal
Styrene-1,4-butadiiene block copolymer	Microcrystal/Glassy state of polystyrene	Microcrystal of poly(1,4-butadiene)
Ethylene oxide-ethylene terphethalate block copolymer	Microcrystal of PET	Microcrystal of PEO
Polymethylene-1,3-cyclopentane)polyethylene block copolymer	Microcrystal of PE	Microcrystal/Glassy state of pmcp

Shape memory polymers can be laminated, coated, foamed, and even straight converted to fibers. There are many possible end uses of these smart textiles. The smart fiber made from the shape memory polymer can be applied as stents, and screws for holding bones together.

Shape memory polymer coated or laminated materials can improve the thermo-physiological comfort of surgical protective garments, bedding and incontinence products because of their temperature adaptive moisture management features.

Films of shape memory polymer can be incorporated in multilayer garments, such as, those that are often used in the protective clothing or leisurewear industry. The shape memory polymer reverts within wide range temperatures. This offers great promise for making clothing with adaptable features. Using a composite film of shape memory polymer as an interliner in multilayer garments, outdoor clothing could have adaptable thermal insulation and be used as protective clothing. A shape memory polymer membrane and insulation materials keep the wearer warm. Molecular pores open and close in response to air or water temperature to increase or minimize heat loss. Apparel could be made with shape memory fiber. Forming the shape at a high temperature provides creases and pleats in such apparel as slacks and skirts. Other applications include fishing yarn, shirt neck bands, cap edges, casual clothing and sportswear. Also, using a composite film of shape memory polymers as an interlining provides apparel systems with variable tog values to protect against a variety of weather conditions [163].

6.13.3 *CHROMIC* MATERIALS

Chromic materials are the general term referring to materials, which their color changes by the external stimulus. Due to color changing properties, chromic materials are also called chameleon materials. This color changing phenomenon is caused by the external stimulus and chromic materials can be classified depending on the external stimulus of induction.

Photochromic, thermochromic, electrochromic, piezochromic, solvatechromic and carsolchromic are chromic materials that change their color by the external stimulus of heat, electricity, pressure, liquid and an electron beam, respectively. Photochromic materials are suitable for sun lens applications. Most photochromic materials are based on organic materials or silver particles. Thermochromic materials change color reversibly with changes in temperature. The liquid crystal type and the molecular rearrangement type are thermochromic systems in textiles. The thermochromic materials can be made as semiconductor compounds, from liquid crystals or metal compounds. The change in color occurs at a predetermined temperature, which can be varied. Electrochromic materials are capable of changing their optical properties (transmittance and/or reflectance) under applied electric potentials. The variation of the optical properties is caused by insertion/extraction of cations in the electrochromic film. Piezochromism is the phenomenon where crystals undergo a major change of color due to mechanical grinding. The induced color reverts to the original color when the fractured crystals are kept in the dark or dissolved in an organic solvent [164].

Solvatechromism is the phenomenon, where color changes when it makes contact with a liquid, for example, water. Materials that respond to water by changing color are also called hydrochromic and this kind of textile material can be used, for example, for swimsuits.

6.13.3.1 APPLICATIONS OF TEXTILES CONTAINING CHROMIC MATERIALS

The majority of applications for chromic materials in the textile sector today are in the fashion and design area, in leisure and sports garments. In workwear and the furnishing sector a variety of studies and investigations are in the process by industrial companies, universities and research centers. Chromic materials are one of the challenging material groups when thinking about future textiles. Color changing textiles are interesting, not only in fashion, where color changing phenomena will exploit for fun all the rainbow colors, but also in useful and significant applications in soldier and weapons camouflage, workwear and in technical and medical textiles. The combination of SMM and thermochromic coating is an interesting area which produces shape and color changes of the textile material at the same time [165].

6.13.4 *DESIGNING* THE SMART TEXTILE SYSTEMS

Comfort is very important in textiles because stresses lead to increased fatigue. The potential of smart textiles is to measure a number of body parameters such as skin temperature, humidity and conductivity and show the level of comfort through the textile sensors. To keep the comfort of textiles, adequate actuators are needed that can heat, cool, insulate, ventilate and regulate moisture. The use of the smart system should not require any additional effort. The weight of a smart textile system should not reduce operation time of the rescue worker.

Other key issues for the design of a smart textile system are [166]:
- Working conditions-relevant parameters: only relevant information should be provided in order to avoid additional workload; this includes indication of danger and need for help.
- Effective alarm generation: the rescue worker or a responsible person should be informed adequately on what needs to be done.
- System maintenance: it must be possible to treat the suit using usual maintenance procedures.
- Cost must be justified
- Robustness
- Energy constraints: energy requirements must be optimized
- Long range transmission: transmission range must be adjusted to the situation of use.

Fighting a fire in a building is different from fighting one in an open field.
- A wearable smart textile system basically comprises following components [36]:
- Sensors to detect body or environmental parameters;
- A data processing unit to collect and process the obtained data;
- An actuator that can give a signal to the wearer;
- An energy supply that enables working of the entire system;
- Interconnections that connect the different components;
- A communication device that establishes a wireless communication link with a nearby base station.

The main layers concerned with smart clothing are the skin layer and two clothing layers. Physically the closest clothing layer for a human user is an underwear layer, which transports perspiration away from the skin area. The function of this layer is to keep the interface between a user and the clothes comfortable and thus improve the overall wearing comfort. The second closest layer is an intermediate clothing layer, which consists of the clothes that are between the underclothes and outdoor clothing. The main purpose of this layer is considered to be an insulation layer for warming up the body. The outermost layer is an outerwear layer, which protects a human against hard weather conditions.

The skin layer is located in close proximity to the skin. In this layer we place components that need direct contact with skin or need to be very close to the skin.

Therefore, the layer consists of different user interface devices and physiological measurement sensors. The number of the additional components in underwear is limited owing to the light structure of the clothing.

An inner clothing layer contains intermediate clothing equipped with electronic devices that do not need direct contact with skin and, on the other hand, do not need to be close to the surrounding environment. These components may also be larger in size and heavier in weight compared to components associated with underclothes. It is often beneficial to fasten components to the inner clothing layer, as they can be easily hidden. Surrounding clothes also protect electronic modules against cold, dirt and hard knocks.

Generally, the majority of electronic components can be placed on the inner clothing layer. These components include various sensors, a central processing unit and communication equipment. Analogous to ordinary clothing, additional heating to warming up a person in cold weather conditions is also associated with this layer. Thus, the inner layer is the most suitable for batteries and power regulating equipment, which are also sources of heat.

The outer clothing layer contains sensors needed for environment measurements, positioning equipment that may need information from the surrounding environment and numerous other accessories. The physical surroundings of smart clothing components measure the environment and the virtual environment accessed by communication technologies. Soldier and weapons camouflage is possible by using chromic materials in outer layer of smart textiles [10].

6.13.5   *DATA* TRANSFER REQUIREMENTS IN SMART TEXTILES

The data transfer requirements can be divided into internal and external. The internal transfer services are divided into local health and security related measurements. Many of the services require or result in external communications between the smart clothing and its environment. Wired data transfer is in many cases a practical and straightforward solution. Thin wires routed through fabric are an inexpensive and high capacity medium for information and power transfer. The embedded wires inside clothing do not affect its appearance. However, wires form inflexible parts of clothing and the detaching and reconnecting of wires decrease user comfort and the usability of clothes. The cold winter environment especially stiffens the plastic shielding of wires. In hard usage and in cold weather conditions, cracking of wires also becomes a problem [167]. The connections between the electrical components placed on different pieces of clothing are another challenge when using wires. During dressing and undressing, connectors should be attached or detached, decreasing the usability of clothing. Connectors should be easily fastened, resulting in the need for new connector technologies.

A potential alternative to plastic shielded wires is to replace them with electrically conductive fibers. Conductive yarns twisted from fibers form a soft cable that naturally integrates in the clothing's structure keeping the system as clothing-like as possible. Fiber yarns provide durable, flexible and washable solutions. Also lightweight optical fibers are used in wearable applications, but their function has been closer to a sensor than a communication medium [168, 169]. The problem of conductive fibers is due to the reliable connections of them. Ordinary wires can be soldered directly to printed circuit boards, but the structure of the fiber yarn is more sensitive to breakage near the solder connections.

Protection materials that prevent the movement of the fiber yarn at the interface of the hard solder and the soft yarn must be used. Optical fibers are commonly used for health monitoring applications and also for lighting purposes.

Low-power wireless connections provide increased flexibility and also enable external data transfer within the personal space. Different existing and emerging WLAN and WPAN types of technologies are general purpose solutions for the external communications, providing both high speed transfer and low costs. For wider area communications and full mobility, cellular data networks are currently the only practical possibility [170].

6.13.6 *OPTICAL* FIBERS IN SMART TEXTILES

Optical fibers are currently being used in textile structures for several different applications. Optic sensors are attracting considerable interest for a number of sensing applications [171]. There is great interest in the multiplexed sensing of smart structures and materials, particularly for the real-time evaluation of physical measurements (e.g., temperature, strain) at critical monitoring points. One of the applications of the optical fibers in textile structures is to create flexible textile-based displays based on fabrics made of optical fibers and classic yarns [172]. The screen matrix is created during weaving, using the texture of the fabric. Integrated into the system is a small electronics interface that controls the LEDs that light groups of fibers. Each group provides light to one given area of the matrix. Specific control of the LEDs then enables various patterns to be displayed in a static or dynamic manner. This flexible textile-based displays are very thin size and ultra-light weight. This leads one to believe that such a device could quickly enable innovative solutions for numerous applications. Bending in optical fibers is a major concern since this causes signal attenuation at bending points. Integrating optical fibers into a woven perform requires bending because of the crimping that occurs as a result of weave interlacing. However, standard plastic optical fiber materials like poly methyl methacrylate, polycarbonate and polystyrene are rather stiff compared to standard textile fibers and therefore their integration into textiles usually leads to stiffen of the woven fabric and the textile touch is getting lost [173]. Alternative fibers with appropriate flexibility and transparency are not commercially available yet.

6.13.7 *CONDUCTIVE* MATERIALS IN SMART TEXTILES

Several conductive materials are in use in smart textiles. Conductive textiles include electrically conductive fibers fabrics and articles made from them. Flexible electrically conducting and semiconducting materials, such as, conductive polymers, conductive fibers, threads, yarns, coatings and ink are playing an important role in realizing lightweight, wireless and wearable interactive electronic textiles. Generally, conductive fibers can be divided into two categories such as naturally conductive fibers and treated conductive fibers.

Naturally, conductive fibers can be produced purely from inherently conductive materials, such as, metals, metal alloys, carbon sources, and conjugated polymers [174].

However, it is important to point out that nanofibers produced from polymers do not in general show quantum effects usually associated with nanotechnology, with spatially confined matter. The reason simply is first of all that the nature of the electronic states of organic materials including polymer materials that control optical and electronic properties does not resemble the one known for semiconductors or conductors.

Electronic states that are not localized but rather extend throughout the bulk material are characteristics of such nonorganic materials, with the consequence that modifications first of all of the absolute size and secondly of the geometry of a body made from them have strong effects on properties particularly as the sizes approach the few tens to a few nanometers scale. Organic materials, on the other hand, display predominantly localized states for electronic excitations, electronic transport with the states being defined by molecular groups such as chromophore groups or complete molecules. The consequence is that the electronic states are not affected as the dimensions of the element such as a fiber element are reduced down into the nanometer scale [175].

Furthermore, both amorphous polymers and partially crystalline polymers have structures anyway even in the bulk, in macroscopic bodies that are restricted to the nanometer scale. The short-range order of amorphous polymers, as represented by the pair correlation function, does not extend beyond about 2 nm and the thickness of crystalline lamellae is typically in the range of a few nanometers or a few tens of nanometer, respectively. So, the general conclusion is that the reduction of the diameter of fibers made from polymers or organic materials for that matter will affect neither optical nor electronic properties to a significant degree or the intrinsic structure. This will, of course, be different if we are concerned with fibers composed of metals, metal oxides, semiconductors as accessible via the precursor route. In such cases, one is well within the range of quantum effects and the properties of such fibers have to be discussed along the lines spelt out in textbooks on quantum effects [176].

Now, staying with the subject of polymer fibers the discussion given above should certainly not lead to the conclusion that nanofibers and nonwovens composed of nanofibers do not display unique properties and functions of interest both in the areas of basic science but also technical applications, just the contrary.

Taking the reduction of the fiber diameter into the nm scale as a first example it is readily obvious that the specific surface area increases dramatically as the fiber diameter approaches this range. In fact, it increases with the inverse of the fiber radius.

It is obvious that the specific surface increases from about 0.1 m^2/g for fibers with a thickness of about 50 micrometers (diameter of a human hair) to about 300 m^2/g for fibers with a thickness of 10 nm.

Secondly, the strength of fibers also scales inversely with the fiber diameter, thus increasing also very strongly with decreasing fiber diameter, following the Griffith criteria. The reason is that the strength tends to be controlled by surface flaws, the probability of which will decrease along a unit length of the fiber as the surface area decreases. So, a decrease of the fiber diameter from about 50 micrometer to about 10 nm is expected to increase the strength from about 300 N/mm 2 by a factor of about 1000 and more. Thirdly, the pores of nonwoven membranes composed of nanofibers reach the nm range as the fibers get smaller and smaller. A reduction of the fiber diameter from say about 50 micrometer, for which pores with diameter of about 500 micrometer are expected to about 100 nm will cause the pore diameter to decrease to about 1 micrometer. In a similar way, a further reduction to fiber diameters to about 10 nm will cause the average pores to display pore diameters around 100 nm. This will certainly show up in the selectivity of the filters with respect to solid particles, aerosols, etc., to be discussed further below [175].

The reduction in pore diameter is, furthermore, connected with strong modifications of the dynamics of gases and fluids within the nonwoven, located within or flowing through the nonwoven. Thermal insulations, for instance, in nonwovens containing a gas is controlled for larger pores-larger than the average free path length of the molecules – by just this free path length. However, as the pores get smaller the collision of the molecules with the pore walls – fiber surface for fiber -based nonwovens – takes over the control of thermal insulation with an increase of the thermal insulation that can amount to several orders of magnitude. This aspect will also be discussed below [177].

Finally, the flow of gases or fluids around a fiber changes very strongly in nature as the fiber diameter goes down to the nanometer range with a transition of the flow regime from the conventional one to the so-called Knudsen regime, to be discussed below in more detail.

All these effects are classical ones yet have major impact on nanofiber properties and applications. These examples show that nanofibers and nonwovens composed of them display unique properties of functions already based on classical phenomena. This suggests that such fibers/nonwovens can be used with great benefits in various types of applications. The spectrum of applications that can be envisioned for electrospun nanofibers is extremely broad due to their unique intrinsic structure, surface properties and functions [175].

Highly conductive flexible textiles can be prepared by weaving thin wires of various metals such as brass and aluminum. These textiles have been developed for higher degrees of conductivity.

Metal conductive fibers are very thin filaments with diameters ranging from 1 to 80 μm produced from conductive metals such as ferrous alloys, nickel, stainless steel, titanium, aluminum and copper. Since they are different from polymeric fibers, they may be hard to process and have problems of long-term stability. These highly conductive fibers are expensive, brittle, heavier and lower processability than most textile fibers.

Treated conductive fibers can be produced by the combination of two or more materials, such as, nonconductive and conductive materials. These conductive textiles can be produced in various ways, such as, by impregnating textile substrates with conductive carbon or metal powders, patterned printing, and so forth. Conducting polymers, such as, polyacetylene, polypyrrole, polythiophene and polyaniline offer an interesting alternative. Among them, polypyrrole has been widely investigated owing to its easy preparation, good electrical conductivity, good environmental stability in ambient conditions and because it poses few toxicological problems. PPy is formed by the oxidation of pyrrole or substituted pyrrole monomers. Electrical conductivity in PPy involves the movement of positively charged carriers or electrons along polymer chains and the hopping of these carriers between chains. The conductivity of PPy can reach the range 102 S cm-1, which is next only to PA and PAn. With inherently versatile molecular structures, PPys are capable of undergoing many interactions [178, 179].

The conductive fibers obtained through special treatments such as mixing, blending, or coating are also known as CPCs, can have a combination of the electrical and mechanical properties of the treated materials. Fibers containing metal, metal oxides and metal salts are a proper alternative for metal fibers. Polymer fibers may be coated with a conductive layer such as polypyrrole, copper or gold. The conductivity will be maintained as long as the layer is intact and adhering to the fiber. Chemical plating and dispersing metallic particles at a high concentration in a resin are two general methods of coating fibers with conductive metals [180].

The brittleness of PPy has limited the practical applications of it. The processability and mechanical properties of PPy can be improved by incorporating some polymers into PPy [181]. However, the incorporation of a sufficient amount of filler generally causes a significant deterioration in the mechanical properties of the conducting polymer, in order to exceed the percolation threshold of conductivity [182]. Another route to overcoming this deficiency is by coating the conducting polymer on flexible textile substrates to obtain a smooth and uniform electrically conductive coating that is relatively stable and can be easily handled [183].

Thus, PPy-based composites may overcome the deficiency in the mechanical properties of PPy, without adversely affecting the excellent physical properties of the substrate material, such as, its mechanical strength and flexibility. The resulting products combine the usefulness of a textile substrate with electrical properties that are similar to metals or semiconductors.

Due to electron-transport characteristics of Conjugated polymers or ICPs, they are regarded as semi conductors or even sometimes conductors. Due to their high conductivity, lower weight, and environmental stability, they have a very important place in the field of smart and interactive textiles [184].

The conductivity of materials is often affected by several parameters which may be exploitable mechanisms for use as a sensor. Extension, heating, wetting and absorption of chemical compounds in general may increase or decrease conductivity. Swelling or shrinkage of composite fibers of carbon nanotubes alters the distance between the nanoparticles in the fibers, causing the conductivity to change.

Fibers containing conductive carbon are produced with several methods such as loading the whole fibers with a high concentration of carbon, incorporating the carbon into the core of a sheath–core bicomponent fiber, incorporating the carbon into one component of a side–side or modified side–side bicomponent fiber, suffusing the carbon into the surface of a fiber.

Nanoparticles such as carbon nanotubes can be added to the matrix for achieving conductivity. Semi-conducting metal oxides are often nearly colorless, so their use as conducting elements in fibers has been considered likely to lead to fewer problems with visibility than the use of conducting carbon. The oxide particles can be embedded in surfaces, or incorporated into sheath–core fibers, or react chemically with the material on the surface layer of fibers.

Conductive fibers can also be produced by coating fibers with metal salts such as copper sulfide and copper iodide. Metallic coatings produce highly conductive fibers; however, adhesion and corrosion resistance can present problems. It is also possible to coat and impregnate conventional fibers with conductive polymers, or to produce fibers from conductive polymers alone or in blends with other polymers.

Conductive fibers/yarns can be produced in filament or staple lengths and can be spun with traditional nonconductive fibers to create yarns that possess varying wearable electronics and photonics degrees of conductivity. Also, conductive yarns can be created by wrapping a nonconductive yarn with metallic copper, silver or gold foil and be used to produce electrically conductive textiles.

Conductive threads can be sewn to develop smart electronic textiles. Through processes such as electrodeless plating, evaporative deposition, sputtering, coating with a conductive polymer, filling or loading fibers and carbonizing, a conductive coating can be applied to the surface of fibers, yarns or fabrics. Electrodeless plating produces a uniform conductive coating, but is expensive. Evaporative deposition can produce a wide range of thicknesses of coating for varying levels of conductivity. Sputtering can achieve a uniform coating with good adhesion. Textiles coated with a conductive polymer, such as, polyaniline and polypyrrole, are more conductive than metal and have good adhesion, but are difficult to process using conventional methods.

Adding metals to traditional printing inks creates conductive inks that can be printed onto various substrates to create electrically active patterns. The printed circuits on flexible textiles result in improvements in durability, reliability and circuit speeds and in a reduction in the size of the circuits. The printed conductive textiles exhibit good electrical properties after printing and abrading. The inks withstand bending without losing conductivity. However, after 20 washing cycles, the conductivity decreases considerably. Therefore, in order to improve washability, a protective polyurethane layer is put on top of the printed samples, which resulted in the good conductivity of the fabrics, even after washing. Currently, digital printing technologies promote the application of conductive inks on textiles [185].

6.13.7.1   APPLICATIONS OF CONDUCTIVE SMART TEXTILE

Electrically conductive textiles make it possible to produce interactive electronic textiles. They can be used for communication, entertainment, health care, safety, homeland security, computation, thermal purposes, protective clothing, wearable electronics and fashion. The application of conductive smart textile in combination with electronic advices is very widespread. In location and positioning, they can be used for child monitoring, geriatric monitoring, integrated GPS monitoring, livestock monitoring, asset tracking, etc.

In infotainment, they can be used for integrated compact disc players, MP3 players, cell phones and pagers, electronic game panels, digital cameras, and video devices, etc.

In health and biophysical monitoring, they can be used for cardiovascular monitoring, monitoring the vital signs of infants, monitoring clinical trials, health and fitness, home healthcare, hospitals, medical centers, assisted-living units, etc.

They can be used for soldiers and personal support of them in the battlefield, space programs, protective textiles and public safety (fire-fighting, law enforcement), automotive, exposure-indicating textiles, etc.

They can be also used to show the environmental response such as color change, density change, heating change, etc. Fashion, gaming, residential interior design, commercial interior design and retail sites are other application of conductive smart textiles.

6.13.8   SMART FABRICS FOR HEALTH CARE

The continuous monitoring of vital signs of some patients and elderly people is an emerging concept of health care to provide assistance to patients as soon as possible either online or offline. A wearable smart textile can provide continuous remote monitoring of the health status of the patient. Wearable sensing systems will allow the user to perform everyday activities without discomfort. The simultaneous recording of vital signs would allow parameter extrapolation and intersignal elaboration, contributing to the generation of alert messages and synoptic patient tables. In spite of this, a smart fabric is capable of recording body kinematic maps with no discomfort for several fields of application such as rehabilitation and sports [186].

6.13.9 *ELECTRONIC* SMART TEXTILES

The components of an electronic smart textile that provide several functions are sensors unit, network unit, processing unit, actuator unit and power unit. On the smart textile, several of these functions are combined to form services. Providing information, communication or assistance are possible services. Because mobility is now a fundamental aspect of many services and devices, these smart textiles can be used for health applications such as monitoring of vital signs of high-risk patients and elderly people, therapy and rehabilitee, knowledge applications such as instructions and navigation and entertainment applications such as audio and video devices. For communication between the different components of smart textile applications, both wired and wireless technologies are applicable. An applied solution for data transferring is often a compromise based on application requirements, operational environment, available and known technologies, and costs [146].

6.13.10 *ELECTRICAL* CIRCUITS IN SMART TEXTILE STRUCTURES

In order to form flexible circuit boards, printing of circuit patterns is carried out on polymeric substrates such as films. Fabric based circuits potentially offer additional benefits of higher flexibility in bending and shear, higher tear resistance, as well as better fatigue resistance in case of repeated deformation. Different processes that have been described in literature for the fabrication of fabric based circuits include embroidery of conductive threads on fabric substrates, weaving and knitting of conductive threads along with nonconductive threads, printing or deposition and chemical patterning of conductive elements on textile substrates.

The insulating fabric could be woven, nonwoven, or knitted [187]. The conductive threads can be embroidered in any shape on the insulating fabric irrespective of the constituent yarn path in a fabric. One of the primary disadvantages of embroidery as a means of circuit formation is that it does not allow formation of multilayered circuits involving conductive threads traversing through different layers as is possible in the case of woven circuits. Conductive threads can be either woven or knitted into a fabric structure along with nonconductive threads to form an electrical circuit. One of the limitations of using weaving for making electrical circuits is that the conductive threads have to be placed at predetermined locations in the warp direction while forming the warp beam or from a creel during set up of the machine.

Different kinds of conductive threads can be supplied in the weft or filling direction and inserted using the weft selectors provided on a weaving machine. Some modifications to the yarn supply system of the machine may be needed in order to process the conductive threads that are more rigid [188].

In most conventional weft knitting machines, like a flatbed machine, the conductive threads can be knitted in the fabric only in one direction, i.e., the course (or cross) direction. In order to keep the conductive element in a knit structure straight, one can insert a conductive thread in the course direction such that the conductive thread is embedded into the fabric between two courses formed from nonconductive threads.

Processes that have been employed to form a patterned conductive path on fabric surfaces include deposition of polymeric or nonpolymeric conducting materials and subsequent etching, reducing, or physical removal of the conductive materials from certain regions. Thus, the conductive material that is not removed forms a patterned electrical circuit or a region of higher conductivity. The biggest problem associated with patterning of circuits from thin conductive films (polymeric or metallic) deposited on fabric substrates is that use of an etching agent for forming a circuit pattern leads to nonuniform etching, as some of the etching liquid is absorbed by the threads of the underlying substrate fabric [189]. Another problem with deposition of conductive films on fabric substrates is that bending the fabric may lead to discontinuities in conductivity at certain points.

There are different device attachment methods like raised wire connectors, solders, snap connectors, and ribbon cable connectors in electronic smart textiles. Soldering produces reliable electrical connections to conductive threads of an electronic textile fabric but has the disadvantage of not being compatible with several conductive threads or materials like stainless steel. Moreover, soldering of electronic devices to threads that are insulated is a more complex process involving an initial step of removal of insulation from the conductive threads in the regions where the device attachments are desired and insulation of the soldered region after completion of the soldering process. The main advantage of employing snap connectors is the ease of attachment or removal of electronic devices from these connectors, whereas the main disadvantages are the large size of the device and the weak physical connection formed between the snap connectors and the devices. Ribbon cable connectors employ insulation displacement in order to form an interconnection with insulated conductor elements integrated into the textiles. A v-shaped contact cuts through the insulation to form a connection to the conductor. Firstly, the ribbon cable connector is attached to the conductive threads in an e-textile fabric and subsequent electronic devices and printed circuit boards are attached to the ribbon cable connector. One of the advantages of employing ribbon cable connectors for device attachment is the ease of attachment and removal of the electronic devices to form the electronic textiles [190, 191].

6.13.11 *NEW* TEXTILE MATERIALS FOR ENVIRONMENTAL PROTECTION

From prehistoric times till now, air pollution from hazardous chemical and biological particles is an essential threat to humans' health. Together with the development of civilization and escalation of the conflicts between nations, the risk of loss of health and even life due to polluted air increases considerably. Therefore the continuous development of the new materials used for protection of human respiratory tracts against hazardous particles is observed. The fibrous materials play a special role in this subject. Davies in his work 'Air Filtration has presented an interesting review of the earliest literature considering problems connected with filtering polluted air [192].

For centuries, miners have used special clothes to protect nose and mouth against dust. Bernardino amazzini, who lived on the turn of the seventeenth century, in his work 'De morbis artificum' indicated the need for protection of the respiratory tracts against dusts of workers laboring in various professions listed by him. Brise Fradin developed in 1814 the first device, which provided durable protection of the respiratory tracts. It was composed of a container filled with cotton fibers which was connected by a duct with the user's mouth. The first filtration respiratory mask was designed at the beginning of the nineteenth century with the aim of protecting the users against diseases transmitted by the breathing system. In these times, firemen began to use masks specially designed for them. The first construction of such a 'mask' was primitive: a leather helmet was connected with a hose which supplied air from the ground level.

The construction was based on the observation that during fire, fewer amounts of toxic substances were at the ground level than at the level of the fireman's mouth. In addition, a layer of fibers protected the lower air inlet. John Tyndall, in 1868 designed a mask which consisted of some layers of differentiated structure. A clay layer separated the first two layers of dry cotton fibers. Between the two next cotton fiber layers can be inserted charcoal, and the last two cotton fiber layers were separated by a layer of wool fibers saturated with glycerin. The history of the development of filtration materials over the nineteenth century has been described in a work elaborated by Feldhaus [193].

The twentieth century left a lasting impression of the First World War, during which toxic gases were used for the first time. This was the reason that after 1914, the further history of the development of filtration materials was connected with absorbers of toxic substances manufactured with the use of charcoal and fibrous materials. The next discovery, which changed the approach to the designing of filtration materials, was done in 1930. Hansen, in his filter applied a mixture of fibers and resin as filtration materials. This caused an electrostatic field being created inside the material. The action of electrostatic forces on dust particles significantly increases the filtration efficiency of the materials manufactured.

The brief historical sketch presented above indicates that textile fibers were one of the material components, which protect the respiratory tracts, and have been applied from the dawn of history. From the beginning they had been used intuitively, without understanding the mechanism of filtration. The first attempts of scientific description of the filtration mechanism were presented by Albrecht [194], Kaufman [195], Langmuir [196], and recently by Brown [6] who characterized the four basic physical phenomena of mechanical deposition in the following way:

- direct interception occurs when a particle follows a streamline and is captured as a result of coming into contact with the fiber;
- inertial impaction is realized when the deposition is effected by the deviation of a particle from the streamline caused by its own inertia; in diffusive deposition the combined action of airflow and Brownian motion brings a particle into contact with the fiber; gravitational settling resulting from gravitation forces.

Illustration of the above mechanisms of filtration is presented in Fig. 6.14.

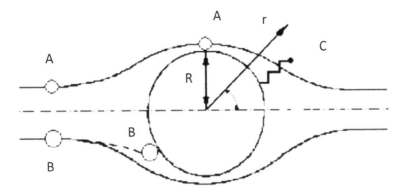

FIGURE 6.14 Particle capture mechanism: A – particle captured by interception; B – particle captured by inertial impaction; C – particle captured by diffusive deposition.

The analysis of equations determining the filter efficiency governed by the mechanisms specified above indicates that the most important parameters deciding about filtering efficiency are the thickness of filters, the diameter of fibers and the porosity of the filter. Identification of these phenomena was the basis for development of new technologies for filtering materials composed from ultra thin fibers. These technologies mainly are based on manufacturing the nonwovens directly from the dissolved or melted polymers using melt-blown technique, flash-spinning and electrospinning.

Additionally theoretical consideration also indicates that the activity of fibers on particles significantly increases if an electrostatic field is formed inside the nonwoven. This is the reason that nonwovens are additional modified. Three following groups of fibrous electrostatic materials used can be distinguished, based upon their ability to generate an electrostatic charge:

- materials in which the charge is generated by corona discharge after fiber or web formation,
- materials in which the charge is generated by induction during spinning in an electrostatic field, and
- materials in which the charge is induced as the result of the triboelectric effect.

6.14 REVIEW OF TECHNIQUES FOR MANUFACTURING FIBROUS FILTERING MATERIALS

Current environmental problems are caused by human activities in the last 150 years. They are having serious negative impacts on us and this is likely to continue for a very long time. Effective solutions are urgently needed to protect our environment. Due to their high specific surface area, electrospun nanofibers are expected to be used to collect pollutants via physical blocking or chemical adsorption.

The first nonwovens using melted organic polymers were manufactured in the 1950s, using a method similar to air-blowing of the polymer melt. Application of this latter method enabled super-thin fibers to be obtained with a diameter smaller than 5 nm. The melt-blown technique of manufacturing nonwovens from super-fine fibers was developed by Wente at the Naval Research Laboratory in USA, [197]. Buntin, a worker at Exxon Research and Engineering introduced the melt-blown technique for processing PP into the industry [198]. Recently, the Nonwoven Technologies Inc., USA has announced the possibility of manufacturing melt-blown PP nonwovens composed from nano-size fibers of a diameter equal to 300 nm. To enhance the filtration efficiency, the melt -blown nonwovens are subjected to the process of activation, mainly using the corona discharge method [105].

An overview of flash-spinning technologies is presented by Wehman. Flashspun nonwovens made from fibers with very low linear density, which can be obtained using splittable fibers as a raw material for production of conventional webs [199].

Subsequently, webs can be subjected to the classical needle punching or spunlace process during which sacrificial polymer is removed and fibers of low linear density are obtained. The flash-spinning process can be also accomplished using such bicomponent melt-blown technology which is based on spinning two incompatible polymers together and forming a web which is then subjected to the splitting process.

Induction of electric charges is another mechanism used in filtering material technology. Induction consists of electric charge generation in a conductor placed in an electric field. Therefore, fine-fibers made from conductive solutions or melts, charged during electrostatic extrusion, belong to this group. Formation of nano-fibers by the electrospinning method results from the reaction of a polymer solution drop subjected to an external electric field. This method enables manufacturing fibers with transversal dimensions of nanometers. Gilbert in 1600 made the first observation concerning the behavior of an electroconductive fluid under the action of an electrical field. He pointed out that a spherical drop of water on a dry surface is drowning up, taking the shape of a cone when a piece of rubbed amber is held above it. One of the first investigations into the phenomena of interaction of an electric field with a fluid drop was carried out by Zeleny [200]. He used the apparatus presented in Fig. 6.15 in his experiment.

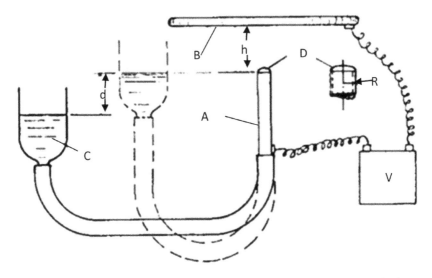

FIGURE 6.15 Scheme showing the idea of an one-plate apparatus for electrospinning.

The apparatus include an open-end capillary tube of metal or glass. The conductive fluid is delivered to this tube using the reservoir C. A plate B is mounted opposite to the capillary tube in a distance of h. The capillary tube and the plate are maintained at a given potential difference V using a high voltage generator. Formhals [201] used this kind of technology for spinning thin polymer filaments.

The electric charges, which diffuse in the liquid, forced by the electric field, cause a strong deformation of the liquid surface in order to minimize the system's total energy. The electric forces exceed the forces of surface tension in the regions of the maximum field strength and charge density, and the liquid forms a cone at the nozzle outlet. A thin stream of liquid particles is torn off from the end of the cone. Taylor [202] proved that for a given type of fluid, a critical value of the applied voltage exists, at which the drop of fluid, flowing from the capillary tube, is transformed into a cone under the influence of the electric field, and loss its stability. The critical value of this potential depends on the surface tension of the fluid and of the initial radius of drop. Zeleny's and Taylor's investigations have been an inspiration for many researchers who carried out observations of the behavior of different kinds of polymers in the electric field. These observations were the basis for the development of manufacturing technologies for a new generation of fibers with very small transversal dimensions. Schmidt demonstrated the possibility of application of electrospun polycarbonate fibers to enhance the dust filtration efficiency [203].

In the 1980s, the Carl Freudenberg company used the electrospinning technology first commercially. Trouilhet and Weghmann presented a wide range of applications of electrospun webs especially in the filtration area. In that time the electrospinning method for manufacturing filtering materials did not find common application. The revival of this technology has been observed for the last five-four years. In 2000 Donaldson Inc., USA realized dust filters with a thin layer of nanofibers [204, 205].

A basic set for electrospinning consists form three major components, such as: a high voltage generator, a metal or glass capillary tube, and a collecting plate electrode, similar to the set designed by Zeleny. Such type of set is characterized by low productivity, usually less than 1 mLh^{-1}. To solve this problem, the array of multiply capillary tubes should be developed. Experiments carried out indicate that due to the interference between the electrical fields developed around such system an uniform electrical field strength cannot be ensured at the tip of each tube. For such a system, high probability of the tube clogging appears. To avoid such problems during the electrospinning process, some authors proposed to spin the fibers directly form the polymer solution surface. A new method with high productivity was developed by Jirsak at the Technical University of Liberec [193]. The proposed invention was commercialized by Elmarco Company. The idea is very simple. The set is composed from two electrodes. The bottom electrode formed in the shape of a roll is immersed in the solution of a polymer, as shown in Fig. 6.16.

A thin layer of polymer solution covers the rotating electrode, and multiple jets are formed due to the action of the electrical field. The Elmarco Company offers a wide assortment of spun-bonded nonwovens covered by nanofiber membranes made of polyamide, polyurethane and polyvinyl alcohol. A further approach related to spinning directly from the solution surface was invented by Yarin and Zussman [206]. The proposed system is composed from two layers: a bottom layer in the form of ferromagnetic suspension and an upper layer in the form of polymer solution. The two-layer system is subjected to the magnetic field provided by a permanent magnet. The scheme of this apparatus is presented in Fig. 6.17.

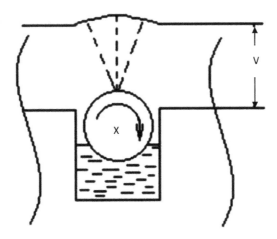

FIGURE 6.16 The idea of the electrospinning method developed by Jirsak.

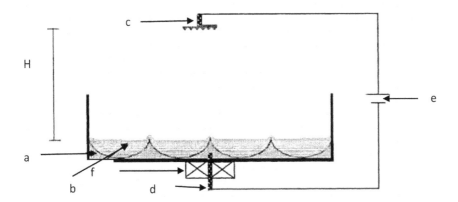

FIGURE 6.17 The idea of manufacturing electrospun nonwovens directly from the surface of a polymer solution: a - ferromagnetic suspension, b - polymer solution, c - upper electrode, d - lower electrode, e - high voltage generator, f - permanent magnet.

Vertical spikes of magnetic suspension appear as the result of action of the magnetic field, what causes the perturbation of the free surface of the polymer solution. Under the action of the electrical field, perturbations of the free surface become the sites of jetting directed upward.

6.14.1   *BASIC* RESEARCH ON ELECTROSPUN NANOFIBERS IN FILTRATION

Usually, the particle filtration occurs via multiple collection mechanism such as sieving, direct interception, inertial impaction, diffusion, and electrostatic collection. In practice, sieving is not an important mechanism in most air filtration application. Additionally, commercial nanofibers are electrically neutral. So, the remaining important mechanisms in mechanical filtration are direct interception, inertial impaction, and diffusion. The reasonable approximations of filtering media performance have been made using single-fiber filtration theory.

6.14.2   *MEMBRANES* FILTRATION

Membrane filtration is a mechanical filtration technique which uses an absolute barrier to the passage of particulate material as any technology currently available in water treatment. The term "membrane" covers a wide range of processes, including those used for gas/gas, gas/liquid, liquid/liquid, gas/solid, and liquid/solid separations. Membrane production is a large-scale operation. There are two basic types of filters: depth filters and membrane filters.

Depth filters have a significant physical depth and the particles to be maintained are captured throughout the depth of the filter. Depth filters often have a flexuous three-dimensional structure, with multiple channels and heavy branching so that there is a large pathway through which the liquid must flow and by which the filter can retain particles. Depth filters have the advantages of low cost, high throughput, large particle retention capacity, and the ability to retain a variety of particle sizes. However, they can endure from entrainment of the filter medium, uncertainty regarding effective pore size, some ambiguity regarding the overall integrity of the filter, and the risk of particles being mobilized when the pressure differential across the filter is large.

The second type of filter is the membrane filter, in which depth is not considered momentous. The membrane filter uses a relatively thin material with a well-defined maximum pore size and the particle retaining effect takes place almost entirely at the surface. Membranes offer the advantage of having well-defined effective pore sizes, can be integrity tested more easily than depth filters, and can achieve more filtration of much smaller particles. They tend to be more expensive than depth filters and usually cannot achieve the throughput of a depth filter. Filtration technology has developed a well defined terminology that has been well addressed by commercial suppliers.

The term membrane has been defined in a number of ways. The most appealing definitions to us are the following:

"A selective separation barrier for one or several components in solution or suspension":

A thin layer of material that is capable of separating materials as a function of their physical and chemical properties when a driving force is applied across the membrane.

Membranes are important materials which form part of our daily lives. Their long history and use in biological systems has been extensively studied throughout the scientific field. Membranes have proven themselves as promising separation candidates due to advantages offered by their high stability, efficiency, low energy requirement and ease of operation. Membranes with good thermal and mechanical stability combined with good solvent resistance are important for industrial processes [207].

The concept of membrane processes is relatively simple but nevertheless often unknown. Membranes might be described as conventional filters but with much finer mesh or much smaller pores to enable the separation of tiny particles, even molecules. In general, one can divide membranes into two groups: porous and nonporous. The former group is similar to classical filtration with pressure as the driving force; the separation of a mixture is achieved by the rejection of at least one component by the membrane and passing of the other components through the membrane (see Fig. 6.18). However, it is important to note that nonporous membranes do not operate on a size exclusion mechanism.

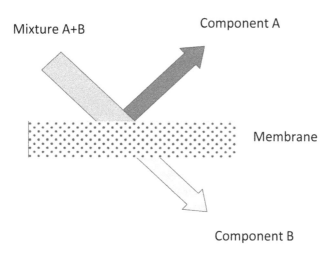

FIGURE 6.18 Basic principle of porous membrane processes.

Membrane separation processes can be used for a wide range of applications and can often offer significant advantages over conventional separation such as distillation and adsorption since the separation is based on a physical mechanism. Compared to conventional processes, therefore, no chemical, biological, or thermal change of the component is involved for most membrane processes. Hence membrane separation is particularly attractive to the processing of food, beverage, and bioproducts where the processed products can be sensitive to temperature (vs. distillation) and solvents (vs. extraction).

Synthetic membranes show a large variety in their structural forms. The material used in their production determines their function and their driving forces. Typically the driving force is pressure across the membrane barrier (see Table 6.4) [208–210]. Formation of a pressure gradient across the membrane allows separation in a bolter-like manner. Some other forms of separation that exist include charge effects and solution diffusion. In this separation, the smaller particles are allowed to pass through as permeates whereas the larger molecules (macromolecules) are retained. The retention or permeation of these species is ordained by the pore architecture as well as pore sizes of the membrane employed. Therefore based on the pore sizes, these pressure driven membranes can be divided RO, NF, UF and MF, are already applied on an industrial scale to food and bioproduct processing [211–213].

TABLE 6.4 Driving Forces and Their Membrane Processes

Driving force	Membrane process
Pressure difference	Microfiltration, Ultrafiltration, Nanofiltration, Reverse osmosis
Chemical potential difference	Pervaporation, Pertraction, Dialysis, Gas separation, Vapor permeation, Liquid Membranes
Electrical potential difference	Electrodialysis, Membrane electrophoresis, Membrane electrolysis
Temperature difference	Membrane distillation

A. MF membranes

MF membranes have the largest pore sizes and thus use less pressure. They involve removing chemical and biological species with diameters ranging between 100 to 10,000 nm and components smaller than this, pass through as permeates. MF is primarily used to separate particles

B. UF membranes

UF membranes operate within the parameters of the micro and nanofiltration membranes. Therefore UF membranes have smaller pores as compared to MF membranes. They involve retaining macromolecules and colloids from solution which range between 2–100 nm and operating pressures between 1 and 10 bar., for example, large organic molecules and proteins. UF is used to separate colloids such as proteins from small molecules such as sugars and salts[210].

C.   NF membranes

NF membranes are distinguished by their pore sizes of between 0.5–2 nm and operating pressures between 5 and 40 bar. They are mainly used for the removal of small organic molecules and di- and multivalent ions. Additionally, NF membranes have surface charges that make them suitable for retaining ionic pollutants from solution. NF is used to achieve separation between sugars, other organic molecules, and multivalent salts on the one hand from monovalent salts and water on the other. Nanofiltration, however, does not remove dissolved compounds [210].

D.   RO membranes

RO membranes are dense semipermeable membranes mainly used for desalination of sea water [38]. Contrary to MF and UF membranes, RO membranes have no distinct pores. As a result, high pressures are applied to increase the permeability of the membranes [210]. The properties of the various types of membranes are summarized in Table 6.5.

TABLE 6.5   Summary of Properties of Pressure Driven Membranes

	MF	UF	NF	RO
Permeability(L/h. m^2.bar)	1000	10–1000	1.5–30	0.05–1.5
Pressure (bar)	0.1–2	0.1–5	3–20	5–1120
Pore size (nm)	100–10000	2–100	0.5–2	<0.5
Separation Mechanism	Sieving	Sieving	Sieving, charge effects	Solution diffusion
Applications	Removal of bacteria	Removal of bacteria, fungi, virus	Removal of multivalen-tions	Desalination

The NF membrane is a type of pressure-driven membrane with properties in between RO and UF membranes. NF offers several advantages such as low operation pressure, high flux, high retention of multivalent anion salts and an organic molecular above 300, relatively low investment and low operation and maintenance costs. Because of these advantages, the applications of NF worldwide have increased [214]. In recent times, research in the application of nanofiltration techniques has been extended from separation of aqueous solutions to separation of organic solvents to homogeneous catalysis, separation of ionic liquids, food processing, etc. [215].

Figure 6.19 presents a classification on the applicability of different membrane separation processes based on particle or molecular sizes. RO process is often used for desalination and pure water production, but it is the UF and MF that are widely used in food and bioprocessing.

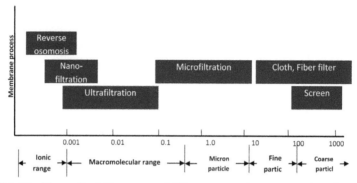

FIGURE 6.19   The applicability ranges of different separation processes based on sizes.

While MF membranes target on the microorganism removal, and hence are given the absolute rating, namely, the diameter of the largest pore on the membrane surface, UF/NF membranes are characterized by the nominal rating due to their early applications of purifying biological solutions. The nominal rating is defined as the molecular weight cut-off (MWCO) that is the smallest molecular weight of species, of which the membrane has more than 90% rejection (see later for definitions). The separation mechanism in MF/UF/NF is mainly the size exclusion, which is indicated in the nominal ratings of the membranes. The other separation mechanism includes the electrostatic interactions between solutes and membranes, which depends on the surface and physiochemical properties of solutes and membranes [211]. Also, The principal types of membrane are shown schematically in Fig. 6.20 and are described briefly below. The membrane process characteristics are shown in Fig. 6.21.

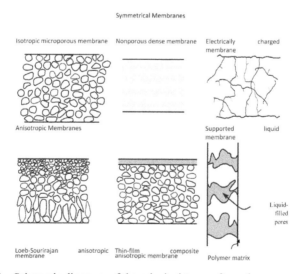

FIGURE 6.20   Schematic diagrams of the principal types of membranes.

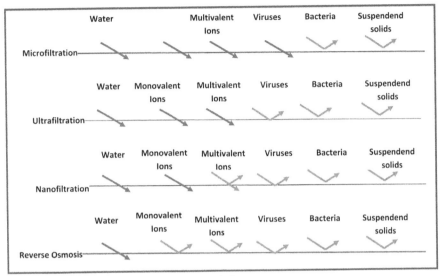

FIGURE 6.21   Membrane process characteristics.

6.14.3   THE RELATIONSHIP BETWEEN NANOTECHNOLOGY AND FILTRATION

Nowadays, nanomaterials have become the most interested topic of materials research and development due to their unique structural properties (unique chemical, biological, and physical properties as compared to larger particles of the same material) that cover their efficient uses in various fields, such as, ion exchange and separation, catalysis, biomolecular isolation and purification as well as in chemical sensing [216]. However, the understanding of the potential risks (health and environmental effects) posed by nanomaterials hasn't increased as rapidly as research has regarding possible applications.

One of the ways to enhance their functional properties is to increase their specific surface area by the creation of a large number of nanostructured elements or by the synthesis of a highly porous material.

Classically, porous matter is seen as material containing three-dimensional voids, representing translational repetition, while no regularity is necessary for a material to be termed "porous." In general, the pores can be classified into two types: open pores which connect to the surface of the material, and closed pores which are isolated from the outside. If the material exhibits mainly open pores, which can be easily transpired, then one can consider its use in functional applications such as adsorption, catalysis and sensing. In turn, the closed pores can be used in sonic and thermal insulation, or lightweight structural applications. The use of porous materials offers also new opportunities in such areas as coverage chemistry, guest–host synthesis and molecular manipulations and reactions for manufacture of nanoparticles, nanowires and other quantum nanostructures. The International Union of Pure and Applied Chemistry (IUPAC) (Fig. 6.22) defines porosity scales as follows:

- Microporous materials 0–2-nm pores
- Mesoporous materials 2–50-nm pores
- Macroporous materials >50-nm pores

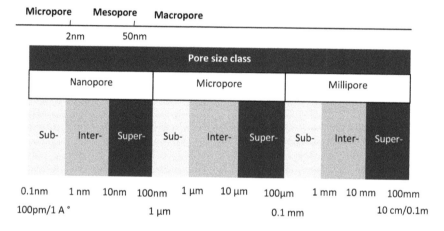

FIGURE 6.22 New pore size classification as compared with the current IUPAC nomenclature.

This definition, it should be noted, is somewhat in conflict with the definition of nanoscale objects, which typically have large relative porosities (>0.4), and pore diameters between 1 and 100 nm. In order to classify porous materials according to the size of their pores the sorption analysis is one of the tools often used. This tool is based on the fact that pores of different sizes lead to totally different characteristics in sorption isotherms. The correlation between the vapor pressure and the pore size can be written as the Kelvin equation:

$$r_p\left(\frac{p}{p_0}\right) = \frac{2\gamma V_L}{RT \ln \left(p/p_0\right)} + t\left(\frac{p}{p_0}\right) \tag{1}$$

Therefore, the isotherms of microporous materials show a steep increase at very low pressures (relative pressures near zero) and reach a plateau quickly. Mesoporous materials are characterized by a so called capillary doping step and a hysteresis (a discrepancy between adsorption and desorption). Macroporous materials show a single or multiple adsorption steps near the pressure of the standard bulk condensed state (relative pressure approaches one) [216].

Nanoporous materials exuberate in nature, both in biological systems and in natural minerals. Some nanoporous materials have been used industrially for a long time. Recent progress in characterization and manipulation on the nanoscale has led to noticeable progression in understanding and making a variety of nanoporous materials: from the merely opportunistic to directed design. This is most strikingly the case in the creation of a wide variety of membranes where control over pore size is increasing dramatically, often to atomic levels of perfection, as is the ability to modify physical and chemical characteristics of the materials that make up the pores [217].

The available range of membrane materials includes polymeric, carbon, silica, zeolite and other ceramics, as well as composites. Each type of membrane can have a different porous structure, as illustrated in Fig. 6.23. Membranes can be thought of as having a fixed (immovable) network of pores in which the molecule travels, with the exception of most polymeric membranes [218, 219]. Polymeric membranes are composed of an amorphous mix of polymer chains whose interactions involve mostly Van der Waals forces. However, some polymers manifest a behavior that is consistent with the idea of existence of opened pores within their matrix. This is especially true for high free volume, high permeability polymers, as has been proved by computer modeling, low activation energy of diffusion, negative activation energy of permeation, solubility controlled permeation [220, 221]. Although polymeric membranes have often been viewed as nonporous, in the modeling framework discussed here it is convenient to consider them nonetheless as porous. Glassy polymers have pores that can be considered as 'frozen' over short times scales, while rubbery polymers have dynamic fluctuating pores (or more correctly free volume elements) that move, shrink, expand and disappear [222].

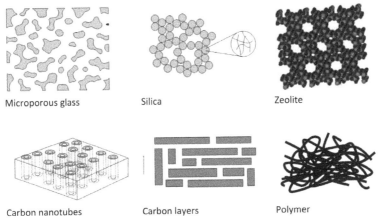

Microporous glass   Silica   Zeolite

Carbon nanotubes   Carbon layers   Polymer

FIGURE 6.23 Porous structure within various types of membranes.

Three nanotechnologies that are often used in the filtering processes and show great potential for applications in remediation are:

1. Nanofiltration (and its sibling technologies: reverse osmosis, ultrafiltration, and microfiltration), is a fully developed, commercially available membrane technology with a large number of vendors. Nanofiltration relies on the ability of membranes to discriminate between the physical size of particles or species in a mixture or solution and is primarily used for water pretreatment, treatment, and purification). There are almost 600 companies in worldwide which offering membrane systems.

2. Electrospinning is a process used by the nanofiltration process, in which fibers are stretched and elongated down to a diameter of about 10 nm. The modified nanofibers that are produced are particularly useful in the filtration process as an ultra-concentrated filter with a very large surface area. Studies have found that electrospun nanofibers can capture metallic ions and are continually effective through refiltration.

3. Surface modified membrane is a term used for membranes with altered makeup and configuration, though the basic properties of their underlying materials remain intact.

6.14.4 TYPES OF MEMBRANES

As it mentioned, membranes have achieved a momentous place in chemical technology and are used in a broad range of applications. The key property that is exploited is the ability of a membrane to control the permeation rate of a chemical species through the membrane. In essence, a membrane is nothing more than a discrete, thin interface that moderates the permeation of chemical species in contact with it. This interface may be molecularly homogeneous, that is completely uniform in composition and structure or it may be chemically or physically heterogeneous for example, containing holes or pores of finite dimensions or consisting of some form of layered structure. A normal filter meets this definition of a membrane, but, generally, the term filter is usually limited to structures that separate particulate suspensions larger than 1–10 µm [223].

The preparation of synthetic membranes is however a more recent invention which has received a great audience due to its applications [224]. Membrane technology like most other methods has undergone a developmental stage, which has validated the technique as a cost-effective treatment option for water. The level of performance of the membrane technologies is still developing and it is stimulated by the use of additives to improve the mechanical and thermal properties, as well as the permeability, selectivity, rejection and fouling of the membranes [225]. Membranes can be fabricated to possess different morphologies. However, most membranes that have found practical use are mainly of asymmetric structure. Separation in membrane processes takes place as a result of differences in the transport rates of different species through the membrane structure, which is usually polymeric or ceramic [226].

The versatility of membrane filtration has allowed their use in many processes where their properties are suitable in the feed stream. Although membrane separation does not provide the ultimate solution to water treatment, it can be economically connected to conventional treatment technologies by modifying and improving certain properties [227].

The performance of any polymeric membrane in a given process is highly dependent on both the chemical structure of the matrix and the physical arrangement of the membrane [228]. Moreover, the structural impeccability of a membrane is very important since it determines its permeation and selectivity efficiency. As such, polymer membranes should be seen as much more than just sieving filters, but as intrinsic complex structures which can either be homogenous (isotropic) or heterogeneous (anisotropic), porous or dense, liquid or solid, organic or inorganic [228, 229].

6.14.4.1 ISOTROPIC MEMBRANES

Isotropic membranes are typically homogeneous/uniform in composition and structure. They are divided into three subgroups, namely: microporous, dense and electrically charged membranes. Isotropic microporous membranes have evenly distributed pores (Fig. 6.24a). Their pore diameters range between 0.01–10 µm and operate by the sieving mechanism. The microporous membranes are mainly prepared by the phase inversion method albeit other methods can be used. Conversely, isotropic dense membranes do not have pores and as a result they tend to be thicker than the microporous membranes (Fig. 6.24b). Solutes are carried through the membrane by diffusion under a pressure, concentration or electrical potential gradient. Electrically charged membranes can either be porous or nonporous. However, in most cases they are finely microporous with pore walls containing charged ions (Fig. 6.24c) [20, 28].

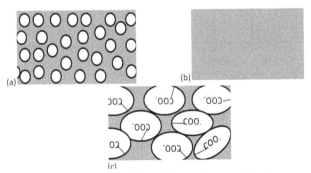

FIGURE 6.24 Schematic diagrams of isotropic membranes: (a) microporous; (b) dense; and (c) electrically charged membranes.

6.14.4.2 ANISOTROPIC MEMBRANES

Anisotropic membranes are often referred to as Loeb-Sourirajan, based on the scientists who first synthesized them [230, 231]. They are the most widely used membranes in industries. The transport rate of a species through a membrane is inversely proportional to the membrane thickness. The membrane should be as thin as possible due to high transport rates are eligible in membrane separation processes for economic reasons. Contractual film fabrication technology limits manufacture of mechanically strong, defect-free films to thicknesses of about 20 μm. The development of novel membrane fabrication techniques to produce anisotropic membrane structures is one of the major breakthroughs of membrane technology. Anisotropic membranes consist of an extremely thin surface layer supported on a much thicker, porous substructure. The surface layer and its substructure may be formed in a single operation or separately [223]. They are represented by nonuniform structures which consist of a thin active skin layer and a highly porous support layer. The active layer enjoins the efficiency of the membrane, whereas the porous support layer influences the mechanical stability of the membrane. Anisotropic membranes can be classified into two groups, namely: (i) integrally skinned membranes where the active layer is formed from the same substance as the supporting layer, (ii) composite membranes where the polymer of the active layer differs from that of the supporting sublayer [231]. In composite membranes, the layers are usually made from different polymers. The separation properties and permeation rates of the membrane are determined particularly by the surface layer and the substructure functions as a mechanical support. The advantages of the higher fluxes provided by anisotropic membranes are so great that almost all commercial processes use such membranes [223].

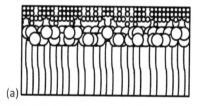
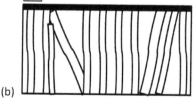

(a)

(b)

FIGURE 6.25 Schematic diagrams of anisotropic membranes: (a) Loeb-Sourirajan and (b) thin film composite membranes.

6.14.4.3 POROUS MEMBRANE

In Knudsen diffusion (Fig. 6.26a), the pore size forces the penetrant molecules to collide more frequently with the pore wall than with other incisive species [232]. Except for some special applications as membrane reactors, Knudsen-selective membranes are not commercially attractive because of their low selectivity [233].

In surface diffusion mechanism (Fig. 6.26b), the pervasive molecules adsorb on the surface of the pores so move from one site to another of lower concentration. Capillary condensation (Fig. 6.26c) impresses the rate of diffusion across the membrane. It occurs when the pore size and the interactions of the penetrant with the pore walls induce penetrant condensation in the pore [234]. Molecular-sieve membranes in Fig. 6.26d have gotten more attention because of their higher productivities and selectivity than solution-diffusion membranes. Molecular sieving membranes are means to polymeric membranes. They have ultra microporous (<7Å) with sufficiently small pores to barricade some molecules, while allowing others to pass through. Although they have several advantages such as permeation performance, chemical and thermal stability, they are still difficult to process because of some properties like fragile. Also they are expensive to fabricate.

Porous membrane   Nonporous membrane

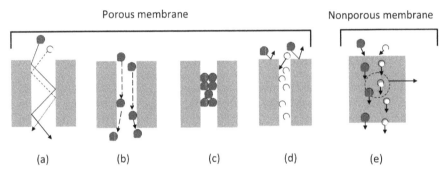

(a)   (b)   (c)   (d)   (e)

FIGURE 6.26 Schematic representation of membrane-based gas separations. (a) Knudsen-flow separation, (b) surface-diffusion, (c) capillary condensation, (d) molecular-sieving separation, and (e) solution-diffusion mechanism.

6.14.4.4 NONPOROUS (DENSE) MEMBRANE

Nonporous, dense membranes consist of a dense film through which permeants are transported by diffusion under the driving force of a pressure, concentration, or electrical potential gradient. The separation of various components of a mixture is related directly to their relative transport rate within the membrane, which is determined by their diffusivity and solubility in the membrane material. Thus, nonporous, dense membranes can separate permeants of similar size if the permeant concentrations in the membrane material differ substantially. Reverse osmosis membranes use dense membranes to perform the separation. Usually these membranes have an anisotropic structure to improve the flux [223].

The mechanism of separation by nonporous membranes is different from that by porous membranes. The transport through nonporous polymeric membranes is usually described by a solution–diffusion mechanism (Fig. 6.26e). The most current commercial polymeric membranes operate according to the solution–diffusion mechanism. The solution–diffusion mechanism has three steps: (1) the absorption or adsorption at the upstream boundary, (2) activated diffusion through the membrane, and (3) desorption or evaporation on the other side. This solution–diffusion mechanism is driven by a difference in the thermodynamic activities existing at the upstream and downstream faces of the membrane as well as the intermolecular forces acting between the permeating molecules and those making up the membrane material.

The concentration gradient causes the diffusion in the direction of decreasing activity. Differences in the permeability in dense membranes are caused not only by diffusivity differences of the various species but also by differences in the physicochemical interactions of the species within the polymer. The solution–diffusion model assumes that the pressure within a membrane is uniform and that the chemical potential gradient across the membrane is expressed only as a concentration gradient. This mechanism controls permeation in polymeric membranes for separations.

6.14.5 INTRODUCTION TO MORPHOLOGY AND POROSITY

Union of Pure and Applied Chemists defines morphology as the "shape, optical appearance, or form of phase domains in substances, such as, high polymers, polymer blends, composites, and crystals." Since this is a very broad and diffuse definition, two classes of morphology are set apart in this work. Shape and bulk morphology are distinguished, because both are very useful in the description of the porous networks. The former concerns the particle size, shape and pore structure, the latter classifies the polymers by the molecular architecture of the pore walls. Polymers have the advantage that they can be prepared in almost any micro or macroscopic size and shape. This allows extensive tuning of the shape morphology to the desired application, while keeping the bulk morphological parameters unchanged and vice versa (Tables 6.6 and 6.7).

TABLE 6.6 Examples of Size-Dependent Shape Morphology

Size range	Shape morphology
nanometer	Polymer brush
	Micelle
	Microgel
	Pores
Micrometer	Powders
	polyHIPE (high internal phase emulsion) pores
macroscopic	Beads
	Discs
	Membranes

TABLE 6.7 Examples of Cross-Link-Dependent Bulk Morphology

Content of cross-links	Bulk morphology
None	Soluble polymer
	Supported polymer brush
Low	Swellable polymer gels
High	Polymer networks
Extra-high	Hypercross-linked
	Polymers

Classically, porous matter is seen as a material that has voids through and through. The voids show a translational repetition in 3-D space, while no regularity is necessary for a material to be termed "porous." A common and relatively simple porous system is one type of dispersion classically described in colloid science, namely foam or, better, solid foam. In correlation with this, the most typical way to think about a porous material is as a material with gas-solid interfaces as the most dominant characteristic. This already indicates that classical colloid and interface science as the creation of interfaces due to nucleation phenomena (in this case nucleation of wholes), decreasing interface energy, and stabilization of interfaces is of elemental importance in the formation process of nanoporous materials [217, 235–237].

These factors are often omitted because the final products are stable (they are metastable). This metastability is due to the rigid character of the void-surrounding network, which is covalently cross-linked in most cases. However, it must be noticed that most of the porous materials are not stable by thermodynamic means. As soon as kinetic energy boundaries are overcome, materials start to breakdown [217].

Porous materials have been extensively exploited for use in a broad range of applications: for example, as membranes for separation and purification [238], as high surface-area adsorbants, as solid supports for sensors [239] and catalysts, as materials with low dielectric constants in the fabrication of microelectronic devices [240], and as scaffolds to guide the growth of tissues in bioengineering [241].

Porous materials occur widely and have many important applications. They can, for example, offer a convenient method of imposing fine structure on adsorbed materials.

They can be used as substrates to support catalysts and can act as highly selective sieves or cages that only allow access to molecules up to a certain size.

Food is often finely structured. Many biologically active materials are porous, as are many construction and engineering materials. Porous geological materials are of great interest; high porosity rock may contain water, oil or gas; low porosity rock may act as a cap to porous rock, and is of importance for active waste sealing.

6.14.6   POROSITY

Porosity φ is the fraction of the total soil volume that is taken up by the pore space. Thus it is a single-value quantification of the amount of space available to fluid within a specific body of soil. Being simply a fraction of total volume, φ can range between 0 and 1, typically falling between 0.3 and 0.7 for soils. With the assumption that soil is a continuum, adopted here as in much of soil science literature, porosity can be considered a function of position.

6.14.7   POROSITY IN NATURAL SOILS

The porosity of a soil depends on several factors, including (1) packing density, (2) the breadth of the particle size distribution (polydisperse vs. monodisperse), (3) the shape of particles, and (4) cementing. Mathematically considering an idealized soil of packed uniform spheres, φ must fall between 0.26 and 0.48, depending on the packing. Spheres randomly thrown together will have φ near the middle of this range, typically 0.30 to 0.35. Sand with grains nearly uniform in size (monodisperse) packs to about the same porosity as spheres. In polydisperse sand, the fitting of small grains within the pores between large ones can reduce φ, conceivably below the 0.26 uniform-sphere minimum. Figure 27 illustrates this concept. The particular sort of arrangement required to reduce φ to 0.26 or less is highly improbable, however, so φ also typically falls within the 0.30–0.35 for polydisperse sands. Particles more irregular in shape tend to have larger gaps between their nontouching surfaces, thus forming media of greater porosity. In porous rock such as sand-stone, cementation or welding of particles not only creates pores that are different in shape from those of particulate media, but also reduces the porosity as solid material takes up space that would otherwise be pore space. Porosity in such a case can easily be less than 0.3, even approaching 0. Cementing material can also have the opposite effect. In many soils, clay and organic substances cement particles together into aggregates. An individual aggregate might have 0.35 porosity within but the medium as a whole has additional pore space in the form of gaps between aggregates, so that φ can be 0.5 or greater. Observed porosities can be as great as 0.8 to 0.9 in a peat (extremely high organic matter) soil.

Porosity is often conceptually partitioned into two components, most commonly called textural and structural porosity. The textural component is the value the porosity would have if the arrangement of the particles were random, as described above for granular material without cementing. That is, the textural porosity might be about 0.3 in a granular medium. The structural component represents nonrandom structural influences, including macropores and is arithmetically defined as the difference between the textural porosity and the total porosity.

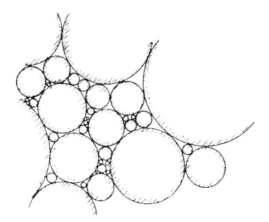

FIGURE 6.27 Dense packing of polydisperse spheres.

The texture of the medium relates in a general way to the pore-size distribution, as large particles give rise to large pores between them, and therefore is a major influence on the soil water retention curve. Additionally, the structure of the medium, especially the pervasive-ness of aggregation, shrinkage cracks, worm-holes, etc. substantially influences water retention.

6.14.8 MEASUREMENT OF POROSITY

The technology of thin sections or of tomographic imaging can produce a visualization of pore space and solid material in a cross-sectional plane. The summed area of pore space divided by total area gives the areal porosity over that plane. An analogous procedure can be followed along a line through the sample, to yield a linear porosity. If the medium is isotropic, either of these would numerically equal the volumetric porosity as defined above.

The volume of water contained in a saturated sample of known volume can indicate porosity. The mass of saturated material less the oven-dry mass of the solids, divided by the density of water, gives the volume of water. This divided by the original sample volume gives porosity.

An analogous method is to determine the volume of gas in the pore space of a completely dry sample. Sampling and drying of the soil must be conducted so as not to compress the soil or otherwise alter its porosity. A pycnometer can measure the air volume in the pore space. A gas-tight chamber encloses the sample so that the internal gas-occupied volume can be perturbed by a known amount while the gas pressure is measured. This is typically done with a small piston attached by a tube connection. Boyle's law indicates the total gas volume from the change in pressure resulting from the volume change. This total gas volume minus the volume within the piston, connectors, gaps at the chamber walls, and any other space not occupied by soil, yields the total pore volume to be divided by the sample volume.

To avoid having to saturate with water or air, one can calculate porosity from measurements of particle density and bulk density. From the definitions of bulk density as the solid mass per total volume of soil and particle density as the solid mass per solid volume, their ratio ρ_b/ρ_p is the complement of φ, so that:

$$\varnothing = 1 - \rho_b/\rho_p \qquad (2)$$

Often the critical source of error is in the determination of total soil volume, which is harder to measure than the mass. This measurement can be based on the dimensions of a minimally disturbed sample in a regular geometric shape, usually a cylinder. Significant error can result from irregularities in the actual shape and from unavoidable compaction. Alternatively, the measured volume can be that of the excavation from which the soil sample originated. This can be done using measurements of a regular geometric shape, with the same problems as with measurements on an extracted sample. Additional methods, such as, the balloon or sand-fill methods, have other sources of error.

6.14.9 *PORES* AND PORE-SIZE DISTRIBUTION: THE NATURE OF A PORE

Because soil does not contain discrete objects with obvious boundaries that could be called individual pores, the precise delineation of a pore unavoidably requires artificial, subjectively established distinctions. This contrasts with soil particles, which are easily defined, being discrete material objects with obvious boundaries. The arbitrary criterion required to partition pore space into individual pores is often not explicitly stated when pores or their sizes are discussed. Because of this inherent arbitrariness, some scientists argue that the concepts of pore and pore size should be avoided. Much valuable theory of the behavior of the soil-water-air system, however, has been built on these concepts, defined using widely, if not universally, accepted criteria.

6.14.10 *POROUS* MATERIALS

Porous materials are solid forms of matter permeated by interconnected or non-interconnected pores (voids) of different kinds: channels, cavities or interstices; that can be divided into several classes.

According to the nomenclature suggested by the IUPAC, porous materials are usually classified into three different categories depending on the lateral dimensions of their pores: microporous (<2 nm), mesoporous (between 2 and 50 nm) and macroporous (>50 nm) [242].

Liquid and gaseous molecules have been known to exhibit characteristic transport behaviors in each type of porous material. For example, mass transport can be obtained via viscous flow and molecular diffusion in a macroporous material, through surface diffusion and capillary flow in a mesoporous material and by activated diffusion in a microporous material.

Pores from the nanoscopic to the macroscopic scale are generated depending on the method. A summary of selected methods that can be applied to styrene-codivinylbenzene polymers is given in Table 6.8 [243].

TABLE 8   Overview of methods of generating porosity during polymer synthesis.

Method	Porogene	Porosity Accessibility	Typical size
Foaming	Gas, solvent, super-critical solvent	open/closed	100 nm–1 mm
Phase separation	Solvent	Open	1 μm–1 mm
High internal phase emulsion polymer	Emulsion	Open	10 μm–100 μm
Soft templating by	Molecules (solvent)	Micelles	Bicontinuous microemulsion
Hard templating by	Colloidal crystals	Porous solids	Open

The internal structural architecture of the void space potentially controls the physical and chemical properties, such as, reactivity, thermal and electric conductivity, as well as the kinetics of numerous transport processes. The characterization of porous materials, therefore, has been of great practical interest in numerous areas including catalysis, adsorption, purification, separation, etc., where the essential aspects for such applications are pore accessibility, narrow pore size distribution (PSD), relatively high specific surface area and easily tunable pore sizes.

Ordered porous materials are judged to be much more interesting because of the control over pore sizes and pore shapes. Their disordered counterparts exhibit high polydispersity in pore sizes, and the shapes of the pores are irregular. Ordered porous materials seem to be much more homogeneous. In many cases a material possesses more than one porosity. This could be for microporous materials: an additional meso- or macroporosity caused by random grain packing

For mesoporous materials: an additional macroporosity caused by random grain packing, or an additional microporosity in the continuous network. For macroporous materials: an additional meso- and microporosity, these factors should be taken into consideration when materials are classified according to their homogeneity. A material possessing just one type of pore, even when the pores are disordered, might be more homogenous than one having just a fraction of nicely ordered pores.

Ordered porous solids contain a regularly arranged pore system and it is desired to design materials with different cylindrical, window-like, spherical or slit-like pore shapes and sizes [244].

6.14.11 PROPERTIES OF POROUS MATERIALS

There are a number of important properties of porous materials such as:
- Porosity
- Specific surface area
- Permeability
- Breakthrough capillary pressure
- Diffusion properties of liquids in pores
- Pore size distribution
- Radial density function

6.14.12 MACROPOROUS MATERIALS AND THEIR USES

Macroporous materials are formed from the packing of monodisperse spheres (polystyrene or silica) into a three-dimensional ordered arrangement, to form face-centered cubic (FCC) or hexagonal close-packed (HCP) structures. The spaces between the packed spheres create a macroporous structure.

Glass or rubbery polymer that includes a large number of macropores (50 nm–1 µm in diameter) that persist when the polymer is immersed in solvents or in the dry state.

Macroporous polymers are often network polymers produced in bead form. However, linear polymers can also be prepared in the form of macroporous polymer beads. They swell only slightly in solvents.

Macroporous polymers can be used, for instance, as precursors for ion-exchange polymers, as adsorbents, as supports for catalysts or reagents, and as stationary phases in size-exclusion chromatography columns.

Macroporous materials have many applications in the field of engineering due to their large effective surface area, and can be used for purposes such as filters, catalysts, supports, heat exchangers, and fuel cells. Although microporous and mesoporous materials also possess large surface areas, their small pore diameters (less than 10 nm) do not allow larger molecules to pass through them. Hence, macroporous materials with larger pore diameters are preferred and are of more practical use. Through colloidal crystal templating techniques, three-dimensionally ordered macroporous materials with uniform pore size can be successfully synthesized, thus improving the efficiency of transport of materials through the pores. Furthermore, photonic crystals possessing optical band-gaps can be synthesized from these macroporous structures by infiltrating the macroporous material with a precursor fluid, followed by removal of the original spheres through calcination.

Photonic crystals are materials in which the dielectric constants vary periodically in space. Due to the alternating dielectric properties, photonic crystals are hence able to control the propagation of photons, by creating a frequency (bandgap) in which light is not able to propagate. Photonic crystals themselves have great potential use in the engineering field. Due to their ability to localize photons, photonic crystals can be used as wave guides in optical fibers, which would prove very valuable in optical communications for the transfer of information. With the advent of information technology and the need for faster, quicker and more efficient data transmission, the importance and potential of photonic crystals are ever more apparent.

6.14.13   *MESOPOROUS* MATERIALS

Mesoporous solids consist of inorganic or inorganic/organic hybrid units of long-range order with amorphous walls, tunable textural and structural properties with highly controllable pore geometry and narrow pore size distribution in the 2–50 nm range [245].

The pores can have different shapes such as spherical or cylindrical and be arranged in varying structures, see Figure 28. Some structures have pores that are larger than 50 nm in one dimension, for example, the two first structures in Fig. 6.28, but there the width of the pore is in the mesorange and the material is still considered to be mesoporous.

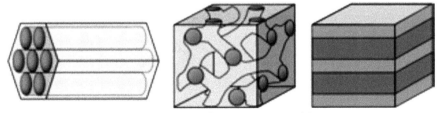

FIGURE 6.28   Different pore structures of mesoporous materials.

Mesoporous materials can have a wide range of compositions but mainly consists of oxides such as SiO_2, TiO_2, ZnO_2, Fe_2O or combinations of metal oxides, but also mesoporous carbon can be synthesized [246–249]. Most commonly is to use a micellar solution and grow oxide walls around the micelles. Both organic metal precursors such as alkoxides [250, 251] as well as inorganic salts such as metal chloride salts [247] can be used. Alternatively a mesoporous template can be used to grow another type of mesoporous material inside it. This is often used for synthesizing, for example, mesoporous carbon [249, 252, 253].

6.14.14. *MICROPOROUS* **MATERIALS**

Porous materials are networks of solid material which contain void spaces. These materials can be further classified depending on the size of the pores present in the material. Microporous solids are materials that contain permanent cavities with diameters of less than 2 nm. Mesoporous materials contain pores ranging from 2 nm to 50 nm and macroporous materials contain pores of greater than 50 nm [254]. The field of microporous materials contains several classes which are well known [255], including naturally occurring zeolites, activated carbons and silica. Synthetic microporous solids have recently emerged as a potentially important class of materials.

These include Metal Organic Frameworks (MOFs), Microporous Organic Polymers (MOPs) including Covalent Organic Frameworks (COFs) and Polymers of Intrinsic Microporosity (PIMs) [256]. It is the very large surface areas and very small pore sizes of these materials which make them of specific interest. These two factors permit microporous materials to be useful in applications such as heterogeneous catalysis, separation chemistry, and potential uses in hydrogen or other gas storage [257, 258]. Most synthetic strategies to prepare microporous materials consist of linking together smaller units with di-topic or poly topic functionalities in order to form extended networks much like is displayed in the general diagram of Fig. 6.29.

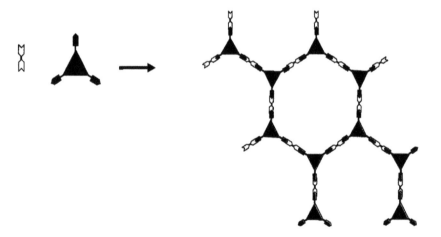

FIGURE 6.29   General schematic showing the linking of polytopic building blocks to form synthetic networks.

Whether a network formed is crystalline or amorphous is generally governed by whether the covalent bonds being formed involve reversible chemistry or not. Crystalline networks are typically formed under reversible reaction conditions that allow error corrections during network formation and produce thermodynamically stable networks. These types of reactions are commonly condensation reactions [259, 260]. On the other hand, when irreversible reactions are employed such as cross couplings, the networks tend to form in a disordered manner [261, 262], resulting in amorphous materials. This is because a carbon-carbon bond is formed irreversibly under conditions such as a Sonagashira or Yamamoto couplings resulting in amorphous networks. This is of course unless some other templating measure is taking into account while considering the reaction conditions [263].

Certain factors such as temperature, solvent and solvent-to-head-space ratio play an important role in the formation of a crystalline framework [260]. Certain solvents can be employed in order to form ordered networks by means of their ability to dissolve the monomer building blocks. If the concentration of a monomer in solution is controlled by a solvent in which it is slightly soluble, then a network is more likely to form under thermodynamic, instead of kinetic control [262]. Solvents could also be used based on their molecular size to act as templates for pores to form around [263]. While this idea of MOP/COF templating is generally understood in qualitative terms (which solvents produce crystalline networks) there has been no research into the quantitative effects (what solvent ratio is required to produce a well-structured network).

Finally, if a material is to exhibit microporosity, it must be composed of somewhat rigid building blocks in order to impart rigidity within the network and provide directionality for the formation of the network. This rigidity prevents collapse of the network upon itself and results in free volume which becomes the pores within a framework.

6.14.15 NANOPOROUS POLYMERS

In the past few decades, nanomaterials have received substantial attention and efforts from academic and industrial world, due to the distinct properties at the nanoscale. Nanoporous materials as a subset of nanomaterials possess a set of unique properties: large specific surface volume ratio, high interior surface area, exclusive size sieving and shape selectivity, nanoscale space confinement, and specific gas/fluid permeability. Moreover, pore-filled nanoporous materials can offer synergistic properties that can never be reached by pure compounds. As a result, nanoporous materials are of scientific and technological importance and also considerable interest in a broad range of applications that include templating, sorting, sensing, isolating and releasing.

Nanoporous materials can be classified by pore geometry (size, shape, and order) or distinguished by type of bulk materials. Nanoporous materials are considered uniform if the pore size distribution is relatively narrow and the pore shape is relatively homogenous. The pores can be cylindrical, conical, slit-like, or irregular in shape. They can be well ordered with an alignment as opposed to a random network of tortuous pores. Nanoporous materials cover a wide variety of materials, which can be generally divided into inorganic, organic and composite materials. The majority of investigated nanoporous materials have been inorganic, including oxides, carbon, silicon, silicate, and metal. On the other side, polymers have been identified as materials that offer low cost, less toxicity, easy fabrication process, diverse chemical functionality, and extensive mechanical properties [264, 265].

Naturally, the success of inorganic materials to form nanoporous materials has promoted the development of analogous polymers. More importantly, advances in polymer synthesis and novel processing techniques have led to various nanoporous polymers.

6.14.16   *NANOPOROUS* MATERIALS CONNECTION

A whole variety of nanoporous materials in nature can be found in many different functions. The most common task for nanoporous materials in nature is to make inorganic material much lighter while preserving or improving the high structural stability of these compounds. Often, by filling the voids between inorganic matters the desired properties of the hybrid materials exceed the performances of the pure compounds by several orders of magnitude. Nanoporous materials in nature are organic-inorganic hybrids. Naturally occurring materials exhibit synergistic properties. Neither the organic material filling the void nor the inorganic network materials are able to achieve comparable performances by themselves [266–268].

It is seen that complex mechanisms are involved in the formation of these hierarchical materials. Similar to the structure motives on different length-scales cells, vesicles, supra-molecular structures, and biomolecules are involved in the structuring process of inorganic matter occurring in nature. This process is commonly known as biomineralization [267].

It is often not seen in this relation, but it will be shown later that ordered porous materials, and therefore artificial materials, are constructed according to very similar principles. A completely different area where nanoporous materials are highly important is in the lungs, where foam with a high surface area permits sufficient transfer of oxygen to the blood. Even the most recent developments in nanoporous materials, such as, their application as photonic materials are already present in nature, the color of butterfly wings, for instance, originates from photonic effects [269, 270].

It can be concluded that nature applies the concept of nanoporous materials (either filled or unfilled) as a powerful tool for constructing all kinds of materials with advanced properties. So it is not surprising how much research has recently been devoted to porous materials in different areas such as chemistry, physics, and engineering. The current interests in nanoporous materials are now far behind their size-sieving properties.

6.14.17   *CLASSIFICATION* BY NETWORK MATERIAL

One of the most important goals in the field of nanoporous materials is to achieve any possible chemical composition in the network materials "hosting" the pores. It makes sense to divide the materials into two categories: (a) inorganic materials and (b) organic materials. Among the inorganic materials, the larger group, it could be found:

(i) Inorganic oxide-type materials. This is the field of the most commonly known porous silica, porous titania, and porous zirconia materials.

(ii) A category of its own is given for nanoporous carbon materials. In this category are the highly important active carbons and some examples for ordered mesoporous carbon materials.

(iii) Other binary compounds such as sulfides, nitrides. Into this category also fall the famous AlPO materials.

(iv) There are already some examples in addition to carbon where just one element (for instance, a metal)could be prepared in a nanoporous state. The most appropriate member of this class of materials is likely to be nanoporous silicon, with its luminescent properties [271, 272]. There are far fewer examples of nanoporous organic materials, such as, polymers [273].

6.14.18   *SUMMARY* OF CLASSIFICATIONS

Three main criteria could be defined:

a) Size of pores
b) Type of network material
c) State of order: ordered or disordered materials

6.14.19   *ORIGIN* AND CLASSIFICATION OF PORES IN SOLID MATERIALS

Solid materials have a cohesive structure which depends on the interaction between the primary particles. The cohesive structure leads indispensably to a void space which is not occupied by the composite particles such as atoms, ions, and line particles. Consequently, the state and population of such voids strongly depends on the interparticle forces. The interparticle forces are different from one system to another; chemical bonding, van der Waals force, electrostatic force, magnetic force, surface tension of adsorbed films on the primary particles, and so on. Even the single crystalline solid which is composed of atoms or ions has intrinsic voids and defects. Therefore, pores in solids are classified into intraparticle pores and interparticle pores (Table 6.9) [274].

TABLE 6.9   Classification of Pores from Origin, Pore width w, and Accessibility

Origin and Structure		
Intraparticle pore	Intrinsic intraparticle pore	Structurally intrinsic type injected intrinsic type Pure type Pillared type
	Extrinsic intraparticle pore	
Interparticle pore	Rigid interparticle pore (Agglomerated)	
	Flexible interparticle pore (Aggregated)	
Pore width		
Macropore	$w < 50$ nm	
Mesopore	2 nm $< w < 50$ nm	
Micropore	$W < 2$ nm	
	Supermicropore, $0.7 < w < 2$ nm	
	Ultramicropore, $w < 0.7$ nm	
	(Ultrapore, $w < 0.35$ nm in this review)	
Accessibility to surroundings		
Open pore	Communicating with external surface	
Closed pore	No communicating with surroundings	
Latent pore	Ultrapore and closed pore	

6.14.20   TRAPARTICLE PORES

6.14.20.1   INTRINSIC IN TRAPARTICLE PORE

Zeolites are the most representative porous solids whose pores arise from the intrinsic crystalline structure. Zeolites have a general composition of Al, Si, and 0, where Al-O and Si-0 tetrahedral units cannot occupy the space perfectly and therefore produce cavities. Zeolites have intrinsic pores of different connectivities according to their crystal structures [275]. These pores may be named intrinsic crystalize pores. The carbon nanotube has also an intrinsic crystalline pore [276].

Although all crystalline solids other than zeolites have more or less intrinsic crystalline pores, these are not so available for adsorption or diffusion due to their isolated state and extremely small size.

There are other types of pores in a single solid particle. α-FeOOH is a precursor material for magnetic tapes, a main component of surface deposits and atmospheric corrosion products of iron-based alloys, and a mineral. The α-FeOOH microcrystal is of thin elongated plate [277].

These new created intrinsic intraparticle pores should have their own name different from the intrinsic intraparticle pore. The latter is called a structurally intrinsic intraparticle pore, while the former is called an injected intrinsic intraparticle pore.

6.14.20.2   EXTRINSIC INTRAPARTICLE PORE

When a foreign substance is impregnated in the parent material in advance this is removed by a modification procedure [278]. This type of pores should be called extrinsic intraparticle pores. Strictly speaking, as the material does not contain other components, extrinsic pure intraparticle pore is recommended. However, the extrinsic intrapore can be regarded as the interparticle pore in some cases.

It can be introduced a pore-forming agent into the structure of solids to produce voids or fissures which work as pores. This concept has been applied to layered compounds in which the interlayer bonding is very weak; some inserting substance swells the interlayer space. Graphite is a representative layered compound; the graphitic layers are weakly bound to each other by the van der Waals force [279]. If it heated in the presence of intercalants such as K atoms, the intercalants are inserted between the interlayer spaces to form a long periodic structure. K-intercalated graphite can adsorb a great amount of H_2 gas, while the original graphite cannot [280]. The interlayer space opened by intercalation is generally too narrow to be accessed by larger molecules. Intercalation produces not only pores, but also changes the electronic properties. Montmorillonite is a representative layered clay compound, which swells in solution to intricate hydrated ions or even surfactant molecules [281, 282]. Then some pillar materials such as metal hydroxides are intricate in the swollen interlayer space under wet conditions. As the pillar compound is not removed upon drying, the swollen structure can be preserved even under dry conditions.

The size of pillars can be more than several nm, being different from the above intercalants. As the graphite intercalation compounds and pillar ones need the help of foreign substances, they should be distinguished from the intrinsic intraparticle pore system. Their pores belong to extrinsic intraparticle pores. As the intercalation can be included in the pillar formation, it is better to say that both the pillared and intercalated compounds have pillared in traparticle pores [283].

6.14.21   INTER-PARTICLE PORES

Primary particles stick together to form a secondary particle, depending on their chemical composition, shape and size. In colloid chemistry, there are two gathering types of primary particles. One is aggregation and the other is agglomeration.

The aggregated particles are loosely bound to each other and the assemblage can be readily broken down. Heating or compressing the assemblage of primary particles brings about the more tightly bound agglomerate [284].

There are various interaction forces among primary particles, such as, chemical bonding, van der Waals force, magnetic force, electrostatic force, and surface tension of the thick adsorbed layer on the particle surface. Sintering at the neck part of primary particles produces stable agglomerates having pores. The aggregate bound by the surface tension of adsorbed water film has flexible pores. Thus, interparticle pores have wide varieties in stability, capacity, shape, and size, which depend on the packing of primary particles. They play an important role in nature and technology regardless of insufficient understanding. The interparticle pores can be divided into rigid and flexible pores. The stability depends on the surroundings. Almost all interparticle pores in agglomerates are rigid, whereas those in aggregates are flexible. Almost all sintered porous solids have rigid pores due to strong chemical bonding among the particles. The rigid interparticle porous solids have been widely used and have been investigated as adsorbents or catalysts. Silica gel is a representative of interparticle porous solids. Ultrafine spherical silica particles form the secondary particles, leading to porous solids [285, 286].

6.14.22 *STRUCTURE* OF PORES

The pore state and structure mainly depend on the origin. The pores communicating with the external surface are named open pores, which are accessible for molecules or ions in the surroundings. When the porous solids are insufficiently heated, some parts of pores near the outer shell are collapsed inducing closed pores without communication to the surroundings. Closed pores also remain by insufficient evolution of gaseous substance. The closed pore is not associated with adsorption and permeability of molecules, but it influences the mechanical properties of solid materials, the new concept of latent pores is necessary for the best description of the pore system. This is because the communication to the surroundings often depends on the probe size, in particular, in the case of molecular resolution porosimetry. The open pore with a pore width smaller than the probe molecular size must be regarded as a closed pore. Such effectively closed pores and chemically closed pores should be designated the latent pores [287]. The combined analysis of molecular resolution porosimetry and SAXS offers an effective method for separate determination of open and closed (or latent) pores, which will be described later. The porosity is defined as the ratio of the pore volume to the total solid volume [288].

The geometrical structure of pores is of great concern, but the three-dimensional description of pores is not established in less-crystalline porous solids. Only intrinsic crystalline intraparticle pores offer a good description of the structure. The hysteresis analysis of molecular adsorption isotherms and electron microscopic observation estimate the pore geometry such as cylinder (cylinder closed at one end or cylinder open at both ends), cone shape, slit shape, interstice between closed-packing spheres and ink bottle. However, these models concern with only the unit structures. The higher order structure of these unit pores such as the network structure should be taken into account. The simplest classification of the higher order structures is one-, two- and three-dimensional pores. Some zeolites and aluminophosphates have one dimensional pores and activated carbons have basically two-dimensional slit-shaped pores with complicated network structures [289].

The IUPAC has tried to establish a classification of pores according to the pore width (the shortest pore diameter), because the geometry determination of pores is still very difficult and molecular adsorption can lead to the reliable parameter of the pore width. The pores are divided into three categories: macropores, mesopores, and micropores, as mentioned above. The fact that nanopores are often used instead of micropores should be noted.

These sizes can be determined from the aspect of N, adsorption at 77 K, and hence N_2 molecules are adsorbed by different mechanisms – multilayer adsorption, capillary condensation, and micropore filling for macropores, mesopores, and micropores, respectively. The critical widths of 50 and 2 nm are chosen from empirical and physical reasons. The pore width of 50 nm corresponds to the relative pressure of 0.96 for the N_2 adsorption isotherm. Adsorption experiments above that are considerably difficult and applicability of the capillary condensation theory is not sufficiently examined. The smaller critical width of 2 nm corresponds to the relative pressure of 0.39 through the Kelvin equation, where an unstable behavior of the N, adsorbed layer (tensile strength effect) is observed. The capillary condensation theory cannot be applied to pores having a smaller width than 2 nm. The micropores have two subgroups, namely ultra-micropores (0.7 nm) and super-micropores (0.7 nm$< w <$ 2 nm). The statistical thickness of the adsorbed N2 layer on solid surfaces is 0.354 nm. The maximum size of ultra-micropores corresponds to the bilayer thickness of nitrogen molecules, and the adsorbed N_2 molecules near the entrance of the pores often block further adsorption. The ultra-micropore assessment by N_2 adsorption has an inevitable and serious problem. The micropores are divided into two groups.

Recently, the molecular statistical theory has been used to examine the limitation of the Kelvin equation and predicts that the critical width between the micropore and the mesopore is 1.3–1.7 nm (corresponding to 4–5 layers of adsorbed N_2), which is smaller than 2 nm [290].

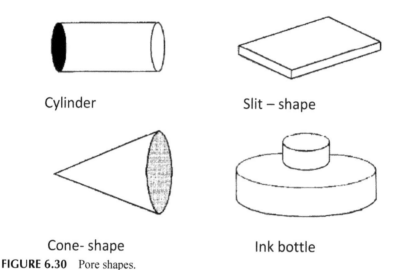

| Cylinder | Slit – shape |

| Cone- shape | Ink bottle |

FIGURE 6.30   Pore shapes.

6.14.23 *POROSITY* AND PORE SIZE MEASUREMENT TECHNIQUES ON POROUS MEDIA

Crushing measure the volume of the porous material, crush it to remove the void space, and remeasure the volume.

- Optically this may involve filling the pores with a material such as black wax or Wood's metal, sectioning and inspecting with a microscope or scanning electron microscope.
- Imbibition weighing before and after filling the pores with a liquid.
- Gas Adsorption measure the change in pressure as a gas is adsorbed by the sample.
- Mercury Intrusion Measure the volume of mercury forced into the sample as a function of pressure.
- Thermoporosimetry fill the pores with a liquid, freeze it, then measure the heat evolved as the sample is warmed, until all the liquid is melted.
- NMR Cryoporometry fill the pores with a liquid, freeze it, then measure the amplitude of the NMR signal from the liquid component as the sample is warmed, until all the liquid is melted.
- SANS scatter neutrons from the pores, then the smaller the dimensions of the variations in density distribution, the larger the angle through which the neutrons will be scattered.

Many of these methods give results that quite frequently differ from one another. This is often because they are in fact measuring different things some measurements are directly on the pores themselves. Others (such as mercury intrusion) are in effect measuring the necks that give access to the pores [291].

6.14.24 *CARBON* NANOTUBES-POLYMER MEMBRANE

Iijima discovered carbon nanotubes in 1991 and it was really a revolution in nanoscience because of their distinguished properties. CNTs have the unique electrical properties and extremely high thermal conductivity [292, 293] and high elastic modulus (>1TPa), large elastic strain – upto 5%, and large breaking strain – up to 20%. Their excellent mechanical properties could lead to many applications [294]. For example, with their amazing strength and stiffness, plus the advantage of lightness, perspective future applications of CNTs are in aerospace engineering and virtual biodevices [295].

CNTs have been studied worldwide by scientists and engineers since their discovery, but a robust, theoretically precise and efficient prediction of the mechanical properties of CNTs has not yet been found. The problem is, when the size of an object is small to nanoscale, their many physical properties cannot be modeled and analyzed by using constitutive laws from traditional continuum theories, since the complex atomistic processes affect the results of their macroscopic behavior. Atomistic simulations can give more precise modeled results of the underlying physical properties. Due to atomistic simulations of a whole CNT are computationally infeasible at present, a new atomistic and continuum mixing modeling method is needed to solve the problem, which requires crossing the length and time scales. The research here is to develop a proper technique of spanning multiscales from atomic to macroscopic space, in which the constitutive laws are derived from empirical atomistic potentials which deal with individual interactions between single atoms at the microlevel, whereas Cosserat continuum theories are adopted for a shell model through the application of the Cauchy-Born rule to give the properties which represent the averaged behavior of large volumes of atoms at the macrolevel [296, 297]. Since experiments of CNTs are relatively expensive at present, and often unexpected manual errors could be involved, it will be very helpful to have a mature theoretical method for the study of mechanical properties of CNTs. Thus, if this research is successful, it could also be a reference for the research of all sorts of research at the nanoscale, and the results can be of interest to aerospace, biomedical engineering [298].

Subsequent investigations have shown that CNTs integrate amazing rigid and tough properties, such as, exceptionally high elastic properties, large elastic strain, and fracture strain sustaining capability, which seem inconsistent and impossible in the previous materials. CNTs are the strongest fibers known. The Young's Modulus of SWNT is around 1TPa, which is 5 times greater than steel (200 GPa) while the density is only 1.2~1.4 g/cm^3. This means that materials made of nanotubes are lighter and more durable.

Beside their well-known extra-high mechanical properties, SWNTs offer either metallic or semiconductor characteristics based on the chiral structure of fullerene. They possess superior thermal and electrical properties so SWNTs are regarded as the most promising reinforcement material for the next generation of high performance structural and multifunctional composites, and evoke great interest in polymer based composites research. The SWNTs/polymer composites are theoretically predicted to have both exceptional mechanical and functional properties, which carbon fibers cannot offer [299].

6.14.24.1 CARBON NANOTUBES

Nanotubular materials are important "building blocks" of nanotechnology, in particular, the synthesis and applications of CNTs [276, 300, 301]. One application area has been the use of carbon nanotubes for molecular separations, owing to some of their unique properties. One such important property, extremely fast mass transport of molecules within carbon nanotubes associated with their low friction inner nanotube surfaces, has been demonstrated via computational and experimental studies [302, 303]. Furthermore, the behavior of adsorbate molecules in nano-confinement is fundamentally different than in the bulk phase, which could lead to the design of new sorbents [304].

Finally, their one-dimensional geometry could allow for alignment in desirable orientations for given separation devices to optimize the mass transport. Despite possessing such attractive properties, several intrinsic limitations of carbon nanotubes inhibit their application in large scale separation processes: the high cost of CNT synthesis and membrane formation (by microfabrication processes), as well as their lack of surface functionality, which significantly limits their molecular selectivity [305]. Although outer-surface modification of carbon nanotubes has been developed for nearly two decades, interior modification via covalent chemistry is still challenging due to the low reactivity of the inner-surface. Specifically, forming covalent bonds at inner walls of carbon nanotubes requires a transformation from sp^2 to sp^3 hybridization. The formation of sp^3 carbon is energetically unfavorable for concave surfaces [306].

Membrane is a potentially effective way to apply nanotubular materials in industrial-scale molecular transport and separation processes. Polymeric membranes are already prominent for separations applications due to their low fabrication and operation costs. However, the main challenge for using polymer membranes for future high-performance separations is to overcome the tradeoff between permeability and selectivity. A combination of the potentially high throughput and selectivity of nanotube materials with the process ability and mechanical strength of polymers may allow for the fabrication of scalable, high-performance membranes [307, 308].

6.14.24.2   STRUCTURE OF CARBON NANOTUBES

Two types of nanotubes exist in nature: MWNTs, which were discovered by Iijima in 1991[276] and SWNTs, which were discovered by Bethune et al. [309, 310] in 1993.

Single-wall nanotube has only one single layer with diameters in the range of 0.6–1 nm and densities of 1.33–1.40 g/cm³[311] MWNTs are simply composed of concentric SWNTs with an inner diameter is from 1.5 to 15 nm and the outer diameter is from 2.5 nm to 30 nm [312]. SWNTs have better defined shapes of cylinder than MWNT, thus MWNTs have more possibilities of structure defects and their nanostructure is less stable. Their specific mechanical and electronic properties make them useful for future high strength/modulus materials and nanodevices. They exhibit low density, large elastic limit without breaking (of up to 20–30% strain before failure), exceptional elastic stiffness, greater than 1000GPa and their extreme strength which is more than 20 times higher than a high-strength steel alloy. Besides, they also posses superior thermal and elastic properties: thermal stability up to 2800°C in vacuum and up to 750°C in air, thermal conductivity about twice as high as diamond, electric current carrying capacity 1000 times higher than copper wire [313]. The properties of CNTs strongly depend on the size and the chirality and dramatically change when SWCNTs or MWCNTs are considered [314].

CNTs are formed from pure carbon bonds. Pure carbons only have two covalent bonds: sp^2 and sp^3. The former constitutes graphite and the latter constitutes diamond. The sp^2 hybridization, composed of one s orbital and two p orbitals, is a strong bond within a plane but weak between planes. When more bonds come together, they form six-fold structures, like honeycomb pattern, which is a plane structure, the same structure as graphite [315].

Graphite is stacked layer by layer so it is only stable for one single sheet. Wrapping these layers into cylinders and joining the edges, a tube of graphite is formed, called nanotube [316].

Atomic structure of nanotubes can be described in terms of tube chirality, or helicity, which is defined by the chiral vector, and the chiral angle, θ. Figure 6.31 shows visualized cutting a graphite sheet along the dotted lines and rolling the tube so that the tip of the chiral vector touches its tail. The chiral vector, often known as the roll-up vector, can be described by the following equation [317]:

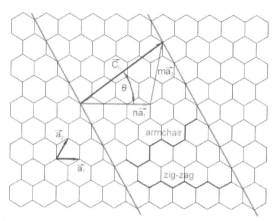

FIGURE 6.31 Schematic diagram showing how graphite sheet is 'rolled' to form CNT

$$C_h = na_1 + ma_2 \qquad (3)$$

As shown in Figure 6.31, the integers (n, m) are the number of steps along the carbon bonds of the hexagonal lattice. Chiral angle determines the amount of "twist" in the tube. Two limiting cases exist where the chiral angle is at 0° and 30°. These limiting cases are referred to as ziz-zag (0°) and armchair (30°), based on the geometry of the carbon bonds around the circumference of the nanotube. The difference in armchair and zig-zag nanotube structures is shown in Figure 6.32. In terms of the roll-up vector, the ziz-zag nanotube is (n, 0) and the armchair nanotube is (n, n). The roll-up vector of the nanotube also defines the nanotube diameter since the interatomic spacing of the carbon atoms is known [299].

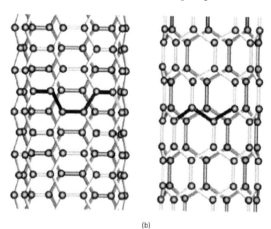

(a) (b)

FIGURE 6.32 Illustrations of the atomic structure (a) an armchair and (b) a ziz-zag nanotube.

Chiral vector C_h can be an integer multiple a_1 of a_2, which are two basis vectors of the graphite cell. Then we have $C_h = a_1 + a_2$, with integer n and m, and the constructed CNT is called a (n, m) CNT, as shown in Fig. 6.33. It can be proved that for armchair CNTs n=m, and for zigzag CNTs m=0. In Fig. 6.33, the structure is designed to be a (4, 0) zigzag SWCNT.

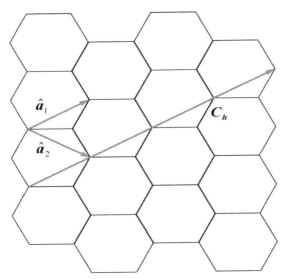

FIGURE 6.33   Basis vectors and chiral vector.

MWCNT can be considered as the structure of a bundle of concentric SWCNTs with different diameters. The length and diameter of MWCNTs are different from those of SWCNTs, which means, their properties differ significantly. MWCNTs can be modeled as a collection of SWCNTs, provided the interlayer interactions are modeled by Van der Waals forces in the simulation. A SWCNT can be modeled as a hollow cylinder by rolling a graphite sheet as presented in Fig. 6.34.

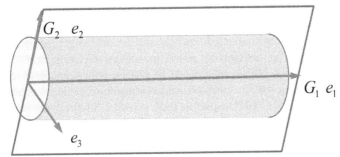

FIGURE 6.34   Illustration of a graphite sheet rolling to SWCNT.

If a planar graphite sheet is considered to be an undeformed configuration, and the SWCNT is defined as the current configuration, then the relationship between the SWCNT and the graphite sheet can be shown to be:

$$e_1 = G_1, \quad e_2 = R\sin\frac{G_2}{R}, \quad e_3 = R\cos\frac{G_2}{R} - R \tag{4}$$

The relationship between the integer's n, m and the radius of SWCNT is given by:

$$R = a\sqrt{m^2 + mn + n^2}/2\pi \tag{5}$$

where $a = \sqrt{3}a_0$, and a_0 is the length of a nonstretched C-C bond which is 0.142 nm [318].

As a graphite sheet can be 'rolled' into a SWCNT, we can 'unroll' the SWCNT to a plane graphite sheet. Since a SWCNT can be considered as a rectangular strip of hexagonal graphite monolayer rolling up to a cylindrical tube, the general idea is that it can be modeled as a cylindrical shell, a cylinder surface, or it can pull-back to be modeled as a plane sheet deforming into curved surface in three-dimensional space. A MWCNT can be modeled as a combination of a series of concentric SWCNTs with inter-layer inter-atomic reactions. Provided the continuum shell theory captures the deformation at the macrolevel, the inner microstructure can be described by finding the appropriate form of the potential function which is related to the position of the atoms at the atomistic level. Therefore, the SWCNT can be considered as a generalized continuum with microstructure [298].

6.14.24.3   CNT COMPOSITES

CNT composite materials cause significant development in nanoscience and nanotechnology. Their remarkable properties offer the potential for fabricating composites with substantially enhanced physical properties including conductivity, strength, elasticity, and toughness. Effective utilization of CNT in composite applications is dependent on the homogeneous distribution of CNTs throughout the matrix. Polymer-based nanocomposites are being developed for electronics applications such as thin-film capacitors in integrated circuits and solid polymer electrolytes for batteries. Research is being conducted throughout the world targeting the application of carbon nanotubes as materials for use in transistors, fuel cells, big TV screens, ultra-sensitive sensors, high-resolution AFM probes, super-capacitor, transparent conducting film, drug carrier, catalysts, and composite material. Nowadays, there are more reports on the fluid transport through porous CNTs/polymer membrane.

6.14.24.4 STRUCTURAL DEVELOPMENT IN POLYMER/CNT FIBERS

The inherent properties of CNT assume that the structure is well preserved (large-aspect-ratio and without defects). The first step toward effective reinforcement of polymers using nano-fillers is to achieve a uniform dispersion of the fillers within the hosting matrix, and this is also related to the as-synthesized nano-carbon structure. Secondly, effective interfacial interaction and stress transfer between CNT and polymer is essential for improved mechanical properties of the fiber composite. Finally, similar to polymer molecules, the excellent intrinsic mechanical properties of CNT can be fully exploited only if an ideal uniaxial orientation is achieved. Therefore, during the fabrication of polymer/CNT fibers, four key areas need to be addressed and understood in order to successfully control the microstructural development in these composites. These are: (i) CNT pristine structure, (ii) CNT dispersion, (iii) polymer–CNT interfacial interaction and (iv) orientation of the filler and matrix molecules (Fig. 6.35) [319].

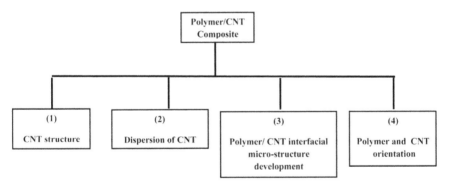

FIGURE 6.35 Four major factors affecting the microstructural development in polymer/CNT composite fiber during processing.

Achieving homogenous dispersion of CNTs in the polymer matrix through strong interfacial interactions is crucial to the successful development of CNT/polymer nanocomposite [320]. As a result, various chemical or physical modifications can be applied to CNTs to improve its dispersion and compatibility with polymer matrix. Among these approaches acid treatment is considered most convenient, in which hydroxyl and carboxyl groups generated would concentrate on the ends of the CNT and at defect sites, making them more reactive and thus better dispersed [321, 322].

The incorporation of functionalized CNTs into composite membranes is mostly carried out on flat sheet membranes [323, 324]. For considering the potential influences of CNTs on the physicochemical properties of dope solution [325] and change of membrane formation route originated from various additives [326], it is necessary to study the effects of CNTs on the morphology and performance.

6.14.24.5 GENERAL FABRICATION PROCEDURES FOR POLYMER/CNT FIBERS

In general, when discussing polymer/CNT composites, two major classes come to mind. First, the CNT nano-fillers are dispersed within a polymer at a specified concentration, and the entire mixture is fabricated into a composite. Secondly, as grown CNT are processed into fibers or films, and this macroscopic CNT material is then embedded into a polymer matrix [327]. The four major fiber-spinning methods (Fig. 6.36) used for polymer/CNT composites from both the solution and melt include dry-spinning [328], wet-spinning [329], dry-jet wet spinning (gel-spinning), and electrospinning [330]. An ancient solid-state spinning approach has been used for fabricating 100% CNT fibers from both forests and aero gels. Irrespective of the processing technique, in order to develop high-quality fibers many parameters need to be well controlled.

All spinning procedures generally involve:
(i)   fiber formation;
(ii)   coagulation/gelation/solidification;
(iii) drawing/alignment.

For all of these processes, the even dispersion of the CNT within the polymer solution or melt is very important. However, in terms of achieving excellent axial mechanical properties, alignment and orientation of the polymer chains and the CNT in the composite is necessary. Fiber alignment is accomplished in post-processing such as drawing/annealing and is key to increasing crystallinity, tensile strength, and stiffness [331].

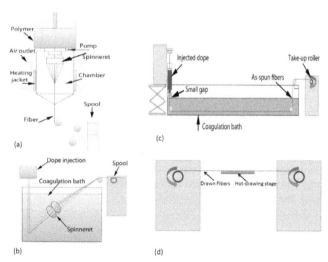

FIGURE 6.36 Schematics representing the various fiber processing methods (a) dry-spinning; (b) wet-spinning; (c) dry-jet wet or gel-spinning; and (d) post-processing by hot-stage drawing.

6.14.25 *FILTER* APPLICATIONS

Nonwovens composed of fibers made from glass, paper or polymers are highly porous membranes – the total porosity typically being of the order of 80 to 95%, which can be used to remove solid particles, dust particles, aerosols, fine fluid droplets from a stream either composed of a gas or a fluid. Water filtration is a topic that is of enormous importance worldwide. Air filtration is highly important for a broad range of industrial applications including power plants, and the same holds for fuel filtration – a must in modern car engines – or coalescence filtration of gasoline for airplanes. Typical high-efficiency filter requirements are that the filters should capture all fluid or solid particles, respectively, surpassing a specified size and that the capture probability should be as high as possible, say in the range of 99 or 99.9% [332–335].

6.14.25.1 *ANTIMICROBIAL AIR FILTER*

It is well known that HVAC air filters usually operated in dark, damp, and ambient temperature conditions, which is susceptible for bacterial, mold, and fungal attacks, resulting in unpredictable deterioration and bad odor. To solve this problem, functionalization of the surface of filtering media with antimicrobial agents for long-lasting durable antimicrobial functionality is of current interest. In 2007, Jeong and Youk [336] explored the electrospun PUCs nanofiber mats with different amounts of quaternary ammonium groups in antimicrobial air filter. They found that PUCs exhibited very strong antimicrobial activities against Staphylococcus aurous and Escherichia coli. Ramakrishna and co-workers [337] induced the silver nanoparticles based on different electrospun polymer [CA; PAN and PVC] nanofiber for antimicrobial functionality owing to the remarkable antimicrobial ability of silver ions and silver compounds.

6.14.25.2 *BASIC PROCESSES CONTROLLING FILTER EFFICIENCIES*

It is helpful at this stage to recall some basic processes controlling filter efficiencies in general, that is, to a first approximation independent of the fiber diameters [338, 339]. What is known for conventional nonwovens composed of fibers with diameters well in the 10–100 micrometer range is that the filter efficiency is controlled by various types of capture processes happening within the nonwoven as the gas/fluids carrying particles pass through their pores. These basic processes are depicted in Fig. 6.37.

The first process to be considered is the interception. Particles following the gas stream around the fiber as depicted in Fig. 6.37a are intercepted by the fiber surfaces if the particles pass the fibers at a distance not larger than the particle diameter. It is obvious that larger particles tend to enhance the probability for such an interception.

The second process of importance is the impaction, as shown schematically in Fig. 37b. Particles do not follow in this case the deflection of the gas stream due to the neighborhood of the fiber as a solid object but because of inertia effects follow the original path. This in turn causes the particle to impact on the fiber surface. Impaction tends to grow in importance as the flow velocity of the gas increases.

Finally, diffusion plays a role in controlling the capture efficiency. Here, particles carried by stream lines that pass the fiber at sufficient distance not to cause a direct interception nevertheless come into contact with the fiber surface because of diffusional motions, as depicted in Fig. 6.37c. Diffusion tends to be of importance for smaller particles and low flow velocities.

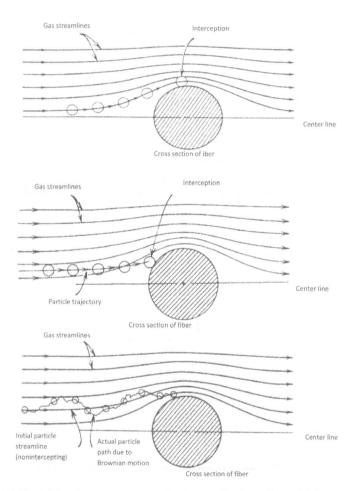

FIGURE 6.37 Molecular processes contributing to filtering effects: (a) interception, (b) impaction, (c) diffusion.

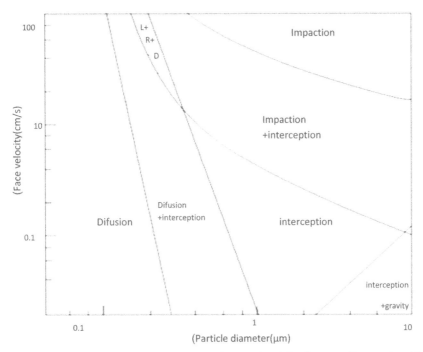

FIGURE 6.38   Survey on the regimes in a flow rate/particle diameter diagram in which either the diffusion, the interception or finally the impaction dominate diffusion.

Figure 6.38 gives a survey on the regimes in a flow-rate/particle – diameter diagram in which either the diffusion, the interception or finally the impaction dominate.

So, at low particle sizes diffusion more or less dominates the control of the filter efficiency, particularly for small flow velocities, whereas the impaction dominates for large flow velocities and particle sizes, with the three processes contributing in different ratios at intermediate particle sizes and flow velocities.

6.15   CATALYST AND ENZYME CARRIERS

A carrier for catalyst in chemistry and biology is used to preserve high catalysis activity, increase the stability, and simplify the reaction process. An inert porous material with a large surface area and high permeability to reactants could be a promising candidate for efficient catalyst carriers. Using an electrospun nanofiber mat as catalyst carrier, the extremely large surface could provide a huge number of active sites, thus enhancing the catalytic capability. The well-interconnected small pores in the nanofiber mat warrant effective interactions between the reactant and catalyst, which is valuable for continuous flow chemical reactions or biological processes.

The catalyst can also be grafted onto the electrospun nanofiber surface via coating or surface modification [340].

6.15.1 CATALYSIS

It is well known that nanostructured materials have opened new possibilities for creating and mastering nanoobjects for novel advanced catalytic materials. In general, catalysis is a molecular phenomenon and the reaction occurs on an active site. A crucial step in catalysis is how to remove and recycle the catalyst after the reaction. The immobilization of catalysts in materials with large surface area advances an interesting solution to this problem. Taking the large surface area and high porosities, electrospun nanofibers, as a novel catalysts or supports for catalysts, have been widely investigated in catalytic field.

6.15.2 ELECTROCHEMICAL CATALYSTS

In 2009, Su and Lei [341] investigated the electrochemical catalytical properties based on Pd/polyamide (Pd/PA6) electrospun nanofiber mats for the oxidation of ethanol in alkaline medium in which Pd/PA6 was directly used as electrocatalytic electrodes. High activity and stable performance based on Pd/PA6 have been obtained.

Simultaneously, Kim and co-workers [342] also explored the electrochemical catalytic properties based on electrospun Pt and PtRh nanowires for dehydrogenative oxidation of cyclohexane to benzene. In contrast to the conventional Pt nanoparticle catalysts (e.g., carbon/Pt or Pt black), Pt and PtRh electrospun nanowires electrocatalysts exhibited higher catalytic activities with the same metal loading amount. Furthermore, PtRh nanowires performed the best catalytic activities with a maximum power density of ca. 23 mWcm^{-2}. Such higher electrocatalytic performances than Pt nanoparticles are attributed to the inherent physicochemical and electrical properties of 1D nanostructures as shown in Fig. 6.39.

1. Nanowire catalysts could provide facile pathways for the electron transfer by reducing the number of interfaces between the electrocatalysts, whereas the nanoparticles are likely to impose more impedance for electrons to transfer particle to particle
2. Adding Rh to Pt to form the PtRh alloy can facilitate the adsorption/desorption properties of benzene and cyclohexane along with the modification of the C–H bond breaking ability
3. The rougher morphology of PtRh nanowires comprise of small nanoparticles, which can provide high catalytic area.

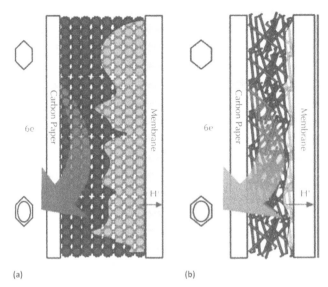

(a)   (b)

FIGURE 6.39 Schematic illustration for cyclohexane electro-oxidation over the nanoparticles (a) and nanowire (b) catalysts.

EPOC is a phenomenon where application of small currents or potentials on catalysts in contact with solid electrolytes leads to pronounced strongly non-Faradaic and reversible changes in catalytic activity and selectivity which is shown in Fig. 6.40.

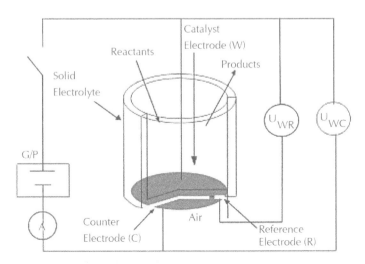

FIGURE 6.40 Fuel cell type reactor.

6.15.3 CATALYSIS

The combination of nanofibers and catalysis seems, at first, to be rather odd. However, considering homogeneous catalysis as a first example it is well known that a huge problem is the removal and recycling of the catalytic agent after the reaction, either from the reaction solution or from the product [343]. Complex separation methods involving in some cases several processing steps have been used for this purpose. The implantation of homogeneous or also heterogeneous catalysts into nanofibers poses an interesting solution for these problems. Now of course, the very nature of homogeneous reactions requires that the catalyst is molecularly dispersed in the same phase as the reaction compounds, so that these compounds come into intimate contact via diffusional motions, thus allowing for the reaction to proceed. The common phase is generally a solution or a melt.

However, considering nanofibers made from polymers in which the catalyst is molecularly dispersed it is well known that smaller but also larger molecules can perform surprisingly rapid diffusional processes in polymer matrices in the amorphous phase, in a partially crystalline phase, above and even below the glass-transition temperature. So, it is highly probable that reaction compounds that are dispersed in a solution or melt surrounding the nanofibers with catalysts dispersed in them can diffuse into the fiber matrix, make contact with the catalyst via diffusion.

Finally, the product molecules diffuse out of the fiber again. In fact, experiments to be discussed in the following have shown that this actually is the case.

One problem that has to be solved is to keep the catalyst within the fiber, despite allowing it to diffuse in the polymer matrix. By choosing the nature of the polymer carrier appropriately to induce specific interactions between carrier and catalysts, by attaching the catalyst via flexible spacers to the polymer backbone, one is able to achieve this goal.

The reaction mixture can circulate around the fibers, as is the case, for example, in the continuously working microreaction technique, or the fibers fixed on a carrier can be immersed repeatedly into a reaction vessel to catalyze the content of the vessel. In addition to the short diffusion distances within nanofibers, the specific pore structures and high surface areas of nanofiber nonwovens allow a rapid access of the reaction components to the catalysts and of the products back into the reaction mixture.

For homogeneous catalysis, systems consisting of core-shell nanofibers combined with proline and $Sc(OTf)_3$ ($TfO = CF_3SO_3$) catalysts were fabricated, for instance, by template methods described above (TUFT (tubes by fiber template process) method). In contrast to conventional catalysis in homogeneous solution or in microemulsions, for which the conversion is 80%, the fiber systems achieved complete conversion in the same or shorter reaction times. The fibers can be used several times without loss of activity. Furthermore, nanofibers were used as carriers for enzymes, where the enzymes were either chemically attached to the electrospun fibers or directly dispersed in the nanofibers during the electrospinning process.

High catalyst activities were reported in this case as well. Current activity will certainly lead to a broad range of catalytic systems.

The use of polymer nanofibers in heterogeneous catalysis was analyzed for nanofibers loaded with monometallic or bimetallic nanoparticles (such as Rh, Pt, Pd, Rh/Pd, and Pd/Pt) has been reported in the literature. These catalyst systems can be applied in hydrogenation reactions, for example. To fabricate such fibrous catalyst systems, polymer nanofibers are typically electrospun from solutions containing metal salts (such as Pd $(OAc)_2$) as precursors. In the next step, the salts incorporated in the fibers are reduced, either purely thermally in air or in the presence of a reducing agent such as H_2 or hydrazine). The Pd nanoparticles formed have diameters in the range of 5–15 nm, depending on the fabrication method. The catalytic properties of these mono- or bimetallic nanofiber catalysts were investigated in several model hydrogenations, which demonstrated that the catalyst systems are highly effective.

6.15.4 ENZYMES

Chemical reactions using enzymes as catalysts have high selectivity and require mild reaction conditions. For easy separation from the reaction solution, enzymes are normally immobilized with a carrier. The immobilization efficiency mainly depends on the porous structure and enzyme-matrix interaction. To immobilize enzyme on electrospun nanofibers, many approaches have been used, including grafting enzyme on fiber surface, physical adsorption, and incorporating enzyme into nanofiber via electrospinning followed by crosslinking reaction.

To graft enzymes on nanofiber surface, the polymer used should possess reactive groups for chemical bonding [344, 345]. In some studies, polymer blends containing at least one reactive polymer were used [346, 347]. The immobilized enzymes normally showed a slightly reduced activity in aqueous environment compared with the un-immobilized native counterpart, but the activity in nonaqueous solution was much higher. For example, α-chymotrypsin was used as a model enzyme to bond chemically on the surface of electrospun PS nanofibers. The enzyme was measured to cover over 27.4% monolayer of the nanofiber surface, and the apparent hydrolytic activity of the enzyme-loaded was 65% of the native enzyme, while the activity in nonaqueous solution was over three orders of magnitude higher than that of its native enzyme under the same condition. In another study using PAN nanofibers to immobilize lipase, the tensile strength of the nanofiber mat was improved after lipase immobilization, and the immobilized lipase retained >90% of its initial reactivity after being stored in buffer at 30°C for 20 days, whereas the free lipase lost 80% of its initial reactivity. Also the immobilized lipase still retained 70% of its specific activity after 10 repeated reaction cycles [348]. In addition, the immobilized enzyme also showed improved pH and thermal stabilities [349]. Ethylenediamine was used to modify PAN nanofiber mat to introduce active and hydrophilic groups, followed by a chitosan coating for improvement of biocompatibility [348].

Enzymes were incorporated into nanofibers via electrospinning, and subsequent crosslinking the enzymes incorporated effectively prevented their leaching. In the presence of PEO or PVA, casein and lipase were electrospun into ultra-thin fibers. After crosslinking with 4,4'-methylenebis(phenyl diisocyanate), the fibers became insoluble, and the lipase encapsulated exhibited 6 times higher hydrolysis activity towards olive oil than that of the films cast from the same solution [350]. The cross-linked enzymes in nanofibers showed very high activity and stability. For example, the immobilized α-chymotrypsin in a shaken buffer solution maintained the same activity for more than 2 weeks [351].

In addition to chemical bonding, the enzymes were also applied onto nanofibers simply via physical adsorption [352]. Polyacrylonitriles-2-methacryloyloxyethyl phosphoryl choline (PANCMPC) nanofiber was reported to have high biocompatibility with enzymes because of the formation of phospholipid microenvironment on the nanofiber surface. Lipase on the nanofibers showed a high immobilization rate, strong specific activity and good activity retention.

6.15.5   PHOTOCATALYSIS

The increasing industrial needs and growing urbanization have led to water scarcity issues around the globe and the wastewater produced has to be treated for reutilization of clean water in daily activities. Among the various techniques, the heterogeneous photocatalysis system is an effective method for treating wastewater and photodegrading organic pollutants. The semiconductor metal oxides have been used as photocatalysts where upon irradiation of sunlight, create electron-hole pairs which in turn produce radicals in different pathways as shown in Figure 6.41[353].

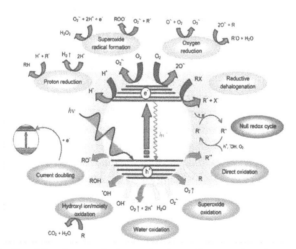

FIGURE 6.41   Possible photocatalytic mechanism of generating free radicals in the conduction band (CB) and valence band (VB) of semiconducting metal oxides.

The photocatalytic mechanism is as follows: upon irradiation semiconductor metal oxides eject an electron from the valance band to the conduction band, thereby leaving behind a hole in the valence band. The generated electrons and holes produce superoxide radicals to degrade the pollutants by reacting with chemisorbed oxygen on the catalyst surface and oxygen in the aqueous solution [354, 355].

Semiconducting metal-oxides such as TiO_2, ZnO, Fe_2O_3, WO_3, Bi_2WO_6, CuO and many more have been widely used as oxidative photocatalysts for effective removal of industrial pollutants and wastewater treatment [356, 357].

Shengyuan et al. [358] synthesized novel rice grain- shaped TiO_2 mesostructures by electrospinning and observed the phase change with increasing temperature from 500°C to 1000°C. A comparison study was performed on the photocatalytic activity of the TiO_2 rice grain and P-25. They observed an enhanced photocatalytic activity on alizarin red dye in rice grain-shaped TiO_2 which was due to its single crystalline nature and larger surface area than P-25. Meng et al. [359] synthesized anatase TiO_2 nanofibers by using a simple electrospinning technique and were able to grow high dense rutile TiO_2 nanorods along the fibers using hydrothermal treatment. The nanofibril-like morphology of TiO_2 nanorods/nanofibers with rutile and anatase phase was able to degrade the rhodamine-6G effectively under UV radiation. Zhang et al. [360] adopted the core/shell technique for synthesizing hollow mesoporous TiO_2 nanofibers with a larger surface area of around 118 m^2/g.

Similarly, Singh et al. [361] was able to show the synthesis of highly mesoporous ZnO nanofibers from electrospinning with high crystallinity and large surface area. The fibers had better interaction with polycyclic aromatic hydrocarbons such as naphthalene and anthracene due to the higher surface to volume ratio of the nanofibers, and thereby had higher rate constants for the UV light photodegradation of the aromatic compounds.

Ganesh et al. [362] demonstrated the superhydrophilic coating of TiO_2 films on glass substrates using electrospinning with the film acting as a self-cleaning coating for photodegradation of alizarin red dye. The self-cleaning property of the TiO_2 can be easily adopted on solid surfaces like stainless steel and also fabrics. Bedford et al. [363] synthesized photocatalytic self-cleaning textile fibers using coaxial electrospinning where cellulose acetate was used as core fiber and TiO_2 nanofiber as shell. They studied the self- cleaning photocatalytic performance of TiO_2 based textile fibers by photodegradation of dyes like keyacid blue and sulforhodamine at moderate exposure of light and were able to degrade it significantly. Neubert et al. fabricated the fibrous conductive catalytic filter membrane consisting of nonconductive polyethylene oxide blended with (±)-camphor-10-sulfonic acid doped conductive polyaniline. They were able to incorporate the TiO_2 nanoparticles on the membrane by electrospraying and achieved the photodegradtion of 2-chloroethyl phenyl sulfide pollutant upon UV exposure.

However, the usage of metal oxides such as TiO_2, ZnO, SnO_2 and Fe_2O_3 for practical applications is still limited because of the fast electron-hole recombination and broad bandgaps which respond only to UV light. Therefore, it is essential to improve the visible light absorption and other shortcomings to ameliorate the photocatalytic property by either by combining with graphene, other semiconducting metal oxides or metal nanoparticles with matching bandgap [364–366].

6.16 MEDICINAL APPLICATIONS FOR ELECTROSPUN NANOFIBERS

6.16.1 NANOTECHNOLOGY AND MEDICINAL APPLICATIONS IN GENERAL

It is obvious that the combination of nanoscience/nanotechnology with medicine makes a lot of sense for many reasons. One major reason is that the nanoscale is a characteristic biological scale, a scale related directly to Life. DNA strains, globular proteins such as ferritins, viruses are all on this scale. For instance, the tobacco mosaic virus is actually a nanotube. The dimensions of bacteria and of cells tend to be already in the micrometer range, but important subunits of these objects such as the membranes of cells have dimensions in the nm scale.

So there are certainly good reasons for addressing various types of medicinal problems on the basis of nanoscience, of nanostructures. Scaffolds used for engineering tissues such as bone or muscle tissues may be composed of nanofibers mimicking the ECM, nanoscalar carriers for drugs to be carried to particular locations within the body to be released locoregionally rather than systemic are examples. Wound healing exploiting fibrillar membranes with a high porosity and pores with diameters in the nanometer scale, thus allowing transport of fluids, gases from and to the wound yet protecting it from bacterial infections, are further examples for the combination of nanoscience and medicine with the focus here on nanofibers.

A highly interesting example along this line certainly concerns inhalation therapy. The concept is to load specific drugs onto nanorods with a given length and diameter rather than onto spherical objects such as aerosols as already done today. The reasoning is that such spherical particles tend to become easily exhaled so that frequently only a minor part of them can become active in the lung. Furthermore, the access to the lung becomes limited with increasing volume of these particles.

However, rather long fibers are known to be able to penetrate deeply into the lung. The reason is that the aerodynamic radius controls this process with the aerodynamic radius of rods being controlled mainly by the diameter and only weakly by the length, as detailed later in more detail. So, inhalation therapy based on nanorods as accessible via nanofibers offers great benefits.

In the following, different areas where nanofibers and nanorods for that matter can contribute to problems encountered in medicine will be discussed.

6.16.2   *INTRODUCTION* TO NANOMEDICAL

Regenerative medicine combines the principles of human biology, materials science, and engineering to restore, maintain or improve a damaged tissue function. Regenerative medicine is divided into cell therapy or "cell transplantation" and "tissue engineering."

The National Institute of Biomedical Imaging and Bioengineering defines tissue engineering as "a rapidly growing area that seeks to create, repair and/or replace tissues and organs by using combinations of cells, biomaterials, and/or biologically active molecules."

Tissue engineering has emerged through a combination of many developments in biology, material science, engineering, manufacturing and medicine. Tissue engineering involves the design and fabrication of three-dimensional substitutes to mimic and restore the structural and functional properties of the original tissue. The term 'tissue engineering' is loosely defined and can be used to describe not only the formation of functional tissue by the use of cells cultured on a scaffold or delivered to a wound site, but also the induction of tissue regeneration by genes and proteins delivered in vivo.

Cell transplantation is performed when only cell replacement is required. However, in tissue engineering, the generated tissue should have similar properties to the native tissue in terms of biochemical activity, mechanical integrity and function. This necessitates providing a similar biological environment as that in the body for the cells to generate the desired tissue. Figure 42 summarizes the important steps in tissue engineering. First, cells are harvested from the patient and are expanded in cell culture medium. After sufficient expansion, the cells are seeded into a porous scaffold along with signaling molecules and growth factors which can promote cell growth and proliferation. The cell-seeded scaffold will be then placed into a bioreactor before being implanted into the patient's body. As it is evident in Fig. 6.42 three major elements of tissue engineering include [367]:

    a)   cells;

    b)   scaffolds; and

    c)   bioreactors.

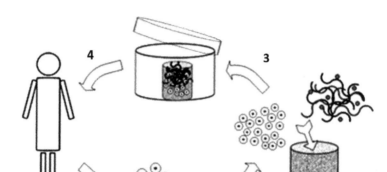

FIGURE 6.42 Schematic diagram summarizing the tissue engineering process.

6.16.2.1 CELLS

Cells are the building block of all tissues. Therefore, choosing the right cell source with no contamination that is compatible with the recipient's immune system is the critical step in tissue engineering. Stem cells are employed as the main source of cells for tissue engineering and are taken from autologous, allogeneic or xenogeneic sources for different applications. Stem cells are divided into the following three groups: embryonic stem cells (ESCs), induced pluripotent stem cells, and adult stem cells. ESCs are isolated from the inner cell mass of preimplantation embryos. These cells are considered pluripotent since they can differentiate into almost any of the specialized cell types. Induced pluripotent stem cells are the adult cells that have been transformed into pluripotent stem cells through programming. Among adult stem cells, Mesenchymal stem cell is widely used as a multipotent source. Mesenchymal stem cell is derived from bone marrow stroma and can differentiate into a variety of cell types in vtiro. Other sources of adult stem cells include the amniotic fluid and placental derived stem cells [368, 369].

6.16.2.2 SCAFFOLDS

Scaffolds are temporary porous structures used to support cells by filling up the space otherwise occupied by the natural ECM and by providing a framework to organize the dissociated cells. A biocompatible and biodegradable material is chosen for tissue engineering scaffolds which have sufficient porosity and pore-interconnectivity to promote cell migration and proliferation, and allow for nutrient and waste exchange. The rate of degradation should be tuned with the rate of cell growth and expansion, so that as the host cells expand and produce their own ECM, the temporary material degenerates with a similar rate. Moreover, the by-products should be confirmed to be nontoxic. Mechanical properties should also match that of the native ECM [370, 371] (Fig. 6.43).

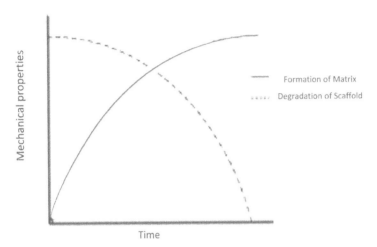

FIGURE 6.43   Mechanical properties of an ideal scaffold.

6.16.2.3   BIOREACTORS

In-vitro tissue engineering requires bioreactors in order to provide sufficient nutrients and oxygen to the cells while removing the toxic materials left by the proliferating cells. Moreover, the essential cell-specific mechanical stimuli are also provided by bioreactors. Each type of cell (cartilage, bone, myocardium, endothelial, etc.) has different requirements in terms of pH, oxygen tension, mechanical stimulation, temperature, etc. As a result, it is necessary to use cell specific bioreactors for generation of different tissues [372].

For the treatment of tissues or organs in malfunction in a human body, one of the challenges to the field of tissue engineering/biomaterials is the design of ideal scaffolds/synthetic matrices that can mimic the structure and biological functions of the natural ECM. Human cells can attach and organize well around fibers with diameters smaller than those of the cells. In this regard, nanoscale fibrous scaffolds can provide an optimal template for cells to seed, migrate, and grow. A successful regeneration of biological tissues and organs calls for the development of fibrous structures with fiber architectures beneficial for cell deposition and cell proliferation [373].

In general, tissue engineering involves the fabrication of three dimensional scaffolds that can support cell in-growth and proliferation. In this context, the generation of scaffolds with tailored, biomimetic geometries (across multiple scales) has become an increasingly active area of research [374].

6.16.3 *TISSUE* ENGINEERING: BACKGROUND INFORMATION

There is without any doubt a constantly increasing need for tissues and organs to replace those that have been damaged by sickness or accidents. It is also without any doubt that this need cannot be covered by allogeneic transplants.

The shortage of donor organs, immunological problems, and possibly contamination of the donor tissue limit the use of organ transplants. This is a good reason for having a closer look at the emerging science of tissue engineering.

Tissue engineering involves the cultivation of various types of tissues to replace such damaged tissues or organs. Cartilage, bone, skin tissue, muscle, blood vessels, lymphatic vessels, lung tissue, and heart tissue are among the target tissues. In vitro, in vivo as well as combination approaches are known. In vitro approaches taken in tissue engineering rely on the seeding of specific cells on highly porous membranes as scaffold. Both homologous and autologous cells have been used for this purpose. Autologous cells, that is, cells that are harvested from an individual for the purpose of being used on that same individual, have the benefit of avoiding an immunologic response in tissue engineering. The bodies of the patients will not reject the engineered tissue because it is their own tissue and the patients will not have to take immunosuppressive drugs.

The concept is that such scaffolds will mimic to a certain extent the extracellular matrix surrounding cells in living tissue. The extracellular matrix is known to have a broad range of tasks to accomplish. It embeds the cells of which the particular tissue is composed, it offers points of contacts to them, provides for the required mechanical properties of the tissue. So, the expectation is that cells seeded on adequate porous scaffolds experience an enhanced proliferation and growth, covering finally the whole scaffold. A further task is to define the three-dimensional shape of the tissue to be engineered. Ideally, such a scaffold may then be reimplanted into the living body provided that an appropriate selection of the nature of the seeded cells was done.

So, as an example, to replace muscles, muscle cells might be chosen for the seeded cells. Yet, frequently rather than choosing specific cell lineages stem, cells such as, for instance, mesenchymal stem cells, to be discussed below in more detail, are seeded for various reasons. The proliferation of such cells is, in this case, just one step, the next involving the differentiation of the cells along specific target cell lines depending of the target tissue. To induce such differentiation various types of biological and chemical signals have been developed. To enhance proliferation it will in general be necessary to include functional compounds such as growth factors, etc., into the scaffold membranes, as discussed below.

Another approach used currently less often (that is, an in vivo approach) in tissue engineering consists in implanting the original scaffold directly into the body to act as nucleation sites for self-healing via seeding of appropriate cells.

Because the tissue engineering technology is based essentially on the seeding of cells into three-dimensional matrices, the material properties of the matrix as well as its architecture will fundamentally influence the biological functionality of the engineered tissue. One has to keep in mind that the carrier matrix has to fulfill a diverse range of requirements with respect to biocompatibility, biodegradability, morphology, sterilizability, porosity, ability to incorporate and release drugs, and mechanical suitability. Also, the scaffold architecture, porosity and relevant pore sizes are very important. In general, a high surface area and an open and interconnected 3D pore system are required for scaffolds.

These factors affect cell binding, orientation, mobility, etc. The pores of scaffolds are, furthermore, very important for cell growth as nutrients diffuse through them. The minimum pore size required is decided by the diameter of cells and therefore varies from one cell type to another. Inappropriate pore size can lead to either no infiltration at all or nonadherence of the cells. Scaffolds with nanoscalar architectures have bigger surface areas, providing benefits for absorbing proteins and presenting more binding sites to cell membrane receptors.

Biological matrices are usually not available in sufficient amounts, and they can be afflicted with biological-infection problems. It is for this reason that during the last decades man-made scaffolds composed of a sizable number of different materials of synthetic or natural, that is biological, origin have been used to construct scaffolds characterized by various types of architectures.

These include powders, foams, gels, porous ceramics and many more. However, powders, foams, and membranes are often not open-pored enough to allow cell growth in the depth of the scaffold; consequently, the formation of a three-dimensional tissue structure is frequently restricted. Even loose gel structures (e.g., of polypeptides) may fail. Furthermore, smooth walls and interfaces, which occur naturally in many membranes and foams, are frequently unfavorable for the adsorption of many cell types.

From a biological viewpoint, almost all of the human tissues and organs are deposited in nanofibrous forms or structures. Examples include: bone, dentin, collagen, cartilage, and skin. All of them are characterized by well-organized hierarchical fibrous structures realigning in nanometer scale. Nanofibers are defined as the fibers whose diameter ranges in the nanometer range. These have a special property of high surface area and increased porosity which makes it favorable for cell interaction and hence it makes it a potential platform for tissue engineering. The high surface area to volume ratio of the nanofibers combined with their microporous structure favors cell adhesion, proliferation, migration, and differentiation, all of which are highly desired properties for tissue engineering applications. There are mainly three techniques involved in synthesizing nanofibers namely electrospinning, self-assembly, and phase separation [375].

6.16.3.1 PHASE SEPARATION

In this technique water-polymer emulsion is formed which is thermodynamically unstable. At low gelation temperature, nanoscale fibers network is formed, whereas high gelation temperature leads to the formation of platelet-like structure. Uniform nanofiber can be produced as the cooling rate is increased. Polymer concentration has a significant effect on the nanofiber properties, as polymer concentration is increased porosity of fiber decreased and mechanical properties of fiber are increased. The final product obtained is mainly porous in nature but due to controlling the key parameters we can obtain a fibrous structure. The key parameters involved are as follows [375].

a) type of polymers and their viscosity;
b) type of solvent and its volatility;
c) quenching temperature; and
d) gelling type.

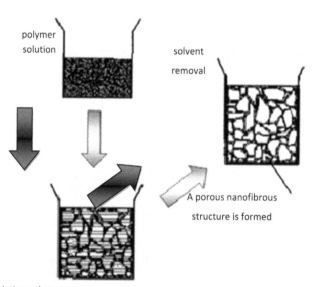

FIGURE 6.44 Nanofibrous structure production through phase separation.

6.16.3.2 SELF ASSEMBLY

It is a powerful approach for fabricating supra molecular architectures. Self-assembly of peptides and proteins is a promising route to the fabrication of a variety of molecular materials including nanoscale fibers and fiber network scaffolds. The main mechanism for a generic self-assembly is the intermolecular forces that bring the smaller unit together [375, 376] (Fig. 6.45).

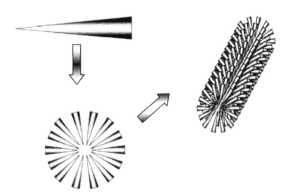

FIGURE 6.45   Schematic presentation of self-assembled nanofiber production.

6.16.3.3   ELECTROSPINNING

It is a term used to describe a class of fibers forming processes for which electrostatic forces are employed to control the production of the fiber. Electrospinning readily leads to the formation of continuous fibers ranging from 0.01 to 10 μm. Electrospinning is a fiber forming processes by which electrostatic forces are employed to control the production of fibers. It is closely related to the more established technology of electrospraying, where the droplets are formed. "Spinning" in this context is a textile term that derives from the early use of spinning wheels to form yarns from natural fiber. In both electrospinning and electrospraying, the role of the electrostatic forces is to supplement or replace the conventional mechanical forces (e.g., hydrostatic, pneumatic) used to form the jet and to reduce the size of the fibers or droplets, hence the term "electrohydrodynamic jetting." Polymer nanofibers fabricated via electrospinning have been proposed for a number of soft tissue prostheses applications such as blood vessel, vascular, breast, etc. In addition, electrospun biocompatible polymer nanofibers can also be deposited as a thin porous film onto a hard tissue prosthetic device designed to be implanted into the human body This method will be discussed in detail, later.

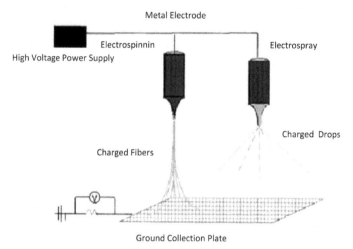

FIGURE 6.46   Electrospinning and electrospraying principles.

6.16.4 NANOFIBERS IN TISSUE ENGINEERING APPLICATIONS

A rapidly growing field of application of polymer nanofibers is their use in tissue engineering. The main areas of intensive research are nerve, blood vessel, skeletal muscle, cartilage, bone and skin tissue engineering.

6.16.4.1 NERVE TISSUE ENGINEERING

Application of electrospun polymeric nanofibers for nerve tissue regeneration is a very significant issue. The most important observation is the elongation and neurite growth of cells parallel to fiber direction, and the effect of fiber diameter is not so significant. They have found that aligned nanofibrous scaffolds are good scaffolds for neural tissue engineering [87].

6.16.4.2 SKIN TISSUE ENGINEERING

Nanofibers exhibit higher cell attachment and spreading, especially when coated with collagen than microfibers. The results, which researchers were obtained, prove the potential of electrospun nanofibers in wound healing and regeneration of skin and oral mucosa [377].

6.16.4.3 BLOOD VESSEL TISSUE ENGINEERING

A number of studies have shown that the biodegradable polymers mimic the natural ECM and show a defined structure replicating the in vivo-like vascular structures and can be ideal tools for blood vessel tissue engineering [378].

6.16.4.4   SKELETAL MUSCLE TISSUE ENGINEERING

Skeletal muscle is responsible for maintenance of structural contours of the body and control of movements. Extreme temperature, sharp traumas or exposure to myotoxic agents are among the reasons of skeletal muscle injury. Tissue engineering is an attractive approach to overcome the problems related to autologous transfer of muscle tissue. It could also be a solution to donor shortage and reduction in surgery time. The studies demonstrate the absence of toxic residuals and satisfactory mechanical properties of the scaffold [379].

6.16.4.5   CARTILAGE TISSUE ENGINEERING

There are three forms of cartilage in the body that vary with respect to structure, chemical composition, mechanical property and phenotypic characteristics of the cells. These are hyaline cartilage, fibrocartilage and elastic cartilage. Cells capable of undergoing chondrogenic differentiation upon treatment with appropriate factors and a 3-D scaffold that provides a suitable environment for chondrogenic cell growth are the two main requirements for successful cartilage tissue engineering. In addition, there are some other conditions to fulfill. First, the matrix should support cartilage-specific matrix production; second, it should allow sufficient cell migration to achieve a good bonding to the adjacent host tissue and finally, the matrix should provide enough mechanical support in order to allow early mobilization of the treated joint [380].

6.16.4.6   BONE TISSUE ENGINEERING

Bone engineering has been studied for a long time to repair fractures and in the last decades, used in preparation of dental and orthopedic devices and bone substitutes. Bone tissue engineering is a more novel technique, which deals with bone restoration or augmentation. The matrix of bone is populated by osteogenic cells, derived of mesenchymal or stromal stem cells that differentiate into active osteoblasts. Several studies have demonstrated that it is possible to culture osteogenic cells on 3-D scaffolds and achieve the formation of bone. They designed a novel 3-D carrier composed of micro and nanofibers and have observed that cells used these nanofibers as bridges to connect to each other and to the microfibers. Furthermore, a higher ability for enhancement of cell attachment and a higher activity was observed in the nano/microfiber combined scaffolds compared to the microfibrous carrier. The fibrous scaffolds improved bone formation [381, 382].

6.16.5   TISSUE ENGINEERING: CELLS AND SCAFFOLDS

At this point in history tissue engineering is largely an Edisonian exercise in which the scaffold provides mechanical support while host-appropriate cells populate the structure and the deposit ECM components specific to the organ targeted for replacement. The current goal is a 'neotissue' that the body can "work with" and eventually adapt to carry out the full range of expected biological activities. The primary constituent of the various ECM's involved is typically collagen; the ratios of collagen type and hierarchical organization define the mechanical properties and organization of the evolving neotissue. In addition, the ECM provides cells with a broad range of chemical signals that regulate cell function [383–385]. Cells have been mainly cultured at the surface of the electrospun materials instead of in the bulk material. 2D monolayer culture models are easy and convenient to set up with good viability of cells in culture. Although, cells on electrospun surfaces have shown 3D matrix adhesion, considerations must be made at a 3D level to truly assess the potential of electrospun biomaterials for tissue engineering by providing cells with the 3D environment found in natural tissues [386, 387].

In recent years, there have been a large number of patients who suffered from the bone defects caused by tumor, trauma or other bone diseases. Generally, autogenetic and allogenetic bones are used as substitutes in treatment of bone defects. However, secondary surgery for procuring autogenetic bone from patient would bring donor site morbidity and allogenetic bone would cause infections or immune response [388, 389]. Thus, it is necessary to find new approach for the bone regeneration. As a promising approach, the tissue engineering develops the viable substitutes capable of repairing or regenerating the functions of the damaged tissue. For the bone tissue engineering, it requires a scaffold system to temporarily support the cells and direct their growth into the corresponding tissue in vivo [390, 391].

There are three basic tissue engineering strategies that are used to restore, maintain, or improve tissue function, and they can be summarized as cell transplantation, scaffold-guided regeneration, and cell loaded scaffold implantation [392, 393].

1. Cell transplantation involves the removal of healthy cells from a biopsy or donor tissue and then injecting the healthy cells directly into the diseased or damaged tissue. However, this technique does not guarantee tissue formation and generally has less than 10% efficiency.

2. Scaffold-guided regeneration involves the use of a biodegradable scaffold implanted directly into the damaged area to promote tissue growth.

3. Cell-loaded scaffold implantation involves the isolation of cells from a patient and a biodegradable scaffold that is seeded with cells and then implanted into the defect location. Prior to implantation, the cells can be subjected to an in vitro environment that mimics the in vivo environment in which the cell/polymer constructs can develop into functional tissue. This in vitro environment is generally the result of a bioreactor, which provides growth factors and other nutrients while also providing mechanical stimuli to facilitate tissue growth. The first phase is the in vitro formation of a tissue construct, by placing the chosen cells and scaffold in a metabolically and mechanically supportive environment with growth media [374, 394]. The key processes occurring during the in vitro and in vivo phases of tissue formation and maturation are:

1. cell proliferation, sorting and differentiation;
2. extracellular matrix production and organization;
3. degradation of the scaffold; and
4. remodeling and potentially growth of the tissue [395].

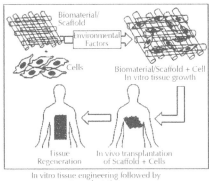 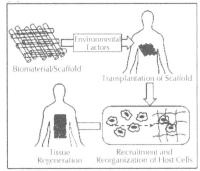

FIGURE 6.47   Tissue engineering steps.

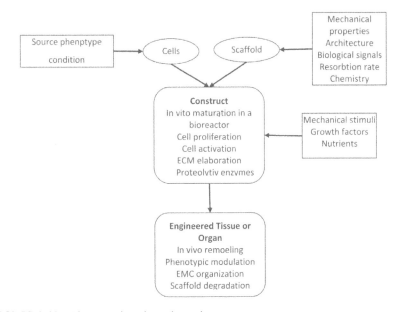

FIGURE 6.48   Tissue engineering schematic.

It is generally accepted that electrospinning has the potential to fabricate scaffolds as it results in a material with sufficient strength, nanostructure, biocompatibility, and economic attractiveness. Structures composed of the thin fibers generated by the electrospinning fall into this category as demonstrated by the widespread use of the process. Electrospinning produces nonwoven meshes containing fibers ranging in diameter from tens of microns to tens of nanometers [396, 397].

A scaffold design based on nanofibers can successfully mimic the structure and components of ECM component in the body and therefore properties of other native tissues. Specifically, the ECM consists of a cross-linked network of collagen and elastin fibrils (mechanical framework), interspersed with glycosaminoglycans (biochemical interactions). In spite of its remarkable diversity due to the presence of various biomacromolecules and their organization, a key feature of native ECM is the nanoscale dimension of its internal components [398].

6.16.6 EXTRACELLULAR MATRIX

6.16.6.1 COLLAGEN

Collagen types II, VI, IX, X, and XI are found in articular cartilage, although type II accounts for 90–95% of the collagen in the matrix. Type II collagen has a high amount of bound carbohydrate groups, allowing more interaction with water than some other types. Types IX and XI, along with type II, form fibrils that interweave to form a mesh. This organization provides tensile strength as well as physically entrapping other macromolecules. Although the exact function of types IX and XI are unknown, type IX has been observed to bind superficially to the fibers and extending into the interfiber space to interact with other type IX molecules, possibly acting to stabilize the mesh structure. Type X is found only near areas of the matrix that are calcified [399, 400].

6.16.6.2 PROTEOGLYCANS

Proteoglycans are composed of about 95% polysaccharide and about 5% protein. The protein core is associated with one or more varieties of glycosaminoglycan chains. GAG chains are unbranched polysaccharides made from disaccharides of an amino sugar and another sugar. At least one component of the disaccharide has a negatively charged sulfate or carboxylate group, so the GAGs tend to repel each other and other anions while attracting cations and facilitating interaction with water. Hyaluronic acid, chondroitin sulfate, keratan sulfate, dermatan sulfate and heparin sulfate are some of the GAGs generally found in articular cartilage [401].

There are both large aggregating monomers and smaller proteoglycans present in articular cartilage. The aggregating proteoglycans, or aggregans, are composed of monomers with keratan sulfate and chondroitin sulfate GAGs attached to the protein core. In most aggregan molecules, link proteins connect many (up to 300) of these monomers to a hyaluronic acid chain. Aggregans fill most of the interfibrillar space of the ECM and are thought to be responsible for much of the resilience and stress distribution in articular cartilage through their ability to attract water. There are no chemical bonds between the proteoglycans and collagen fibers; aggregation prevents diffusion of the proteoglycans out of the matrix during joint loading [402, 403].

The smaller proteoglycans include decorin, biglycan and fibromodulin. They have shorter protein cores and fewer GAG chains than their larger counterparts. Unlike aggregans, these molecules do not affect physical properties of the tissue, but are thought to play a role in cell function and organization of the collagen matrix [402].

6.16.6.3   NONCOLLAGENOUS PROTEINS

In contrast to proteoglycans, glycoproteins have only a small amount of oligosaccharide associated with the protein core. These polypeptides help to stabilize the ECM matrix and aid in chondrocyte-matrix interactions. Both anchoring CII and cartilage oligomeric protein anchor chondrocytes to the surrounding matrix. Other noncollagenous proteins commonly found in most tissues, such as, fibronectin and tenascin, are also observed in articular cartilage and are believed to perform similar functions as the glycoproteins [404].

6.16.6.4   TISSUE FLUID

Tissue fluid is an essential part of hyaline cartilage, comprising up to 80% of the wet weight of the tissue. In addition to water, the fluid contains gases, metabolites and a large amount of cations to balance the negatively charged GAG's in the ECM. It is the exchange of this fluid with the synovial fluid that provides nutrients and oxygen to the avascular cartilage. In addition, the entrapment of this fluid though interaction with ECM components provides the tissue with its ability to resist compression and return to normal shape after deformation [367].

6.16.7   SCAFFOLDS

It is evident that scaffolds mimicking the architecture of the extracellular should offer great advantages for tissue engineering. The extracellular matrix surrounds the cells in tissues and mechanically supports them, as discussed above. This matrix has a structure consisting of a three-dimensional fiber network, which is formed hierarchically by nanoscale multifilaments. An ideal scaffold should replicate the structure and function of the natural extracellular matrix as closely as possible, until the seeded cells have formed a new matrix.

The use of synthetic or natural nanofibers to build porous scaffolds, therefore, seems to be especially promising and electrospinning seems to be the fabrication method of choice for various reasons. Electrospinning first of all allows construction of nanofibers from a broad range of materials of synthetic and natural origin. The range of accessible diameters of the nanofibers is extremely broad covering the range from a few nm up to several micrometers. Secondly, the nonwovens composed of nanofibers and produced by electrospinning have a total porosity of up to 90%, which is highly favorable in view of the requirements defined above. By controlling the diameter of the nanofibers one is able to control directly the average pore sizes within the nonwovens.

When constructing the scaffold by electrospinning the material choice for nanofibers is important. One often chooses degradable polymers designed to degrade slowly in the body, disappearing as the cells begin to regenerate. The degradation rate must therefore match the regeneration rate of tissue in this case.

Biocompatible and biodegradable natural and synthetic polymers such as polyglycolides, polylactides, polycaprolactone, various copolymers, segmented polyurethanes, polyphosphazenes, collagens, gelatin, chitosans, silks, and alginates are used as the carrier materials. Mixtures of gelatin and chitosans or synthetic polymers like PCL and PEO (polyethylene oxide) are also employed, as are PCL modified by grafting, and copolymers coated or grafted with gelatin. The material choice for the applications depends upon the type of scaffold required, nature of the tissues to be regenerated and their regeneration time. The correct material helps in fulfilling the requirement of specific mechanical properties and degradation times for the particular application.

The highly porous nature of the scaffolds is apparent in all cases and the fiber diameter and thus the pore dimensions can be controlled over a sizable range in all cases, as discussed above. Functional compounds such as growth factors can be introduced in large quantities of up to 50% and more into the fibers if required, by adding these compounds to the spinning solution.

A variety of cells (e.g., mesenchymal stem cells, endothelial cells, neural stem cells, keratinocytes, muscle cells, fibroblasts, and osteoblasts) have been seeded onto carrier matrices for the generation of target tissues (such as skin tissue, bone, cartilage, arteries, and nerve tissue). The diameters of the fibers used generally conform to the structural properties of the extracellular matrix and are of the order of 100 nm. However, in some cases, fibers with diameters of less than 100 nm or of the order of 1 nm were used. The observation is that the number of cells located on the scaffold increase in time due to proliferation processes, if the nature of the scaffold is chosen appropriately.

A frequent requirement for the growth of cells in tissue engineering is that the cells are not oriented randomly within the scaffold but are oriented planar or even uniaxial. Tissue engineering of bones or muscles is an example.

In several studies, the proliferation behavior of cells in such fiber structures was compared with that on films cast from the same polymer material. The results showed that the fiber architecture generally affects cell growth positively. For endothelial cells, however, it was reported that a smoother surface can be beneficial for cell adhesion and proliferation. Another conclusion made was that the biocompatibility of a material improves with decreasing fiber diameter. Porosity also seems to have a favorable influence on cell growth. For instance, it was observed that mesenchymal stem cells form branches to the pores on porous nanofibers.

Another important requirement is that the scaffolds are porous enough to allow cells to grow in their depths, while being provided with the necessary nutrients and growth factors. The degree of porosity and the average pore dimensions are significant factors for cell proliferation and the formation of three-dimensional tissues. Depending on the cell type, the optimal pore diameters are 20–100 nm; pore diameters larger than 100 nm are in general not required for optimal cell growth. It was also found that cells can easily migrate to a depth of about 100 nm, but encounter problems at greater depths.

One solution to this problem involves a layer-by-layer tissue-generation procedure. In this approach, cells are uniformly assembled into multilayered three-dimensional (3D) structure with the assistance of electrospun nanofibers. This approach offers lot of flexibility in terms of varying cell seeding density and cell type for each cell layer, the composition for each nanofiber layer, precise control of fiber layer thickness, fiber diameter, and fiber orientation. A further answer to this problem consists in introducing the cells directly during the preparation step of the scaffold via electrospinning. The concept is to combine the spinning process of the fibers with an electrospraying process of the cells.

Combination of electrospinning of fibers and electrospraying of cells Scaffolds based on nanofiber nonwovens offer a lot of further advantages. One important prerequisite for a scaffold is a sufficient mechanical compatibility. Cartilage, for example, is characterized by a Young's modulus of about 130 MPa, a maximum deformation stress of about 20 MPa, and a maximum deformation of 20–120%; the corresponding values for skin tissue are 15–150 MPa, 5–30 MPa, and 35–115%. These ranges of values can be achieved with electrospun nanofibers. For instance, for scaffolds composed of electrospun collagen fibers with diameters of about 100 nm, a Young's modulus of 170 MPa and maximum deformation stress of 3.3 MPa were found. However, the maximum elongation is usually less than 10%. Another important finding is that the fibers can impart mechanical stress to the collective of growing cells. It was reported that the production of extracellular tissue is greater if oriented rather than unoriented matrix fibers are employed. This production can be significantly increased by the application of a periodical mechanical deformation (typically 6%).

Mimicking functional gradients (one of the important characteristic features of living tissue) that is gradients in composition, microstructure and porosity in scaffolds is also possible in a simple way by electrospinning. For example layer-by-layer electrospinning with composition gradients via controlled changes in the composition of the electrospinning solutions provide functional gradient scaffolds. The incorporation of bioactive agents in electrospun fibers will lead to advanced biofunctional tissue engineering scaffolds. The biofunctionalization can alter the efficiency of these fibers for regenerating biological functional tissues. The bioactive agents can be easily incorporated onto fibers just by mixing them in electrospinning solution or by covalent attachment.

Scaffolds fabricated from electrospun nanofibers have definitely several advantages. However, considerable room for optimization remains with respect to architecture, surface properties, porosity, mechanical and biomechanical properties and functional gradient, and also with respect to the seeding of cells in the three-dimensional space and the supply of nutrients to the cells. It is often observed that the cells preferentially grow on the surfaces or that they initially adhere to the carrier fibers, but then detach after differentiation. Toxicity of the organic solvents used for electrospinning is another issue for in vivo applications. The solution is to use either water-soluble polymers for electrospinning with subsequent crosslinking after scaffold formation or to make use of water-based polymeric dispersions. One final remark: early investigations on the co-growth of different types of cells on scaffolds are very promising. For instance, the co-growth of fibroblasts, keratinocytes, and endothelial cells was reported; the astonishing result is that co-growth enhances cell growth. So, a lot can be expected from scaffolds for tissue engineering based on electrospun nanofibers.

Ideal scaffolds probably approximate the structural morphology of the natural collagen found in the target organ. The ideal scaffold must satisfy a number of often conflicting demands: (1) appropriate levels and sizes of porosity allowing for cell migration; (2) sufficient surface area and a variety of surface chemistries that encourage cell adhesion, growth, migration, and differentiation; and (3) a degradation rate that closely matches the regeneration rate of the desired natural tissue [374]. While a broad range of tissue engineering matrices has been fabricated, a few types of synthetic scaffolding show special promise. Synthetic or natural materials can be used that eliminate concerns regarding unfavorable immune responses or disease transmission [396, 397].

Synthetic tissues help stimulate living tissues to repair themselves in various parts of the human body, such as, cartilage, blood vessels, bones and so forth, due to diseases or wear and tear. Victims whose skin are burned or scalded by fire or boiling water may also find an answer in synthetic tissues. The newest generation of synthetic implant materials, also called biomaterials, may even treat diseases such as Parkinson's, arthritis and osteoporosis. The uses of synthetic tissues are numerous. And there are several methods available to create them. One method is to make use of scaffold fabrication technology. Under this technology, synthetic tissues are cultured and placed on the scaffold that is shaped accordingly to, say a tendon or ligament, and then grafted onto the damaged part of the body. Once the new tissues grow over the damaged part of the organ or achieved sufficient structural integrity, the scaffolds would eventually degrade until only the tissues remain. It is also possible to use biocompatible materials that do not degrade, in which case, the scaffolds remain harmlessly in the body.

Synthetic polymers typically allow a greater ability to tailor mechanical properties and degradation rate. Clearly, the electrospinning process can eventually be developed to achieve successful utilization in vivo on a routine basis. Electrospinning offers the ability to fine-tune mechanical properties during the fabrication process, while also controlling the necessary biocompatibility and structure of the tissue engineered grafts [405]. The ability of the electrospinning technique to combine the advantages of synthetic and natural materials makes it particularly attractive, where a high mechanical durability, in terms of high burst strength and compliance (strain per unit load), is required. Advances in processing techniques, morphological characteristics and interesting, biologically relevant modifications are underway [406, 407].

However, there is one big drawback. This is the unstable dynamical behavior of the liquid jet which is formed during the electrospinning process. This instability inhibits the fibers to be aligned in a regular way, which is crucial to satisfy the scaffolds specifications. Many researches have been done to study this jet behavior. When the instability of the jet can be controlled, electrospinning can not only be adapted to produce high quality scaffolds for tissue engineering, but also for many other applications.

6.16.8 WOUND HEALING

An interesting application of electrospun nanofibers is the treatment of large wounds such as burns and abrasions [408–410]. It is found that these types of wounds heal particularly rapidly and without complications if they are covered by a thin web of nanofibers, in particular, of biodegradable polymers. Such nanowebs have suitable pore size to assure the exchange of liquids and gases with the environment, but have dimensions that prevent bacteria from entering. Mats of electro-spun nanofibers generally show very good adhesion to moist wounds. Furthermore, the large specific surface area of up to $100 \text{ m}^2 \text{ g}^{-1}$ is very favorable for the adsorption of liquids and the local release of drugs on the skin, making these materials suitable for application in hemostatic wound closure. Further, multifunctional bioactive nanofibrous wound healing dressings can be made available easily simply by blending with bioactive therapeutic agents (like antiseptics, antifungal, vasodilators, growth factors, etc.) or by coaxial electrospinning. Compared to conventional wound treatment, the advantage is also that scarring is prevented by the use of nanofibers.

The nanofibrillar structure of the nanoweb promotes skin growth, and if a suitable drug is integrated into the fibers, it can be released into the healing wound in a homogeneous and controlled manner. The charging of biodegradable nanofibers with antibiotics was realized with the drugs cefazolin and mefoxin. Generally, different drugs with antiseptic and antibiotic effects, as well as growth and clotting factors, are available for wound healing. Polyurethane is widely used as the nanoweb material because of its excellent barrier properties and oxygen permeability. Electrospun mats of PU nanofibers as wound dressings were successfully tested on pigs. Histological investigations showed that the rate of epithelialization during the healing of wounds treated with nanofiber mats is higher than that of the control group. Another promising and, in contrast to PU, biodegradable material is collagen. The wound healing properties of mats of electrospun fibers of type I collagen can be investigated on wounds in mice. It was found that especially in the early stages of the healing process better healing of the wounds was achieved with the nanofiber mats compared to conventional wound care. Blends of collagen or silk and PEO were also electrospun into fibers and used in wound dressings.

Numerous other biodegradable polymers that can be electrospun can be applied in wound healing, for example, PLA and block-copolymer derivatives, PCL, chitin, and chitosan. Using tetracycline hydrochloride as a model drug, it was shown that the release kinetics can be adjusted by varying the polymer used for the fabrication of the nanofibers. Poly[ethylene-co(vinyl acetate)], PLA, and a 50:50 mixture of the two polymers were investigated. With PEVA, faster drug release was observed than with PLA or the blend. With PLA, burst release occurred, and the release properties of the blend are intermediate between those of the pure polymers. The morphology of the fibers and their interaction with the drug are critical factors. The concentration of the drug in the fibers also affects the release kinetics. The higher the concentration, the more pronounced the burst, evidently because of an enrichment of the drug on the surface.

Handheld electrospinning devices have been developed for the direct application of nanofibers onto wounds. In such a device, a high voltage is generated with the voltage supplied by standard batteries. The device has a modular construction, so that different polymer carriers and drugs can be applied, depending on the type of wound, by exchanging containers within the spinning device.

6.16.9   TRANSPORT AND RELEASE OF DRUGS

Nanostructured systems for the release of drugs (or functional compounds in general) are of great interest for a broad range of applications in medicine, including among others tumor therapy, inhalation and pain therapy [410–412].

Nanoparticles (composed of lipids or biodegradable polymers, for example) have been extensively investigated with respect to the transport and release of drugs. Such nanostructured carriers must fulfill diverse functions. For example, they should protect the drugs from decomposition in the bloodstream, and they should allow the controlled release of the drug over a chosen time period at a release rate that is as constant as possible. They should also be able to permeate certain membranes (e.g., the blood/brain barrier), and they should ensure that the drug is only released in the targeted tissue. It may also be necessary for the drug release to be triggered by a stimulus (either external or internal) and to continue the release only as long as necessary for the treatment. A variety of methods have been used for the fabrication of such nanoparticles, including spraying and sonification, as well as self-organization and phase-separation processes. Such nanoparticles are primarily used for systemic treatment. Experiments are currently being carried on the targeting and enrichment of particular tissues (vector targeting) by giving the nanoparticles specific surface structures (e.g., sugar molecules on the surface).

A very promising approach is based on the use of anisometric nanostructures that is, of nanorods, nanotubes, and nanofibers for the transport and release of drugs. In the focus of such an approach will, in general, be a locoregional therapy rather than a systemic therapy. In a locoregional therapy the drug carriers are localized at the site where the drug is supposed to be applied. Such anisometric carriers can be fabricated by electrospinning with simultaneous incorporation of the drugs via the spinning solution. Another approach envisions the preparation of core-shell objects via coaxial electrospinning where the drug is incorporated in the core region of the fibers with the shell being composed of a polymer.

Nanofibers with incorporated super paramagnetic Fe_3O nanoparticles serve as an example the carrier should be possible with the application of an external magnetic field.

An interesting property of super paramagnetic systems is that they can be heated by periodically modulated magnetic fields. This feature allows drug release to be induced by an external stimulus.

A broad set of in vitro experiments on the release kinetics of functional molecules has been performed among others by fluorescence microscopy. The experiments often have demonstrated that the release occurs as a burst, that is, in a process that is definitely nonlinear with respect to time. It was, however, found that the release kinetics, including the linearity of the release over time and the release time period, can be influenced by the use of core-shell fibers, in which the core immobilizes the drugs and the shell controls their diffusion out of the fibers.

In addition to low molecular weight drugs, macromolecules such as proteins, enzymes, growth factors and DNA are also of interest for incorporation in transport and release systems. Several experimental studies on this topic have been carried out. The incorporation of plasmidic DNA into PLA -b -PEG -b- PLA block copolymers and its subsequent release was investigated, and it was shown that the released DNA was still fully functional. Bovine serum albumin (BSA) and lysozyme were also electrospun into polymer nanofibers, and their activities after release were analyzed, again yielding positive results. In the case of BSA, is was shown that the use of core-shell fibers fabricated by the chemical vapor deposition (CVD) of poly (p-xylylene) PPX onto electrospun nanofibers affords almost linear release over time.

Further investigations deal with the incorporation and release of growth factors for applications in tissue engineering. In the following, some specific applications of nanofibers in drug release are described in more detail.

6.16.10 APPLICATION IN TUMOR THERAPY

Nanofibers composed of biodegradable polymers were investigated with respect to their use in local chemotherapy via surgical implantation. A selection of approaches will be discussed in the following. The water-insoluble antitumor drug paclitaxel (as well as the antituberculosis drug rifampin) was electrospun into PLA nanofibers. In some cases, a cationic, anionic, or neutral surfactant was added, which influenced the degree of charging of the nanofibers. Analysis of the release kinetics in the presence of proteinase K revealed that the drug release is nearly ideally linear over time. The release is clearly a consequence of the degradation of the polymer by proteinase. Analogous release kinetics were found when the degree of charging was increased to 50%. Similar investigations were also carried out with the hydrophilic drug doxorubicin.

To obtain nanofibers with linear release kinetics for water-soluble drugs like doxorubicin, water-oil emulsions were electrospun, in which the drug was contained in the aqueous phase and a PLA – co-PGA copolymer (PGA: polyglycolic acid) in chloroform was contained in the oil phase. These electrospun fibers showed bimodal release behavior consisting of burst kinetics for drug release through diffusion from the fibers, followed by linear kinetics for drug release through enzymatic degradation of the fibers by the proteinase K. In many cases, this type of bimodal behavior may be desired. Furthermore, it was shown that the antitumor drug retained its activity after electrospinning and subsequent release. The drug taxol was also studied with respect to its release from nanofibers. These few examples show that nanofibers may in fact be used as drug carrier and release agents in tumor therapy.

6.16.11 INHALATION THERAPY

Finally, a unique application for anisometric drug carriers, inhalation therapy, will be discussed. The general goal is to administer various types of drugs via the lung. One key argument is that the surface of the lung is, in fact, very large, of the order of 150 m^2, so that this kind of administration should be very effective.

Indications for such treatments are tumors, metastases, pulmonary hypertension, and asthma. But these systems are also under consideration for the administration of insulin and other drugs through the lung.

Further advantages of anisometric over spherical particles as drug carriers for inhalation therapy are that a significantly larger percentage of anisometric particles remain in the lung after exhalation and that the placement of the drug carriers in the lung can be controlled very sensitively via the aerodynamic radius. To produce rod-shaped carriers with a given aerodynamic diameter, nanofibers were electrospun from appropriate carrier polymers that were subsequently cut to a given length either by mechanical means or by laser cutting.

Further progress in inhalation therapy will mainly depend first of all on finding biocompatible polymer systems that do not irritate the lung tissue and on the development dispensers able to dispense such rod-like particles.

6.17 AN INTRODUCTION TO CONDUCTIVITY BASED SENSORS FOR PROTECTION AND HEALTHCARE

Smart textiles are an emerging area in textiles. They allow monitoring on a permanent base without affecting the comfort of the person wearing them. They will generate a real breakthrough in the area of protection and healthcare. Indeed increase of risks can be detected in the earliest possible phase, allowing a fast and adequate reaction. Consequently, it will become an important tool in view of prevention.

However, many problems need to be solved before such smart systems will be actually on the market. At this moment, the materials are not always good enough in several aspects, current data processing techniques do not allow full extraction of the information, long-term behavior is poor. Extensive multidisciplinary research is required to solve all these aspects.

6.17.1 SMART TEXTILES FOR HEALTH CARE AND PROTECTION

As stated in previous chapter, the potential of smart textiles for health care is still largely unexploited.

A particular application area is public health. Researchers warn for world wide epidemia. In the past, particular types of flue have caused enormous casualties. With our society of huge mobility pandemic diseases will spread far quicker than ever before. Smart textile suits can play a role in remote monitoring, diagnosis and advanced protection.

For protection, very smart textile materials can play a role in many aspects. Also the textile can react when necessary, in a passive way or by active control mechanisms. Passive protection systems as are being used today usually have an important impact on comfort, esthetics and freedom to move. Just look at fire suits where the insulation level is so high that the firemen fade because of overheating caused by their own body heat, irrespective of the external fire. Or hip protectors for the elderly that make people look like M. Michelin. So in general smart clothes offer the possibility of adapting itself to the environment, allowing to provide protection only when required, for instance when temperature is too high, when harmful chemicals or microorganisms have been detected and so on.

The smart suit can detect increased risk and react on it in order to prevent accidents to take place. It can protect against hazards and assess the impact of accidents.

Consequently it can provide instant aid as well as long-term support to rehabilitation.

6.17.2 NETWORKS AND ORGANIZATIONS: STRUCTURE, FORM, AND ACTION

Let's look at a scenario for protecting against falling. The suit will help to avoid risky situations. The suit detects a person has an increased risk on falling. It sends out a warning in order to inform the person and his relatives. The suit can supply drugs should this be necessary. It communicates with the house in order to switch on the light when entering a room. It informs objects are lying on the ground.

Integrated artificial muscles help to maintain one's equilibrium. When detecting an actual fall, the suit instantaneously turns into an impact absorbing material.

After the fall, it assesses whether help is needed. It calls for help and sends out information on the situation. It treats wounds and provides a splint should this be necessary. It provides help to rehabilitation, for instance by stimulating the healing process or by keeping the body in shape during immobilization.

In case of fire men heat protection is required only a very small fraction of time. So the self adapting heat protection level considerably contributes to comfort during most of the operations. However, the real threat is the sudden stroke of heat. A smart suit can help to follow up adequately when the risk is rising and it's time to go. Such a suit is equipped with several sensors at different positions on the body and inside the suit. This allows adequate follow up of the status of the person but also of the suit.

As a result the intervention time can be prolonged without loss of safety. Drivers' attention can be monitored and actions can be taken before accidents happen. Here the main challenge is to identify relevant body information from which attention can be calculated in a quantitative way.

6.17.3 CONDUCTIVE FIBERS AND FIBROUS MATERIALS

6.17.3.1 CONDUCTIVE FIBERS

Polymer materials and fibers in particular do not conduct electrical currents. They are considered to be insulating materials. Metallic fibers on the contrary show good conductivity. Conductive polymers have been developed quite some time ago, but unfortunately their conductivity is low as compared to real conductors like cupper. Polyanilin, polythyophene and polypyrole are such polymers. The levels of conductivity are illustrated in Table 6.10.

TABLE 6.10   Volumetric Resistance of Various Conductive Fibers

Fiber	Volumetric resistance (ohm.cm)
Silver	1.63×10^{-3}
Copper	1.72×10^{-6}
Stainless steel	$72. \, 10^{-6}$
Carbon	From 2.2×10^{-4} till 10×10^{-3}
Polymer	10^{-2}–10^{-3}
Pani (panion ™)	10^{-3}
PA charged with nanoparticles	6.5×10^{-4}

Not only the conductivity of so called conductive polymeric fibers is limited, they also have poor mechanical properties and therefore they are usually applied on a textile substrate [1].

Some of the problems with current conductive fibers:
- conductivity of polymers is not so good, as well as long-term stability; they are slightly harmful;
- metal and metallized fibers are expensive; their mechanical properties are quite different from polymeric fibers
- some fibers have dark color (metallic, carbon);
- adding conductive particles may considerably affect on processing and/or fiber properties.

Electro-conductive fibers are used on a large scale for a variety of functions: antistatic applications, electromagnetic shielding, electronic applications, infrared absorption, protective clothing in explosive areas, etc. Their use as a sensor however is a rather new field of application.

6.17.3.2   CONDUCTIVE FIBROUS STRUCTURES

Arranging conductive fibers in a structure like textiles generates a material with a complex behavior in terms of conductivity. Fiber length being limited, the electron flow has to be transferred from one fiber to the other, from one yarn to the other. Contact resistance between fibers plays a determining role here. Contact resistance usually is quite high as compared to the intrinsic conductivity of the material.

Any rearrangement of the fibers in a textile may affect the global conductivity of the structure. It changes the contact resistance, number of contact points, path followed by the current, for some applications this is a source of error, for other it is the base of sensor properties.

6.17.3.3. NANOFIBERS ELECTRICAL CONDUCTIVITY

Electrical conductivity is an important property for sensor devices. Conducting nanofibers can be produced from semiconducting oxides, conducting polymers and nonconductive polymers. Pure oxide nanofibers are normally produced by electrospinning a solution containing oxide sol-gel and polymer, followed by calcining treatment to remove the polymer. The detection of gas molecules using oxide nanofibers is based on the conductivity changes due to the doping effect of analyst gases to the oxide. A few oxide nanofibers have been assessed for detecting different gases, such as, MoO_3 [413] nanofibers for ammonia, WO_3 nanofibers for ammonia [352] and NO_2 [414], TiO_2 nanofibers for NO_2 and H_2 [415]. These sensors exhibited improved sensitivity, faster response and lower detection limit than that of sol-gel based films Conducting polymer is another interesting sensor material. Electrospun polyaniline (PANi)/PS nanofibers containing glucose oxidase have been demonstrated to have a high sensitivity to glucose [416]. PANi/PVP nanofibers also exhibited sensing ability to NO_2. Organic/inorganic semiconductor Schottky nanodiode was fabricated by PANi nanofibers and inorganic n-doped semiconductor. The device has a rapid response and supersensitive to ammonia. Poly(3,4-ethylenedioxythiophene)-poly(styrene sulfonate) (PEDOT:PSS) has also been blended into PVP nanofibers for chemical vapor detection [417].

Besides the conducting and semiconducting materials, insulating polymers were also used to fabricate electrical sensors. In this case, ions or conductive nano-fillers were added to improve the conductivity. When PEO nanofibers were doped with $LiClO_4$, the mat showed low conductivity and was sensitive to moisture, and the nanofiber mat was reported to have much higher sensitivity than its film-type counterpart [416]. Carbon nanotubes/poly(vinylidene fluoride) (PVDF) composite nanofibers showed an increased straining sensing ability (as measured by voltage across the sensor), 35 times higher than that of the film counterpart [418]. In addition, electrospun nanofibers incorporated with carbon black showed sensitivity to VOCs and when the carbon black concentration was near the percolation threshold, the composite fibers changed their resistance in volatile organic compounds. Using different polymer matrices, the sensor can be used to detect toluene, trichloroethylene, methanol, and dichloropentane vapors[419].

Humidity sensor is a very important device for environment tests. KCl doped ZnO [420] and TiO_2 [421] nanofibers have shown higher humidity sensitivity with faster response and recovery time compared to pure ZnO and TiO_2 nanofibers. Ba-TiO_3 nanofibers also exhibited excellent humidity sensing behavior because its complex impedance varied around three orders [421] of magnitude in the whole humidity range [422].

6.17.4 FIBROUS SENSORS

Sensors have been widely used to detect chemicals for environment protection, industrial process control, medical diagnosis, safety, security and defense applications. A good sensor should have a small dimension, low fabrication cost and multiple functions, besides the high sensitivity, selectivity and reliability [423]. High sensitivity and fast response require the sensor device having a large specific surface area and highly porous structure. Several approaches have been used to impart nanofibers with a sensing capability, such as, using a polymeric sensing material to electrospin nanofibers, incorporating sensing molecules into nanofibers, or applying sensing material on nanofiber surface via coating/grafting technique.

6.17.5 ELECTROSPINNING AND SENSORS

Development of electrospun nanomaterials, such as, nanofibers and nanowebs, provided researchers with an opportunity to construct electronic interfaces with components whose sizes are comparable to the size of molecules, potentially leading to a much more efficient interface. Nanometer cross-sections of nanomaterials gives them enhanced surface sensitivity and allows them to use the benefits of size effects, such as, quantization and single molecule sensitivity. The comparatively large surface area and high porosity make electrospun nanomaterials highly attractive candidates for use in a range of devices, including ultrasensitive sensors. This is one of the most desirable properties for improving the sensitivity of sensors because a larger surface area will absorb more analytes and change the sensor's signal more significantly.

6.17.6 TECHNIQUES

Electrospun PAA nanofibers have been grafted with pyrene methanol as the sensing material to detect metal ions Fe^{3+} and Hg^{2+}, and an explosive 2,4-dinitrotuloene in water [424]. Due to the quenching effect of these chemicals to the pyrene moieties, the fluorescent intensity of nanofibers had a linear response to the concentration of quenchers, and the nanofibers showed high sensitivities. Similarly, a PM-grafted poly(methyl methacrylate) PMMA nanofibers showed an order of magnitude higher sensitivity to target analyte DNT than its cast film counterpart [425]. Explosive 2,4,6-trinitrotoluene vapor can be accurately detected using electrospun PAN nanofibers that were coated with a thin layer of conjugated polymer, poly TPA-PBPV [426].

Fluorescence optical sensors were also prepared by a layer-by-layer electrostatic assembly technique to apply a conjugated polymer onto nanofiber surface for detection of methyl viologen and cytochrome c in aqueous solution [427] and porphyrin-doped silica nanofibers were used to trace TNT vapor [428]. All those nanofiber sensors showed high sensitivity and rapid response. Besides fluorescent properties, conjugated polymer embedded electrospun nanofibers were also reported to be able to sense VOCs based on optical absorption properties [429].

A gas sensor using a specific absorption interaction between ammonia and poly (acrylic acid) nanofibers was reported. The weight difference induced by the gas absorption was measured by a QCM. This sensor was capable of detecting ppb level NH_3 in air, and the sensitivity was four times higher than that of the PAA cast film [430]. The absorption of gas also leads to changes in FTIR absorption. PAN nanofibers containing metal oxide nanoparticles, such as, iron oxide and zinc oxide, have been used to detect carbon dioxide [431]. The addition of metal oxide nanoparticles enhanced the gas adsorption and thus improved the sensitivity. In a electrospun carbon nanofiber gas sensor [432], carbon black was blended into the fibers to increase the conductivity, and a porous structure was introduced to the fiber surface by etching under a basic condition. The sensory ability for NO and CO gases was reported to be improved dramatically, and the enhanced sensitivity was attributed to higher surface area and improved electrical conductivity due to the formation of carbon black network. It has been found that electrospun SnO_2 fibers with a smaller grain size showed much better sensitivity to CO and NO gases. This was probably because the fibers containing smaller grains had higher resistance, which was the result of larger number of grain-grain interconnections [159].

6.17.7 FIBROUS TEXTILE STRUCTURES

Conductive fibers are being used as passive sensors for monitoring biopotential, mainly heart rate. Several research projects have been carried out on this topic. The feasibility has clearly been demonstrated, although the sensor needs to be optimized and practical problems need to be solved (Table 6.11).

Table 6.11. Textile Electrodes for Measuring Heart Rate

Name	Application	Level of transformation
Smartex	Health care	Woven /knitted textile sensors
Intellitex	Children's health care	Knitted textile structure, textile antenna
VTAM	Health care	Partly textile structures
Wearable motherboard	Health care, military	Partly textile structures

As explained in paragraph 1.2 and 2.1 several mechanisms cause the resistance to go up or down due to extension of the material. The global effect of these combined mechanisms depends on the type of material and its structures. The conductivity of the textile materials for applications as passive sensor should be as consistent as possible, so their piezo-resistive effect is a source of error. This can be achieved by careful selection of fiber type and proper design of the textile structure. The resistance of such a structure is constant when for instance a cyclic extension is applied as a simulation of the breathing movement (Fig. 6.49).

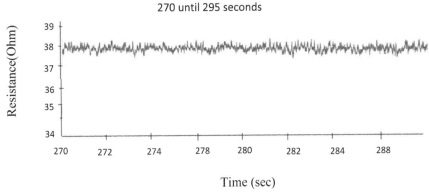

FIGURE 6.49   Textile structure without piezo-resistive effect.

On the contrary the same piezo-resistive effect makes the textile a versatile tool for a broad range of sensor applications where extension is a crucial parameter. This is the case for instance for respiration measurements (expansion/contraction of the chest), all kinds of movements (dance, sports, …) as well as volumetric changes like volume of inhaled air.

6.17.8 SMART TEXTILE STRUCTURES

The applications mentioned in the previous section are rather straightforward. Careful design of the textile structure enables more advanced sensing properties. The basic mechanisms are related to conductivity, changes in conductivity, currents or change in currents and so on.

Any mechanism that affects such parameters is useful. Electrochemistry is an extremely important discipline in this respect.

A set of fibers, yarns or fabrics separated one way or another can be considered as a double electrode system. Such a system can be used to detect water. The presence of water will be reflected in an increase of conductivity between the two electrodes. The increase will be bigger when the water contains salt. The reaction of such a textile sensor (i.e. resistance as a function of time), consisting of two conductive yarns, on wetting with water with different salt concentrations.

Impedance spectroscopy has been used to optimize the test set up. A double set of such a 2-electrode system of which one is coated with a coating that is impermeable to the salt but permeable to water allows separation of the quantity of water and the quantity of salt. Coatings with selective permeability can be the base for a huge number of specific sensors, for instance for a qualitative as well as quantitative analysis of sweat. This basic method is suited for a huge range of applications, provided the right electrode configuration, measuring conditions and textile configuration are selected. Electrode configuration for instance includes diameter of the fibers or yarn electrodes and distance between the electrodes.

Another approach to design conductive fiber based sensors is based on the piezoresistive effect, whereby the separation of conductive (nano) particles is not achieved by fiber extension, but by fiber swelling. In this case as well one basic technology is capable of generating an enormous range of sensing capabilities. Selection of adequate polymeric materials for the fibers or inclusion of swelling components like gels must be adapted to the triggering agent. In addition coatings with selective permeability can be applied to increase selectivity and specificity of the sensor system.

These are just two examples of relatively simple systems with an enormous range of applicability.

6.17.9 STRAIN SENSORS

The main advantage of smart textiles, sensors or actuators, is that textile materials in general are common products that are comfortable materials that are easy to use. Thanks to these properties it becomes possible to wear the sensors and actuator in an imperceptible way.

Of course the smart character of the textile should not affect these advantages. Experience shows that two problems arise. The first is the flexibility that on the one hand it is necessary for achieving a good level of comfort, but on the other hand enables multiple deformation of the material. The other are chemical effects. Laundry for instance combines multiple deformation and chemical effects.

6.17.10 PHYSICAL EFFECTS

Conductive fibers often have mechanical properties that are quite different from those of "regular" textile fibers. This causes them to react differently to deformation, bending, extension. As a result a slow but consistent migration of those fibers occurs. This eventually leads to separation of both components and this effect may become clearly visible after long-term use as for instance breathing sensor (Fig. 6.50).

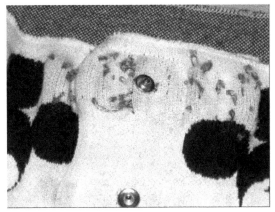

FIGURE 6.50   External loops formed by stainless steel yarns due to repeated extension.

This effect is obviously not welcome because of several reasons:
- it negatively affects the esthetic aspect of the fabric;
- it may affect the sensor function;
- contacts may occur with the skin or the environment, leading to false signals, increased noise, etc.;
- fabric feel may be affected.

Rearrangement of the fibers happens mostly in the initial phase of use. As a result resistance of a fabric will experience its fastest changes at the beginning of deformation tests later on it will stabilize more or less (Fig. 6.51).

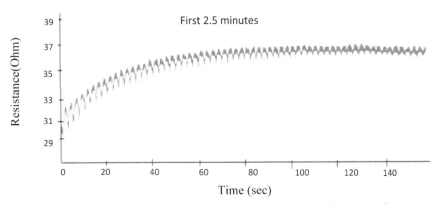

FIGURE 6.51   Change of resistance of a conductive yarn during initial phase of use.

6.17.11   FIBER BREAKAGE

Apart from the quite obvious macroscopic effect described in previous paragraph more complex phenomena influence the sensor function of textile sensors. Stainless steel fibers for instance are rather brittle, so repeated extension and bending will cause them to break. Consequently, the number of fiber to fiber switches will increase with each fiber breakage and contact resistance being the biggest resistance by far, overall resistance of the textile structure will drastically increase. Particularly during laundry deformation is quite intensive, and laundry is of course a very relevant operation so it is a good way to test on impact of defeormation. Measuring changes in length of fibers in an actual textile structure is very difficult, because the fibers are embedded in the textile structure and its unraveling may cause more fibers to break. Fibers may also be crimped considerably so the length measurement in itself gets difficult. An indirect method to evaluate fiber length is yarn strength as this relationship has been demonstrated in numerous studies (Fig. 6.52).

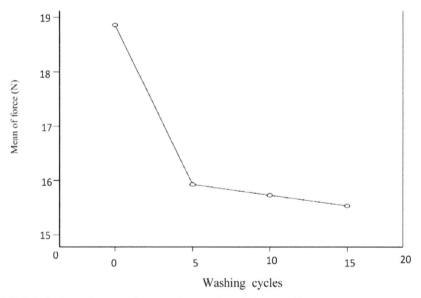

FIGURE 6.52   Influence of repeated extension during washing on yarn strength as a measure of fiber breakage.

Mechanical damage of fibers has also been reported by Tao. This work describes the appearance of cracks at the surface of PANi and PPY coated fibers at extensions from 6% onwards. It is quite clear that all factors that affect the conductivity of the material, also affect its proper functioning in the intelligent textile (Fig. 6.53).

Mechanical damage due to multiple deformations in general is an important problem for all kinds of conductive textile materials. Also interconnections between different components (sensors, actuators, electronics, battery, wires) have been reported in many studies as weak spots, in particular at places where soft (textile) and hard (electronics) elements are connected.

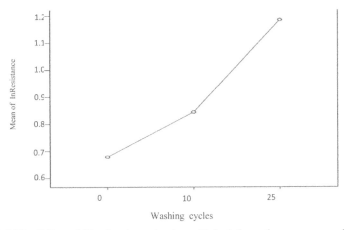

FIGURE 6.53   Effect of fiber breakage due to multiple deformations on yarn resistance.

6.17.12   RESULTING LONG TERM BEHAVIOR OF TEXTILE STRAIN SENSORS

As explained before several factors may affect the proper sensor function of textile strain sensors. To test this a cyclic loading was applied while measuring the resistance of the textile structure. This resistance should go up and down with extension and the amplitude should be high enough for an accurate sensor (Fig. 6.54).

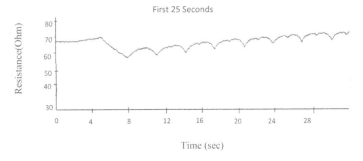

FIGURE 6.54   Variation of resistance due to cyclic loading of the textile strain sensor.

For many textile structures this amplitude slowly goes down turning the textile material into an unreliable sensor (Fig. 6.55).

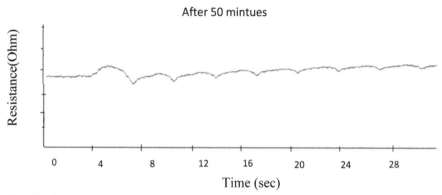

FIGURE 6.55   Loss of sensor capacity due to multiple deformation.

Surprisingly the amplitude temporarily increases after washing. This is probably due to a sort time rearrangement of the fibers after washing following the considerable fiber rearrangement during washing (Fig. 6.56).

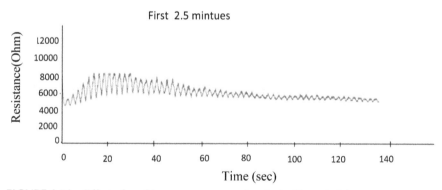

FIGURE 6.56   Effect of washing on sensor capacity of a textile material.

So the sensor sensitivity of some textile structures actually improves due to washing. This can be represented by expressing the amplitude relative to the value of the actual resistance, for instance at the maximum extension. It can be concluded that textile structures behave in a very complex way as sensor. Depending on the actual fiber type and the fabric structure a wide range of responses can be found.

6.18   TEXTILE ELECTRODES

Textile sensors to be used for medical purposes are usually in contact with the skin. This is particularly the case for electrodes used for monitoring heart signals (cardiogram). In order to test the performance of textile structures for this application an actual cardiogram can be recorded; one of the main problems here is the extreme variability of the skin properties. Even for one person skin conductivity changes from moment to moment making objective testing very difficult. Therefore a phantom test set up has been developed. In this method an electrolyte is used to simulate body fluids, separated from the textile electrodes by polymeric membranes mimicking the skin (Fig. 6.57).

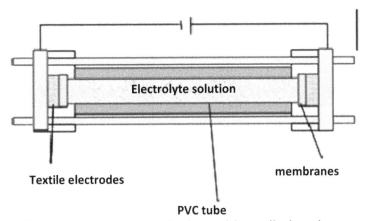

FIGURE 6.57   Impedance spectroscopy for characterizing textile electrodes.

Signal transfer is analyzed using impedance spectroscopy. This method allows to separate the impedance of the separate components of the system. By varying different parameters, their influence on resistivity can be studied quite easily in an accurate and reproducible way. Obviously, heart rate sensors are in permanent contact with the skin. This means that they will be wetted by sweat. Although stainless steel fibers resist corrosion due to for instance NaCl, at the skin a voltage also occurs, causing electrochemical attack. As a consequence the electric resistance increases in time when the material is in contact with artificial sweat.

Smart textile structures are here to stay. They have demonstrated their feasibility both from the point of view of technical specifications as well as regarding their textile character The enormous versatility of textiles in terms of (combinations of) fiber types to be used, processing technologies and textile structure is at the same time an opportunity to be exploited but also a confusing space of possibilities. Different textile materials may show different, even opposite behavior. It is a huge challenge to find the right set of materials for each particular application.

Properties that are beneficial for one application may be disruptive for another one. Technical features may be in contrast to textile characteristics so a balance may have to be looked for. Objective testing is another field of research. No evaluation is possible without an accurate and reliable test method. But the result will be worthwhile, as it will definitely lead to a better quality of our lives.

6.19 INDUSTRIAL APPLICATIONS

Studies have shown that a low fiber diameter allows a filter element with similar operational characteristics but with much higher filtration performance. Beyond this there is also the potential for the incorporation of biologically or chemically active elements either on the surface of the fiber, or intrinsic to the structure of the fiber allowing for active filtration to occur. Active filtration implies that the entrapment method is based on chemical attraction rather than simple physical entanglement. The advantages of this method are a lower resistance to flow across the filter element, and the possibility of selectivity so that particular elements can be removed during filtration.

Using a high specific surface area as a site for chemical reactions shows clear potential as a delivery technology for catalysts. Due to the high surface area to volume ratio there would be minimal waste of catalysts that operate by surface reaction kinetics rather than playing an active role in the chemical reaction. It has been shown by Viswanathamurthi and co-workers [433], and Ding and co-workers that it is possible to include metal salts in the electrospinning solution and perform a thermal post-treatment that removes the polymer carrier and leaves a metal oxide. For more complex metal catalysts the work by He and Gong shows that multiple metal salts can be incorporated, with the final fiber having a mixed composition [434].

Another important industrial application being explored is the use of electrospun fiber as a substrate for the biological element in a bioreactor. Again due to the large surface area to volume ratio and the general strength of these small fibers it is possible to create a substrate that allows a high density of biologically active material such as cells to be packed into a small space while still permitting the flow of nutrient. The higher the density of the biologically active material, the better the efficiency per volume that the bioreactor should be able to achieve. A further advantage for the electrospinning process is the ability to form a surface texture from small pores that could act as anchoring points for cells in the bioreactor [435].

6.19.1 NANOFIBER-REINFORCED COMPOSITES IN BIODEGRADABLE APPLICATION

Natural electrospun nanofibers have also been investigated in nanofiber-reinforced composite. For example, Han and Park [436] explored the potentialities of ultra-fine cellulose electrospun nanofibers (a. v. 560 nm) as reinforcing agent of poly butylene succinate (PBS) for biodegradable composite. Tang and Liu used electrospun cellulose nanofibers (a. v. 500 nm) to reinforce poly vinyl alcohol (PVA) films for biodegradable composite and evaluated composite mechanical properties and visible light transmittance in relation to the fiber content [437].

6.19.2 HOW TO IMPROVE THE INTERFACE BONDING WITHIN THE ELECTROSPUN-REINFORCED COMPOSITES

Although many papers on nanofiber reinforcement have been reported, the interface bonding between the electrospun nanofibers and different polymer matrix is generally poor, resulting in the low nanofiber-reinforcing effect. To solve this problem, Özden and Papila [438] induced cross-linking agent into the electrospun nanofiber reinforced composites. In their study, ethylenediamine (EDA) acted as the supplementary cross-linking agent was sprayed onto the copolymer polystyrene-co-glycidyl methacrylate [P (St-co-GMA)] electrospun nanofiber mat prior to embed into an epoxy resin. They found that the storage modulus of the P(St-co-GMA)/EDA nanofiber-reinforced epoxy resin exhibited about 10 and 2.5 times higher than that of neat and P(St-co-GMA) nanofiber-reinforced epoxy, respectively, even though the weight fraction of the nanofibers was as low as 2 wt. %. They suggested that the superior mechanical properties were attributed to the inherently cross-linked fiber structures and the surface (Fig. 6.58).

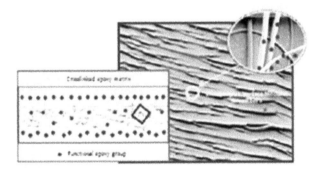

FIGURE 6.58   Schematic illustration of the P(St-co-GMA)/EDA nanofiber-reinforced epoxy resin.

6.19.3 TEXTILE APPLICATIONS

Nanofiber-based nonwovens can be used to strongly modify the properties of conventional textiles composed of much thicker textile fibers designed for clothing applications but also for furniture, various applications in hospitals, technical applications, etc. This can be achieved very effectively, among other approaches, by depositing thin layers of such nonwovens on the textiles. The aim may be among others the protection of the human body against the effect of strong winds, cold temperatures or bacteria. Another aspect may be self-cleaning of the textile directed towards food or drinks such as ketchup or red wine [439, 440].

One interesting aspect along this line is that a thin layer of nanofibers deposited on the surface of a textile is able to increase the wind resistance strongly. The finding is that the wind resistivity of a textile can be increased by five orders of magnitude as the pore diameter goes down by from 100 micrometers to 1 micrometer. This can be achieved by replacing conventional fibers with a diameter of about 10 micrometers by electrospun nanofibers with diameters of the order of 100 nm. To achieve such a high wind resistivity only a small coating level of the order of about 1 g/m² is required. Furthermore, the wind resistance is not influenced to any significant degree for the electrospun nonwoven by the relative humidity of the environment, quite similar to other fabrics, in particular poly(tetrafluoroethylene) microporous membranes.

A positive aspect of the modification of textiles via the deposition of thin layers of nanofiber nonwovens is that the vapor permeability is not negatively influenced.

The electrospun membrane has obviously about the same low water vapor diffusion resistance as PTFE microporous membranes and much lower values as Gore-Tex laminates.

Another property of textiles that can be strongly influenced by depositing nanofiber nonwovens as coating is the thermal isolation. The thermal conductivity characteristic for porous media such as conventional textiles is primarily controlled in air by the diffusion of the gas molecules within the pores. This diffusion in-turn is controlled by the frequent collisions of the gas molecules with each other.

A mean free path length results the order of about 70 nm for air at normal pressure. The thermal conductivity is given for ideal gases by:

$$K = c\lambda\, c_v/3N_A \tag{6}$$

The Avogadro constant causes the thermal conductivity to be of the order of 0.025 W/m K.

Now, when applying a coating composed of a nanofiber-based nonwoven or when considering a textile made purely from nanofiber nonwovens the diffusion of the gas molecules becomes modified as the average size of the pores gets smaller than the length of the mean free path of the gas molecules. This situation is again characterized by the Knudsen number introduced above. The diffusion becomes strongly limited in this case due to the dominance of particle-pore-wall collisions-particle nanofiber collisions and the thermal conductivity is greatly reduced for electrospun nonwovens.

Coating of textiles by antibacterial nanofibers is of significant interest due to the large surface area of electrospun nanofiber nonwovens that in turn cause higher antibacterial efficiency. Here, antibacterial compounds like classical low molecular weight antibacterial agents, silver nanoparticles, or oligomer/polymer ammonium compounds can be incorporated into the electrospinning formulations yielding nanofiber nonwovens that can be coated onto textile surfaces.

However, the application of antibacterial electrospun nonwovens is a true challenge for real textile applications as permanence, adhesion on the textile of the nonwovens, wear resistance, and leaching of the antibacterial compounds are difficult to manage. The combination with other properties of electrospun nonwovens, such as, self-cleaning by super hydrophobic electrospun nonwovens or photocatalytically active electrospun nonwovens containing, for example, TiO_2 may be an interesting approach. Self-cleaning textile fibers were obtained via coaxial electrospinning of cellulose acetate and dispersions of nanocrystalline TiO_2 [363]. Discoloration of model dyes by illumination proved to be feasible.

However, the technical application of coaxial electrospinning in a stable technical electrospinning process still has to be proven.

6.19.4   TEXTILES USING ELECTRONIC APPLICATIONS

Smart textiles, from fibers to fabrics, with integrated special electronics are nowadays used to develop smart clothing. In this paper, some examples for future designs and development are presented, as well as the safety vest that was presented at the LOPE-C conference in Frankfurt with integrated photovoltaic cells and LED lights.

Smart textiles are, by definition, textiles which respond to the changes in the environment as a result of mechanical, thermal, chemical or electromagnetic influences. Interactive textiles represent textiles that have built-in into their structure the elements to control (sensors, switches, communication components, batteries). Most commonly, these elements control the health care functions (pulse, temperature, blood sugar, etc.), enable communication or represent the security and entertainment systems, as well as they allow the power supply thereof.

The development of textiles with electronic components can be subdivided into:
- Simple systems: electronic components are incorporated into pockets sewn-in or attached over soft cables and should be removed before cleaning, for example, LED lighting devices;
- Hybrid systems: the elements are a part of permanent fabrics, woven or embroidered, using conductive yarns;
- Complete integration: the elements are integrated, using the fibers with special properties that act as electronic textiles (sensors, etc.).
- The functions [441, 442] of textiles using electronic applications are:
- Passive functions: as a result of material properties, they can sense the environment (sensors);

- Active functions: as a result of installed sensors, they can act to the environment-actuators, and can work to supply energy (work actively to changes in temperature hot/cold), for protection (inflatable elements for protection against impact, missiles), protection from water-floating clothing, increased active visibility, communication – a cry for help, protection from hazardous substances, chemicals, gas, alarm and protection against radiation, measurement of vital signals (pulse, temperature, etc.), integrated antennas for communication and embedded components for the photovoltaic generation of electricity for the operation-independent of batteries.

Embedded components that measure the heart function and enable monitoring of the user throughout the day, the data being transferred to a medical institution, are already used for medical purposes. After the surgery, patients wear clothing with integrated elements of control, thus allowing the movement and they avoid the pain for the installation of measurement probes for their medical condition control.

For communication[443], the developed elements can be washed and are a part of the clothing, where they operate as touch screens (touch-pad), are flexible, lightweight, durable and allow interactivity.

Smart textiles, which include safety clothing, are materials that allow the installation of a variety of technologies, for example, various electronic components, in clothing. Such fabrics permit the perception of the environment and thus the adaptation to different conditions.

The main functions enabled in smart textiles are integrated sensors[444] that measure vital functions (medical textiles), enable communication, processing and storage of data, acquisition and transmission of energy (including PCM materials-Phase Change Materials) [445].

Electronic components can be fitted directly into textile fibers, for example, conductive materials and conductive textile fibers, diodes, transistors and photovoltaic fibers. Current electronic components prepared on a silicon base are not flexible whereas the new elements developed on the base of organic polymers are.

The fibers that are made from materials which convert light into electricity or electricity is the result of the fiber movement enable the production of electricity. The energy is stored in batteries and when necessary, it powers the built-in OLED lighting. At MIT (Massachusetts Institute of Technology), dyes have been developed for the print of solar cells [446] on different materials, including textiles.

6.19.5 USE OF SMART TEXTILES

Smart textile products have been used in various fields, as smart textiles SFIT (Smart Fabrics and Interactive Textiles), as wearable technology (wearable tech), and as interactive textiles. The areas of application range from the clothing for personal protection, for example, work clothes for special environments, for the protection of the health of workers and for the protection in extreme sports in a variety of environments (hot, cold, wet, dry, etc.). Smart textiles are intended for everyday use, heated/cooled clothing, for entertainment (clubs, concerts, public events) and special effects with fashionable elements such as built-in lighting, changing colors, as well as for communication.

Many of these technologies are being used or planned for use for the elderly who need active assistance or protection in everyday life (control and communication-garments with wearable physiological sensors) in order to reduce the costs of care and treatment. The research is conducted in the areas of direct installation in the clothing, enabling easy maintenance (washing) and increased generation of energy or lower consumption to operate.

Energy can be produced on the basis of photovoltaic cells that can be a part of the garment – as a fashion accessory or using piezo crystals, based on the move-ment of MEMS (Micro-Electronic Mechanical Systems). It also works in the field of storage and use of heat energy that is released in the movement-walking user. For all these cases of energy generation, it is necessary to develop more storage options to be able to use printed batteries that are smaller and can be incorporated into products.

All of the above results in increased consumption and increased amount of post-consumer textile waste. Development can take place in the direction of one time use – large quantities of waste, or multiple uses – problems of maintenance, but less waste.

The ecology [447] of smart textiles is still at the beginning, which means that there is still a lot of work to be done and the need for deciding whether to use better and more expensive organic materials or recycled towards the development of smart textiles. Of course, these problems are also the opportunity for smaller companies which could focus on the development and manufacture of tailor-made ecosmart clothes. This could lead to the development of products with high added value, with added knowledge and the use of novel developed materials.

In today's world, when electronics become a part of clothing, with embedded microcontrollers and variety of sensors (e.g., temperature or light) and LED light-ing, endless combinations for use in both protective and decorative purposes are allowed. Microcontrollers that can be washed, which is their major advantage, allow various connections between the LED elements that can be programmed for any ap-plication, which is extremely important for the use in textiles.

On the market, there are different microcontrollers (e.g., ATMEL ATmega, AT-tiny etc.) which can be used in the products of wearable electronics, for example, LilyPad Arduino. By using these microcontrollers, different applications can be de-veloped, with custom made circuits with arbitrary shape and at reduced price.

In our research, the design of printed circuit boards and programs for different behaviors of LED lighting (gradual or simultaneous switched LED lighting) was performed. The final product represents a warning-decorative LED light arrow that lights up differently.

The research showed that the knowledge in the field of textiles, chemistry, electronics and programming contributes to the manufacture of high quality applications that in addition to textile components also includes the elements of electronics.

6.19.6 EXAMPLES OF SMART TEXTILES

6.19.6.1 PLED DRESS

Clothes are changing every day, not only on the basis of fashion trends, but also to follow the research in the field of technology, new materials and innovations from other fields.

Predicting the future has never been easy, people have predicted flying cars, peace, a disease free world etc., which has not (yet) happened, while nobody foresaw the use of mobile phones, 3D printed food and invisible clothes.

At fashion events, we can see clothing equipped with the LED technology and micromotors that change dimensions and act as a light show. Chalayan presented high-tech dresses that transform on the body and translated them into wearable, commercially viable pieces. The models, when walking, activated the application at the collar and the fabric unraveled to reveal an entirely different look (Figs. 6.59 and 6.60).

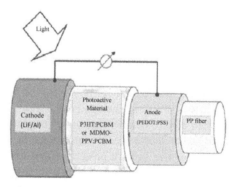

FIGURE 6.59 Schematic drawing of photovoltaic fiber.

FIGURE 6.60 Photovoltaic fibers and single fiber.

For the production of such garments, fabrics with special properties are necessary, for example, photovoltaic fibers [448], which act as photovoltaic cells for generating electrical energy. Figure 59 shows a scheme of such fibers.

Fibers can under the influence of light, wind or rain generate electric potential energy and act as a hybrid photovoltaic-piezoelectric device. Hybrid films are constructed by depositing an organic photovoltaic cell on a commercial PVDF film, while hybrid fibers are developed by depositing an organic solar cell on a piezoelectric polymer fiber. When the hybrid film/fiber is subjected to mechanical vibrations from the wind, rain or tide, the piezoelectric part produces an electrical voltage that is converted to a constant DC voltage by a rectifier.

The photovoltaic part of the hybrid film produces constant DC voltage from the solar energy. Electrical energy can then either be used online or stored in a battery. These materials are now available as an energy harvesting device for the use in various e-textile [449] applications.

Materials can be fitted on textiles as added materials on the surface or integrated as a part of the material itself, for example, photovoltaic fiber [450]. An interesting example of this technology is the printed dress which acts as a screen "Printing Dress." The dress integrates different technologies. It consists of three main parts of the upper corset and a skirt. The corset has fitted four elements LilyPad Arduino, a USB port for connecting to a laptop computer, keyboard, a built-in corset and wires to connect. The skirt is made up of a material with incorporated aluminum wires and hangs over the projector which projects images directly onto the skirt. Each time the user presses a key, it communicates with the processor to display the animation typewritten text on the skirt.

The dress is presented as a prototype of the concept and application of printed electronics in the clothing purposes. The experience will contribute to the creation and development of smart clothing. Nowadays, the dress can be used as a tool to communicate or tweet [451].

6.19.6.2   SAFETY VEST

In the more and more changing climate on the planet, we have to be able to work and travel in the most challenging situations. The weather pattern can change very fast from hot to cold, from draft to floods, so there is a need for the clothing to protect us. The advances in new textiles and printed electronics on flexible substrates can together offer some interesting possibilities.

The OE-A association (Organic Electronic Association, Germany), organizer of LOPE-C, presented an international competition for student projects in order to present the possibilities of using printed electronics, they offered a set of elements that had to be assembled in working demonstrator. Safety clothing – the safety vest presented in Fig. 6.61, with printed solar cells on the back and LED lights was developed at our department. The generated energy was stored in the built-in battery and when necessary, the LED lights that are on the back can be used. The basic concept (cf. Fig. 6.62) is to use polymer solar cells to generate the power that is stored in the built-in batteries and used for better visibility on the roads or in nature. The polymer solar cells (Konarka), batteries and LED (Light Emitting Diode) lights are linked by the special built-in software, the switch, battery and operation button being in the pocket. In the sunlight, solar cells convert light into electricity, the energy is stored in the batteries and when it gets dark, the LED lights, which are integrated on the back, give light.

Previous applications based on reflection, in our case the LED, give light in the dark. LED lights are semiconductor diodes that emit light under the influence of electricity. Photovoltaic cells are the elements that unlike LED lights emit electricity when under light.

Unlike past examples of safety clothing that acted on the principle of reflective elements embedded in clothing, our safety vest in Fig. 6.5 has built-in LED lights that are powered by photovoltaic cells. Figure 6.63 presents the details of the links which are located in the inner side. The basic elements consist of polymer photovoltaic solar cells, LED lighting and batteries. The safety vest represents the beginning of the research in the field of clothing and the added value represented by the elements of printed electronics. Photovoltaic polymer cells printed on a flexible substrate are suitable for the use on textile substrates (Fig. 6.64).

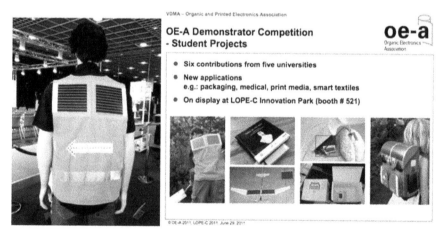

FIGURE 6.61   Safety vest on model of conference LOPE-C 2011.

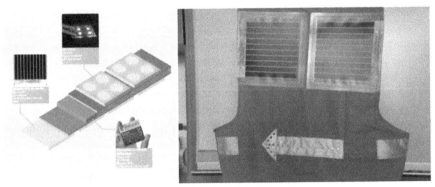

FIGURE 6.62 Basic concept and developed model.

FIGURE 6.63 Integration of polymer photovoltaic solar cells, LED lighting and batteries.

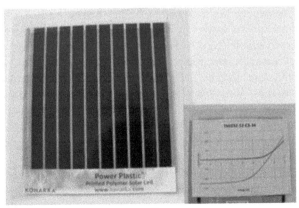

FIGURE 6.64 Power plastic photovoltaic cells.

6.19.6.3   DECORATIVE ARROW WITH CUSTOM MADE PRINTED CIRCUIT BOARD

In addition to the development of high-tech fabrics for the manufacture of sports clothing [452], it is always possible to achieve better sports results with the incorporation of different elements into fibers and fabrics, which is becoming more and more popular. Some of these elements are microcapsules. Back in the early 1980s, NASA developed the technology for embedding microencapsulated phase change materials into textiles for their temperature control [142, 453]. In the printing and graphic arts industry, the microcapsules are used for pharmaceutical and medical purposes, in cosmetic and food industries, for agricultural products, as well as in the chemical, textile and construction industries, biotechnology, photography, electronics and waste management [454] (Fig. 6.65).

In addition to the microcapsules integrated into fibers, there can be different electronic elements, interlacing for the protective or decorative purposes. Electronic components can be fitted directly into textile fibers (hybrid systems) or integrated in pockets connected via flexible cables, for example, batteries connected to LED lights (simple systems)[455]. Different integrated sensors can be used for a variety of medical purposes, for example, for the control of respiration or the measurement of respiratory signals [444]. One study presented the usage of sensors and a printed circuit board embedded in textiles, which enables the detection of the changes in the basic life functions for infants, for example, breathing and heart rate [456], while another study improved the integration of sensors and electronics into textiles enabling the control of the ECG combined with wireless communication [457]. Ultrasonic sensors in combination with a printed circuit board in textiles can be used to detect the obstacles in helping people with impaired vision [458]. The sensors are small, use little power, can be installed internally and can be washed.

Embedded microcontrollers, in combination with a variety of sensors (e.g., temperature or light sensors) and LED lighting, can be programmed and used for any purpose. Their advantage lies in the fact that they can be washed, which is for the use in textile applications extremely important. On the market, there are different microcontrollers available (e.g., ATMEL ATmega, ATtiny etc.), which are used in the products of wearable electronics such as LilyPad Arduino. LilyPad Arduino was designed and developed by Leah Buechley in collaboration with the company SparkFun Electronics. Wearable electronics LilyPad consist of different components (LED light, processor board, light sensor, etc.). LilyPad electronics are well suited for prototypes and unique design products, while their size (processor board is approx. 50 mm in diameter and approx. 3 mm in thickness) and price make them unsuitable for the mass production or very small items.

Some researchers design their own printed circuit boards, while others use Lily-Pad Arduino microcontrollers, for example, for an immediate determination of the pH value of the sweat that is excreted in sporting activities. In that case, Arduino controls the operation of LED lights that change color according to the measured pH value of the sweat [459].

The advantage of self-made decorative-protective applications is that they are made as a separate element which can be integrated into various garments or fashion accessories, that they are affordable, and their size and purpose is adjusted to the application.

Another example of our work presents a design of wearable electronics that consists of LED lights, a processor, circuit board and custom written program for different behaviors of these LED lights (gradually or simultaneously switched LED lights). The final product represents a warning-decorative arrow shaped with LED lights that light up differently.

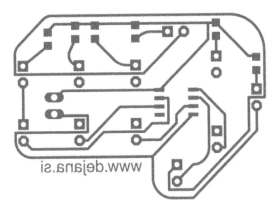

FIGURE 6.65   Printing template of circuit board.

FIGURE 6.66   Printed circuit board.

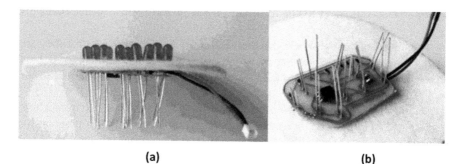

(a)   **(b)**

FIGURE 6.67   Insertion of electronic components: (a) side view, (b) back view.

FIGURE 6.68   Soldering.

For that case, a printed circuit board was designed using the program EAGLE (Easily Applicable Graphical Summary Layout Editor) from CadSoft, in which we made a printed circuit diagram which represents the logical symbols and signs of electronic components and their connections. In Fig. 6.9, the printing template of the circuit board is represented.

The manufacturing of a printed circuit board (cf. Fig. 6.66) was followed by the drilling of holes for the insertion of electronic components (cf. Fig. 6.67) and solder (cf. Fig. 6.68) electronic components into the preprepared printed circuit board.

6.20 ENERGY HARVEST AND STORAGE

Energy is essential for our modern civilization. The rapidly growing global energy demand has not only sped up the consumption of nonrenewable fossil fuels, but also threatened regional stability. In addition to reduce the energy consumption using highly efficient technology, converting other energies into electrical power can considerably assist in alleviating the energy crisis. In this direction, nanotechnology is providing new solutions to solve the problems. It has been found that nanofibrous materials can have significantly higher energy conversion and storage efficiency than their bulk counterparts.

6.20.1 PIEZOELECTRIC NANOFIBERS FOR ENERGY SCAVENGING APPLICATIONS

After decades of developments in the miniaturization of portable and wireless devices, new power sources beyond rechargeable batteries have become important topics for current and future stand-alone devices and systems. Specifically, ideal power sources should be scalable for power demands of various portable devices without the necessity of a recharging process or replacement. Recent work in the field of nanomaterials has shown considerable progress toward self-powered energy sources by scavenging energy from ambient environments (solar, thermal, mechanical vibration, etc.). In particular, the use of piezoelectric generators by nanomaterials as a robust and simple solution for mechanical energy harvesting has attracted lots of attentions. One of the earliest nanogenerators for possible energy scavenging applications from mechanical strain used piezoelectric zinc oxide nanowires. By coupling their semiconducting and piezoelectric properties, mechanical strains can be converted into electricity. In recent years, numerous research groups have demonstrated results in the field of mechanical energy scavenging using nanomaterials with different architectures, including: film-based, nanowire-based and nanofiber-based nanogenerators. Film-based nanogenerators are often made by the spin- on or thin-film deposition methods. Mechanical strains due to the bending, vibration or compression of the thin-film structure can be the source of the energy generation. Nanowire-based nanogenerators are typically made of semiconducting materials such as ZnO, ZnS, GaN or CdS. These piezoelectric nanowires have been demonstrated to build up an electrical potential when mechanically strained by an AFM tip, zig-zag electrodes or a compliant substrate to convert mechanical strains into electricity. The third group of nanogenerators is based on nanofibers often constructed by the electrospinning process with piezoelectric materials such as PZT or PVDF. PZT is a ceramic material exhibiting exceptionally good piezoelectric properties and is only recently used in nanofiber based energy harvesters. Most of PZT- based energy harvesters have film-based structures, including the typical design of cantilever structure with proof mass [460–462].

Two factors have been the key bottlenecks for the applications of PZT nanofibers nanogenerators. First, high temperature annealing (4600 1C) is generally required to enhance the piezoelectric property of PZT. Second, the electrospinning process requires the mixing of PZT with solvent which lowers the density of the PZT in the nanofiber and lowers overall energy conversion efficiency [461].

Compared with those aforementioned nanomaterials, PVDF nanofibers have the unique good combination of material properties in flexibility, lightweight, biocompatibility and availability in ultra-long lengths, various thicknesses and shapes, making them an interesting candidate for energy harvesting applications in wearable and/or implantable devices [462].

6.20.2 SOLAR CELLS

Solar cells use unlimited solar energy for power generation and have been considered as a major solution to current energy crisis. So far, single crystal and polycrystalline silicon based solar cells are dominating the commercial solar cell market. Dye-sensitized solar cells and organic solar cells are still under development (Fig. 6.69).

Conventional DSSCs have a dye-anchored mesoporous TiO_2 nanoparticle thin layer sandwiched between two conducting glass plates in the presence of an electrolyte. Since electrospun nanofibers are one-dimensional material with better electrical conductivity and higher specific surface area than nanoparticles, and large pore size in nanofibrous mat allows increased penetration of viscous polymer gel electrolyte, nanofibers have shown great application potential in DSSCs.

The basic working principle of DSSC is the photoexcitation of dye resulting in electron injection into the conduction band of the metal oxide (which acts as photoanode) and the whole injection into the electrolyte (Fig. 6.5). The metal oxide anode is percolated with an electrolyte whose redox potential supports the separation of bound electron-hole pairs in the metal oxide and photoexcited dye [463]. Electrospun metal oxides (TiO_2, ZnO, SnO_2, and CuO) and its composites with CNTs/graphene have been widely used for DSSCs.

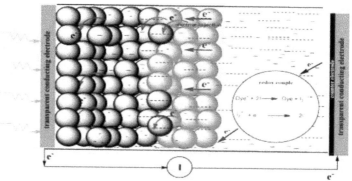

FIGURE 6.69 Schematic diagram of dye sensitized solar cells.

By using a thin layer of TiO_2 nanofibers as working electrode in DSSC, the photocurrent generation using polymer gel electrolytes was over 90% of liquid electrolyte [464]. DSSC devices [465] using polymer electrolytes based on electrospun PVDF-HFP nanofibers could achieve higher conversion efficiency and better long-term stability than conventional liquid electrolyte ones. The largest problem of using inorganic nanofibrous layers in DSSC is the poor adhesion with conductive substrate. The nanofiber layers are easily separated after calcination due to different thermal expansion coefficients. Many approaches have been tested to address this issue. A TiO_2 nanoparticle layer was coated on conductive glass before nanofiber deposition to increase the adhesion between the glass and the TiO_2 layer [466]. Hot pressing pretreatment was applied on ZnO/PVA composite nanofiber mat before calcination [467]. A self-relaxation layer [468] was also spontaneously formed during electrospinning to release the interfacial tensile stress generated in calcination.

To increase the short-circuit current, the electrospun TiO_2 electrode was treated with $TiCl_4$ aqueous solution to form an additional rutile TiO_2 layer on the fiber surface. Such a rutile TiO2 layer increased the fraction volume of active TiO_2 and interfiber connection, resulting in an increased photocurrent [464]. Another method to improve charge generation and transport was through materials doping. QDs were decorated onto TiO_2 nanofibers to yield multiple carrier generation due to the quantum confinement effect [469]. The electron diffusion coefficient and mobility of the niobium-doped anatase TiO_2 nanofibers were an order of magnitude higher than those of the un-doped fibers [470].

Another advantage of electrospun nanofibers is that they can be used to form a transparent mat for counter electrode. This effort has become more and more urgent because of the rapidly growing price of tin and indium, which are typically used for making conductive glass. Copper nanofiber thin layers were prepared using electrospinning for transparent electrode in DSSC[471]. The resulted fibers had ultrahigh aspect ratios of up to 100,000 and fused crossing points with ultra-low junction resistance. The fibrous mat exhibited great flexibility and stretchability as well. Electrospun carbon nanofibers were also explored as counter electrode material and an energy conversion efficiency of 5.5% was achieved in the resultant solar cell[472]. The carbon nanofiber based DSSC had lower fills factor and overall performance, because of the higher total series resistance (15.5 $\Omega \cdot cm^2$) than that (4.8 $\Omega \cdot cm2$) of Pt based traditional conductive substrate.

6.20.3 FUEL CELLS

Fuel cells are electrochemical devices capable of converting hydrogen or hydrogen-rich fuels into electrical current by a metal catalyst. There are many kinds of fuel cells, such as, proton exchange mat fuel cells, direct methanol fuel cells, alkaline fuel cells and solid oxide fuel cells [473]. PEM fuel cells are the most important one among them because of high power density and low operating temperature. Pt nanoparticle catalyst is a main component in fuel cells. The price of Pt has driven up the cell cost and limited the commercialization. Electrospun materials have been prepared as alternative catalyst with high catalytic efficiency, good durability and affordable cost. Binary PtRh and PtRu nanowires were synthesized by electrospinning, and they had better catalytic performance than commercial nanoparticle catalyst because of the one-dimensional features [474]. Pt nanowires also showed higher catalytic activities in a polymer electrolyte membrane fuel cell [342, 474].

Instead of direct use as catalyst, catalyst supporting material is another important application area for electrospun nanofibers. Pt clusters were electrodeposited on a carbon nanofiber mat for methanol oxidation, and the catalytic peak current of the composite catalyst reached 420 mA/mg compared with 185 mA/mg of a commercial Pt catalyst [475, 476]. Pt nanoparticles were immobilized on polyimide-based nanofibers using a hydrolysis process [477] and Pt nanoparticles were also loaded on the carbon nanotube containing polyamic acid nanofibers to achieve high catalytic current with long-term stability [477].

Proton exchange mat is the essential element of PEM fuel cells and normally made of a Nafion film for proton conduction. Because pure Nafion is not suitable for electrospinning due to its low viscosity in solution, it is normally mixed with other polymers to make blend nanofibers. Blend Nafion/PEO nanofibers were embedded in an inert polymer matrix to make a proton conducting mat, and a high proton conductivity of 0.06–0.08 S/cm at 15°C in water and low water swelling of 12–23 wt% at 25°C were achieved. Besides blend electrospinning, Nafion surface coating on polymeric nanofibers [478] is another efficient way to obtain better fuel cell performance and reduce Nafion consumption.

6.20.4   MECHANICAL ENERGY HARVESTERS

Piezoelectric material can directly convert mechanical energies into electrical power and has shown great potential in powering low energy consumption devices. Piezoelectric power generators, also called nanogenerators in some literatures, are becoming an important source for renewable energy [479]. These nanogenerators are normally made of aligned inorganic nanowires and their preparation required preciously controlled conditions [480].

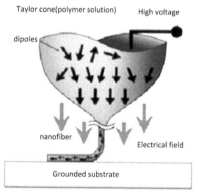

FIGURE 6.70   Scheme of near-field electrospinning process.

Electrospinning can process different piezoelectric materials into nanofibrous structure and electrospun piezoelectric nanofibers have been demonstrated to have energy scavenging capability. PVDF nanofiber was directly deposited across two metal electrodes without extra poling treatment to make single fiber nanogenerator using a near field electrospinning technique [481] (Fig. 6.70). Rigid lead-zirconate-titanate (PZT) material has been made into electrospun nanofibers with highly improved flexibility for fabricating flexible power generator after a soft polymer packaging [461]. The output voltage and power under repeated stress application were 1.63 V and 0.03 μW, respectively. Aligned PVDF nanofibers were collected using a normal electrospinning process and the following poling treatment was used to improve piezo-responsiveness [482]. A hybrid nano system combining this fiber based power generator and a flexible enzymatic biofuel cell was demonstrated as a self-powered UV light sensor.

6.20.5 LITHIUM ION BATTERIES

Lithium ion batteries are compact and rechargeable electrical energy storage devices. They have very high energy density. Like most other chemical batteries, porous structure is an essential requirement for the battery. A sponge-like electrode has high discharge current and capacity, and a porous separator between the electrodes can effectively stop the short circuit, but allows the exchange of ions freely. Solid electrolytes used in lithium ion battery are typically composed of a gel or porous host to retain the liquid electrolyte inside. To have high ion conductivity, the host material, also called separator, should have high permeability to ions. A porous mat with well interconnected pores, suitable mechanical strength and high electrochemical stability could be a potential candidate.

A lithium-ion battery is a family of rechargeable battery types in which lithium ions move from the negative electrode to the positive electrode during discharge and back when charging. Chemistry, performance, cost and safety characteristics vary across lithium-ion battery types. Unlike lithium primary batteries (which are disposable), lithium-ion batteries use an intercalated lithium compound as the electrode material instead of metallic lithium.

Lithium-ion batteries are common in consumer electronics. They are one of the most popular types of rechargeable battery for portable electronics, with one of the best energy-to-weight ratios, high open circuit voltage, low self-discharge rate, no memory effect and a slow loss of charge when not in use. Beyond consumer electronics, lithium-ion batteries are growing in popularity for military, electric vehicle and aerospace applications due to their high energy density.

Lithium-ion batteries were first proposed by M. S. Whittingham at Binghamton University in the 1970s [483]. Whittingham used titanium (II) sulfide as the cathode and lithium metal as the anode. The electrochemical properties of lithium intercalation in graphite were first discovered in 1980 by Rachid Yazami et al.., who showed the reversible intercalation of lithium into graphite in a lithium/polymer electrolyte/ graphite half cell [484]. In 1981, Bell Labs developed a workable graphite anode to provide an alternative to the lithium metal battery. Following cathode research performed by a team led by John Goodenough, in 1991 Sony released the first commercial lithium-ion battery. Their cells used layered oxide chemistry, specifically lithium cobalt oxide. In 1983, Dr Michael Thackeray, Goodenough and co-workers identified manganese spinel as a cathode material [485]. Spinel showed great promise, given low cost, good electronic and lithium ion conductivity and three-dimensional structure, which gives it good structural stability.

Although a pure manganese spinel fades with cycling, this can be overcome with chemical modification of the material. A manganese spinel is currently used in commercial cells.

In 1989, Goodenough showed that cathodes containing polyanions – for example, sulfates – produce higher voltages than oxides due to the inductive effect of the polyanion [486]. In 1996, Goodenough, Akshaya Padhi and co-workers identified lithium iron phosphate ($LiFePO_4$) and other phospho-olivines (lithium metal phosphates with olivine structure) as cathode materials [487]. In 2002, Yet-Ming Chiang and his group at MIT showed a substantial improvement in the performance of lithium batteries by boosting the material's conductivity by doping it with aluminum, niobium and zirconium. The exact mechanism causing the increase became the subject of a debate. In 2004, Chiang again increased performance by using iron phosphate particles of less than 100 nm in diameter. This decreased particle density by almost a hundredfold, increased the cathode's surface area and improved capacity and performance.

In recent years, with the rapid development of nanotechnology, nanomaterials are promising candidates for lithium-ion battery electrodes. As lithium-ion battery electrode materials, nanomaterials have some unique physical and chemical properties, such as, the large surface area, shorter transport length, high reversible capacity and long cycle life. These properties can significantly improve specific capacity and high-rate performance of lithium-ion batteries.

6.20.6 APPLICATIONS OF NANOFIBERS IN LITHIUM-ION BATTERIES

Electrospun polymers, such as, polyvinylidene-difluoride and polyacrylonitrile and their derivatives, can be used as nanofibers in separators of Li-ion batteries, providing a nanoporous structure leading to increased ionic conductivity of a membrane soaked with liquid electrolyte.

Aligned electrospun PVDF fibrous membranes can enhance the tensile strength and modulus of membranes by improving interfiber compaction under hot pressing. Such electrospun membranes may have important applications as battery separators. Uniform electrospun PVDF membrane thickness and fiber diameter can be obtained by using high polymer concentrations and electrospinning at high voltage. This improves mechanical strength and provides the PVDF separator with charge and discharge capacities that exceed commercial polypropylene separators, while resulting in little capacity loss [488, 489]. Electrospun PAN nonwoven fibers show higher porosities with lower Gurley values and increased wettability's compared to conventional separators. Furthermore, cells with separators of PAN nonwovens show enhanced cycle life/performance, higher rate capabilities and smaller diffusion resistance than cells with conventional separators [490]. High surface area conducting nanofibers make electrospun materials attractive as electrodes for batteries. Compared to conventional powder and film electrodes, nanofiber electrodes are expected to have improved electrode properties due to their high surface areas and 1-D nanostructure properties and, therefore should provide higher rates of electron transfer. In addition, their entangled network allows easy access of the ions to the electrode/ electrolyte interface. Nanostructured $LiCoO_2$ fiber electrodes with faster diffusion of Li^+ cations prepared by electrospinning afford a higher initial discharge capacity of 182 mAh/g compared with that of conventional powder and film electrodes (140 mAh/g) [491]. In the present commercial lithium-ion batteries, graphite has been widely used as the intercalating anode for its high reversible capacity, even discharge–charge potential profile and low cost. However, graphite does have some disadvantages and cannot satisfactorily meet the performance requirements of some important applications, especially in the safety and rate performance [492]. Therefore, researchers turn to investigate nanostructured materials as alternative anode material for lithium-ion batteries because of their superior performances. In recent years, nanostructured carbon materials, such as, CNFs [493], have been employed to construct anodes for lithium- ion batteries because these materials have relatively short lithium-ion pathways, large specific surface area and extremely charming surface activity.

However, in order to further improve the performance of carbon materials as anodes for batteries, an effective porous structure that is controllable is needed to provide a desirable surface area and open pore structure, which can achieve larger energy conversion density, higher rate capability and better cycle performance [494]. Therefore, activated or porous CNFs with large specific surface area and controlled pore structure could be an ideal candidate to meet these requirements.

Wei and co-workers [495] reported that the reversible specific capacity of the vapor-grown CNFs at a 0.1C rate was 461 mAh/g, and even at a high cycling rate of 10C the reversible capacity is around 170 mAh/g with a 95% Coulombic efficiency. High-purity CNFs with dome-shaped interiors form the noncatalytic thermal decomposition of acetylene over a copper surface at atmospheric pressure, and electrochemical measurements, indicated that the CNFs deliver good cyclability and a reversible capacity of 260 mAh/g at the high specific current of 100 mA/g. Zou et al. [496] reported that the CNF electrode which was synthesized by cocatalyst deoxidization process by a reaction between $C_2H_5OC_2H_5$, Zn and Fe powder at 650°C for 10 h, shows a capacity of 220 mAh/g and high reversibility with little hysteresis in the insertion/deintercalation reactions of lithium-ion. Wang and co-workers [497] synthesized porous CNFs by an easy pyrolysis of conducting a polymer in an argon atmosphere, with an average pore diameter of 27.98 nm and a specific surface area of about 74.5 $m^2 g^{-1}$. The reversible specific capacity of the CNFs at a 0.5°C rate is about 400 mAh g^{-1} when used as anode material in lithium-ion batteries. Moreover, they exhibit considerably specific capacity even at a high charge-discharge current; that is, the reversible capacities are around 250 and 194 mAh g^{-1} at a rate of 10 and 20°C, respectively.

Porous CNFs with large accessible surface areas and well-developed pore structures were prepared by electrospinning and subsequent thermal and chemical treatments [498]. The electrochemical performance results show that porous CNF anodes have improved lithium-ion storage ability, enhanced charge–discharge kinetics and better cyclic stability compared with nonporous counterparts. Ji et al. [499, 500] reported that Mn oxide-loaded porous CNFs were prepared by electrospinning polyacrylonitrile nanofibers containing different amounts of $Mn(CH_3COO)_2$, followed by thermal treatments in different environments. Electrochemical measurements indicated that the resulting composite nanofibers exhibit high reversible capacity, improved cycling performance and elevated rate capability even at high current rates, when used as anodes for rechargeable lithium-ion batteries without adding any polymer binder or electronic conductor. Zhang and co-workers also synthesized Si [501], copper [502] and nickel [503]/carbon composite nanofibers by the same method for use as anodes in lithium-ion batteries. The resultant composite nanofiber anodes exhibited a large accessible surface area, high reversible capacity and relatively good cycling performance. The novel Sn-C composite with a core-shell morphology of Sn nanoparticles encapsulated in porous multichannel carbon microtubes [504] and bamboo-like hollow CNFs [505] was fabricated by a single-nozzle electrospinning technique. The resulting material exhibited excellent characteristics in terms of reversible capacities, cycling performance and rate capability for applications as an anode material in lithium-ion batteries.

6.20.7 FUNCTIONAL NANOFIBERS AND THEIR APPLICATIONS

Perfluoro polymers have been widely studied in polymer electrolyte because of their chemical and mechanical stability. An electrospun PVDF polymer electrolyte showed high uptake to electrolyte solution (320–350%) and high ion conductivity (1.7×10^{-3} S/cm at 0°C). The fibrous electrolyte also had high electrochemical stability of more than 5 V. The prototype cell (MCMB/PVDF based electrolyte/ $LiCoO_2$) exhibited a very stable charge discharge behavior [435]. When a thin layer of polyethylene (PE) was plasma polymerized onto PVDF nanofiber surface, a role of shutter by melting of the PE layer grafted was rendered to the nanofiber mat and improved the safety of battery [506]. It was also found that the formation of interconnected web structure via heat treatment improved both the mechanical properties and dimensional stability of nanofiber mats [489, 506].

Nanofiber mats from other types of polymers, such as, PAN [507], were also studied as lithium battery separator. The electrospun PAN mat showed high ion conductivity and electrochemical stability. The prototype cell based on the electrospun PAN electrolyte separator with 1 M $LiPF_6$-EC/DMC exhibited an initial discharge capacity of 145 mAh/g, and 94.1% of the initial discharge capacity after 150 cycles at a charge/discharge rate of 0.5 C/0.5 C. To improve ionic conductivity, charge-discharge capability and stability, SiO_2 nanoparticles[508], PMMA [509] and PAN were blended with PVDF-HFP nanofibers for making composite polymer electrolyte.

Besides being used as a separator, some electrospun nanofiber mats have been used as battery electrodes. For example, a carbonated electrospun nanofiber mat was used as anode in lithium ion battery, and the batteries showed a large reversible capacity of 450 mAh/g [510]. Some recent researches are focusing on improving battery performance by incorporating different inorganic nanomaterials into carbon nanofibers as electrodes. A anode material made of Fe_3O_4/C composite nanofibers had much better electrochemical performance with a high reversible capability of 1007 mAh/g at the 80th cycle and excellent rate capability [479]. A Sn/C composite were encapsulated into hollow carbon nanofibers as an anode material for lithium batteries [505] with a high reversible capacity of 737 mAh/g after 200 cycles at 0.5°C (480 mAh/g at 5°C).

6.20.8 SUPERCAPACITORS

As an electrochemical device with a high power density and super high charging-discharging rate, supercapacitors (also known as double-layer capacitor) have been demonstrated with great potential in different emerging applications, such as, power back-up for laptop or mobile phone and power source for hybrid electric vehicle. For a supercapacitor with high capacitive behavior, electrodes made of porous carbon materials are extremely important, and electrospinning technique has been used to prepare carbon nanofibrous mats with high specific surface area and controllable pore size for this purpose. Activated carbon nanofibers prepare from a PAN/DMF solution have shown a maximum specific capacitance of 173 F/g [511]. The polybenzimidazole (PBI) based carbon nanofibers had specific surface areas ranging from 500 to 1220 m^2/g, and the fabricated double-layer capacitor exhibited specific capacitance between 35~202 F/g.

The capacitance of electrospun nanofiber based supercapacitor can be enhanced by using composite carbon electrodes. $ZnCl_2$ [512], carbon nanotubes and nickel [513], have been blended into procurer solutions for electrospinning and then carbonization. The $ZnCl_2$/C composite nanofibers exhibited a capacitance of 140 F/g with a specific surface area of 55 m^2/g when the carbon nanofibers were doped with

5 wt% $ZnCl_2$. The specific capacitance of an electrical double-layer capacitor with electrodes made of carbon nanotubes embedded carbon nanofiber reached as high as 310 F/g.

In a recent publication [514], bicomponent electrospinning was applied to prepare side-by-side nanofibers with PAN on one side and a thermoplastic polymer, PVP on the other side. The resultant carbon fibers after pyrolysis showed improved inter carbon fiber connections and crystallization. The capacitance of the electrochemical cell (Fig. 6.71) made of these interfolded carbon nanofibers was much higher than that of the carbon nanofibers from polymer blends. Apart from carbon, other inorganic nanofiber electrodes have also been examined for supercapacitor applications. For example, RuO_2 has been deposited on a Pt nanofiber mat to function as hybrid electrode and the supercapacitor exhibited a specific capacitance of 409.4 F/g with a capacity loss of only 21.4% from 10 to 1000 mV/s [515]. Electrospun V_2O_5 nanofiber based supercapacitor had the highest specific capacitance of 190 F/g in an aqueous electrolyte and 250 F/g in an organic electrolyte when the nanofibers were annealed at 400°C [516].

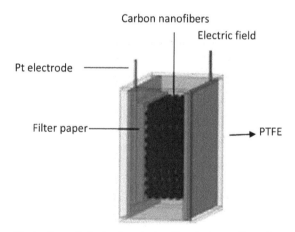

FIGURE 6.71   Illustration of electrochemical cell and the cyclic voltammetry curves of side-by-side.

6.20.9   HYDROGEN GENERATION

Hydrogen is an excellent fuel gas due to its high energy density, environmental cleanness and zero carbon footprints. Back in 1972, Honda and Fujishima discovered a method to produce hydrogen through photocatalytic water-splitting using TiO_2 as the electrode. Since then, intensive efforts have been focused on the fabrication of highly efficient nanostructured electrodes for photocatalytic/photoelectrochemical hydrogen generation. Among various synthesis techniques, electrospinning has drawn much of the scientific attention because of its versatility, simplicity, and cost- effectiveness. Now, the application of electrospun nanofibers in photocatalytic/photoelectrochemical hydrogen generation is presented. In general, the hydrogen generation process involves three steps, namely charge generation, charge separation and redox reactions. Upon light irradiation, electrons and holes are generated from the semiconducting electrode, as similar in photovoltaic solar cells. These charge carriers are then separated to avoid reverse recombination while taking part in water oxidation and reduction reactions, respectively. Electrospinning as a method for nanostructure fabrication is able to effectively influence all these three steps.

It is a well-known physical phenomenon that light absorption by a semiconducting material is highly determined not only by its own physical and chemical properties, such as, the band gaps, but also by its structural configurations. The limitation of pure semiconducting materials can be easily addressed by chemical modification of the electrodes like doping, which can be effectively carried out in electrospinning by simple addition of dopant salts in the electrospinning precursor solution. The introduced dopants are able to form extra energy levels within the forbidden bands, thereby reducing the energy required for valence electrons excitation up into a higher energy level. Through such mechanisms, a larger proportion of the solar spectrum can be used to achieve higher energy conversion efficiency. N-doped TiO_2 grain- like nanostructures have been successfully synthesized via electrospinning. The doped-nitrogen reduced TiO_2 band gap down to 2.83 eV, and increased H_2 generation rate up to 28 μmol/h, which was 12 times more than pristine TiO_2. This superiority was mainly ascribed from the reduced band gap combined with a larger surface area. Compared with its nanoparticle counterparts, nanofibers demonstrate much better mesoporosity, enlarged surface area, tunable fiber morphology and aspect ratios which can significantly enhance the light scattering effect [417]. Chen et al. have reported that the preferred scattering wavelength and intensity was linearly proportional to the fiber diameter and the fiber deposition density. An overlapping between scattering bands and absorption bands was much favorable for an improved photocatalytic activity since the scattered light would have a longer pathway within the material and much enhanced light absorption.

The photogenerated charge carriers have to be separated before recombination occurs. Efficient charge separation can be achieved by inducing some secondary structures within the electrospun nanofibers, or by introducing electron reservoirs or highly conducting materials on the nanofiber surface. Regarding the former approach, Chuangchote et al. had prepared 1-D nanofibrils along the electrospun TiO_2 nanofibers and these nanofibrils were believed to be beneficial for a higher material crystallinity as well as a retarded electron-hole recombination compared with their nanoparticle counterpart. In addition, nanofibers consisting of interconnected.

0-D nanoparticles can also be achieved by directly electrospinning the nanoparticle dispersed solution. Within the nanofibers, TiO_2 nanoparticles were densely packed and closely interconnected, therefore the photogenerated charge carriers were efficiently transported away to different sites for water oxidization and reduction reactions.

Additionally, the nanomaterial heterojunctions with nonaligned band configuration were much more favorable for efficient charge separation due to the presence of built-in band bending, as discussed in the "Photocatalysis" section. Electrospun nanofibers with heterojunction of anatase TiO_2/rutile TiO_2/rutile SnO_2 had been successfully synthesized by Lee et al. Charge carriers were generated in all of the three metal oxides due to the similarity of their band gap width. However, because of the nonaligned energy bands, the electrons were transported to SnO_2 due to its lowest conduction band edge, while holes to rutile TiO_2 due to its highest valence band edge among the three materials. Thus, the charge carriers were efficiently separated and the recombination process was lowered to a minimum level. Another similar heterojunctional configuration was achieved by the same group, but in a component assembly level. A "forest-like" photocatalyst was successfully synthesized using the electrospun anatase TiO_2 as the stems (200–400 nm), hydrothermally grown ZnO nanorods as twigs (100–300 nm) and photo-deposited CuO nanoparticles (10–100 nm) as individual leaves. These nature mimicked structures were favorable due to their hierarchical structures leading to better light absorption, and also enhanced charge separation at the various metal oxide interfaces. Highly conductive materials or electron reservoirs were also commonly used to enhance the charge separation process. Other than the commonly used Pt deposition, Yousef et al. had managed to synthesize graphite protected Ni-doped TiO_2 nanofibers by sintering the nickel acetate tetrahydrate-titanium isopropoxide-PVP electrospun fiber mesh in a hydrogen/argon condition. This architecture led to superior performance not only because of the graphite protection of the catalysts, but also due to the efficient charge transportation induced by the graphite layers.

In the following hydrogen generation process, the photogenerated electrons and holes are consumed in water reduction and oxidation reactions, respectively. Therefore, the number of reactive sites as well as the activity of each individual site plays an important role in determining the H_2 evolution rate. By employing electrospinning, the nanomaterials with higher surface area can be easily achieved which in turn enhance the redox reactions occurring at catalyst interfaces. The electrospun nanofibers can also act as structural support for photocatalysts to increase their chemical stability while retaining the interfacial surface area. The fluoropolymer with the stable C-F bond demonstrated an excellent stability in various conditions, that is, thermal, radiational and chemical treatment. Nanoparticles such as $ZnIn_2S_4$ [517] and ZnS-$AgIn_5S_8$ were successfully grown on electrospun fluoropolymer nanofibers and these materials achieved much enhanced H_2 evolution rate compared with nanoparticles without support.

Additionally, this electrospun fluoropolymer support was used for hydride hydrolysis to produce hydrogen gas. A similar concept was also applied to glucose hydrolysis using electrospun silica fiber as support and iron as catalyst [482].

6.20.10   HYDROGEN STORAGE

Hydrogen has been widely known as an ideal alternative energy to solve energy crisis and global warming. Because high-pressure and cryogenic storage systems can't satisfy the criteria for on-board storage, the key issue for the current hydrogen energy research is hydrogen storage.

It has been reported that carbonaceous materials, such as, active carbon, carbon nanotubes and graphite, are able to store hydrogen because of their large specific surface area, high pore volume and light weight. The hydrogen-storage capabilities of electrospun carbon fibers carbonated from different starting polymers (PAN) [518], PVDF [519], PANi [520] and PCS [521] have also been assessed. Carbon nanofiber can be a better hydrogen storage material than other carbon materials because it has an optimized pore structure with controlled pore size [522].

To increase graphitization during carbonation, Fe(III) acetylacetonate was used as catalyst to prepare graphite nanofibers and the resulted nanofiber (surface areas of 60–253 m^2/g) had the H-storage capacity of 0.14–1.01 wt%. To increase the specific surface area of carbon nanofibers, different inorganic materials normally metal [523] and metal oxides [524] were mixed into polymer solutions for electrospinning. The resultant carbon nanofiber could have a specific surface area of 2900 m^2/g with an H-storage capacity of 3 wt%. Even though the hydrogen adsorption ability increases with the increasing the specific surface area, pore volume is also very important. It has been concluded that the most effect pore width is in the range from 0.6 to 0.7 nm, which is slightly larger than hydrogen molecule (0.4059 nm).

Exciting results have been obtained more recently. PANi fibers were prepared by electrospinning and showed a reversible hydrogen storage capacity of 3–10 wt% at different temperatures [520]. Highly porous carbide-derived carbon fibers with a specific surface area of 3116 m^2/g were prepared after pyrolysis and chlorination of electrospun polyacrbosilane fibers [521]. The fibers have shown a very high hydrogen storage capability of 3.86 wt% at a low pressure (17 bar).

6.20.11   OTHER APPLICATIONS

Early studies on electrospun nanofibers also included reinforcement of polymers. As electrospun nanofiber mats have a large specific surface area and an irregular pore structure, mechanical interlocking among the nanofibers should occur. When a thin electrospun nylon-4, 6 nanofiber mat was added to epoxy, the composite showed transparency to visible light, and both the stiffness and strength were increased considerably compared with the pure epoxy film [525]. Recently, polysulfone [526] and polyetherketone cardo (PEK-C) [527] nanofibers were used to improve the toughness of carbon fiber/epoxy composite. When carbon nanofibers were dispersed into the PEK-C nanofiber phase, a synergistic effect to enhance the toughness and other mechanical properties was also observed [528].

Electrospun polybenzimidazole nanofibers have been used as fillers to reinforce epoxy and rubber [529]. An epoxy containing 15 wt% electrospun PBI nanofibers was found to have higher fracture toughness and modulus than the one containing 17 wt% PBI whiskers. Also the Young's modulus and tear strength of styrene-butadiene rubber containing the PBI nanofibers were higher than those of pure SBR. In addition, electrospun nylon PA $_6$ nanofibers were used to improve the mechanical properties of a BISGMA/TEGDMA dental restorative composite resins [530].

Electrospun nanofibers showed excellent capability to absorb sound. A leading nanofiber technology company, Elmarco, recently patented an electrospun nanofiber material that had unique sound absorption characteristics, with only about one-third of the weight of conventional sound absorption materials. It was able to absorb sounds across a wide range of frequencies, especially those below 1000 Hz.

Electrospun nanofibers have also shown application potential in field-effect transistor. EFT behavior has been observed in camphor sulfonic-acid-doped electrospun PANi/PEO nanofibers [531]. Saturation channel currents were found at low-source-drain voltage with a hole mobility in the depletion regime of 1.4×10^{-4} cm^2/Vs. Electrospun nanofiber mat has also been demonstrated with the application ability for ultrafast identification of latent fingerprints [532].

6.20.12 ELECTRONICS AND ENERGY APPLICATIONS

The use of electrospun fiber in the electronics and energy industry is a growing area, and a number of interesting applications have been pursued. Work by Kim and co-workers has been exploring the use of electrospun fiber as one of the elements in lithium polymer batteries. It was found that electrospun polyvinyl di-fluoride fibers were suitable as a highly porous membrane that showed good high temperature (60°C) performance when used in a lithium polymer battery. The high temperature performance would enable lithium polymer batteries to operate at a higher discharge rate and have a better performance cycle.

A patent by Best and co-workers describes a method for using electrospun fiber to manufacture a flexible battery that may become integral to the flexible electronics industry.

The high surface specific area of electrospun fibers may find application in the production of super capacitors [533]. Super capacitors have a much larger capacitance than those traditionally used in modern electronics although at the time of writing they suffer from lower operating voltages. However, they can store relatively large quantities of energy for short periods of time. As they can be produced with a very high energy density (energy per unit mass of storage medium) they have been proposed as an energy source for electric vehicles. They could also be used as cheap power sources for mobile appliances that are frequently used such as cellphones or portable media players. This would increase the number of times they need to be charged but with low power operation this could only demand daily charging similar to modern devices.

The work by Kim and Yang has explored the possible application of electrospun polyacrylonitrile fibers for super capacitor manufacture. It was found that using a steam activation method on the fiber membrane one could produce a suitable electrode for a capacitor with a high surface area due to the formation of pores on the fiber structure. It was observed that the electrode manufactured by this method had a specific capacitance of 120 F/g while sustaining a discharge current of 1000 mA/g. In this case the specific capacitance did vary by 18% over the range from 1 to 1000 mA/g. Using the method of coaxial electrospinning it would be possible to produce a core-sheath structure that would consist of a conducting core polymer and an insulating sheath. These fibers could then act as insulated nanowires. The same sintering process used in the production of metal oxide fibers can also be used to manufacture conductive gold nanowires when a suitable metal salt is introduced to the polymer solution as described by Pol and co-workers [534]. As yet the steps beyond the synthesis of the raw nanowire, allowing the manufacture of more complex nanoscale electronics, are unknown to the authors.

However, these nanowires may find application in the manufacture of cheap chemical sensors where the nanowires are used to connect the sensor element to the processing circuit. It could also be possible to use the electrospun fiber as the sensor element as well as the transmission medium. The patent by Han and co-workers describes a method of using electrospun fiber to manufacture numerous chemical sensors for different chemical species to be used in a device described as an 'electronic nose'[534].

6.20.13 *ELECTRONIC* DEVICES

It is well known that 1D semiconductor or conductive nanostructures can act as a good model to investigate the dependence of electrical and thermal transport or mechanical properties on dimensionality and size reduction. In addition, they can also be used as interconnects and functional units in fabricating electronic, optoelectronic, electrochemical, and electromechanical devices. Among those 1D nanostructures, electrospun semiconductor or conductive nanomaterials gain special attention for their continuous length and controllable deposited position, which can act as the direct interconnects and functional unit in fabricating diverse devices. Till now, many functional devices have been developed based on electrospun nanomaterials.

6.20.14   BASIC ELECTRONIC RESEARCH ON ELECTROSPUN NANOSTRUCTURES

Same to the semiconductor or conductive 1D nanostructures prepared by other methods, 1D electrospun nanostructures with electrical activities have also received much interest in recent years. In 2003, Wang and Santiago-Avilés [535] firstly investigated the electrical performances (zero magnetic field conductivity and magneto conductivity) based on carbonized carbon nanofibers from PAN nanofibers through a two-probe method. Large negative magneto resistance (−75%) was found at 1.9 K and 9 T. Zhou and MacDiarmid [536] electrically characterized the electrospun camphor sulfonic acid (HCSA)-doped polyaniline/polyethylene oxide composite fibers with the diameters less than 100 nm. They found that as the fibers below 15 nm are electrically insulating through SCM owing to small diameter allow the complete dedoping in air or be smaller than phase-separated grains of PAn and PEO. The I–Vs of asymmetric fibers were rectifying for the formation of Schottky barriers at the nanofiber metallic working electrode contacts. From then on, a number of electric researching data based on diverse electrospun nanomaterials have been reported.

With the development of electrospinning technique, the electric measurement based on single- or aligned-electrospun nanostructures gains special attention. For example, In 2007, Wang and Santiago-Avilés [537] reported the electric performance based on single–porous electrospun SnO_2 ribbon via a two-probe method following a cycle of heating from 300 to 660 K and subsequent cooling from 660 to 300 K. The conductance (G) of the sample is insensitive to the temperature below 380 K and flows an Arrhenius relation with a thermal activation energy of 0.918 ± 0.004 eV from 380 to 660 K; Upon cooling, conductance follows the same Arrhenius relation until 570 K, and another Arrhenius relation with a lower activation energy of 0.259 ± 0.006 eV from 570 to 380 K. The higher Arrhenius relations are attributed to the surface adsorption and desorption of moisture and oxygen; the lower one is attributed to the particle replacement of adsorbed oxygen by moisture. In 2009, Hou and co-workers [538] compared the electric properties based on aligned carbon nanofibers and found that the aligned carbon nanofibers exhibited anisotropic electrical conductivities. The differences between the parallel and perpendicular directions to the carbon bundle axes were more than 20 times.

6.20.15   FIELD-EFFECT TRANSISTOR

Field-effect transistors [539–541], sometimes called as unipolar transistor, as the basic building block in logic circuits, have gained special attention, in which an electric field was applied to control the shape and hence the conductivity of a channel of one type of charge carrier in a semiconductor material. The first electrospun-based

FET was constructed by Pinto and co-workers [539] with polyaniline/polyethylene nanocomposite fibers as channel.

6.20.16   ENERGY GENERATIONS USING NANOFIBER NANOGENERATORS

Ceramic PZT and polymeric PVDF are two piezoelectric materials which have been previously demonstrated as viable nanofiber nanogenerator materials. In these efforts, either near-field electrospinning or the conventional far-field electrospinning process has been the key manufacturing tool to produce nanofibers [542]. For the NFES process, a continuous single nanofiber can be deposited in a controllable manner, while the FFES process can produce dense nanofibers networks on large areas for the nanogenerator demonstrations. In general, a poling process, consisting of both electrical poling and mechanical stretching, is required in the fabrication of materials with piezoelectric properties at moderate temperature. Given the high electrostatic field and polymer jet characteristics of the electrospinning process, electrospinning is ideally suited for producing piezoelectric nanofibers through in-situ electric poling and mechanical stretching. Here, key achievements in nanofiber nanogenerators made of PVDF and PZT are described and discussed.

PVDF has superior piezoelectric properties as compared with other types of polymeric materials due to its polar crystalline structure. In nature, PVDF polymer consists of at least five different structural forms depending on the chain conformation of transormation and gauche linkages. Figure 6.72(a) shows the crystalline structure of the α and β-phase, respectively. While the -phase is known as the most abundant form in nature, β-phase is responsible for most of PVDF's piezoelectric response due to its polar structure with oriented hydrogen and fluoride (CH_2–CF_2) unit cells along with the carbon backbone. In order to obtain the β-phase PVDF, electrical poling and mechanical stretching processes are required during the manufacturing process to align the dipoles in the crystalline PVDF structures as illustrated in Fig. 6.72.

PZT is another good piezoelectric material with its crystalline structure illustrated in Fig. 6.72(b). An electric polarization of PZT can shift up/down of Zr/Ti atom and remain their positions after applying and removing an external electric field for the piezoelectric property. In their bulk or thin film format, PZT can generate higher voltage as compared with other piezoelectric materials for sensing [543, 544], and actuation [544] and energy harvesting applications [545]. As a ceramic material, bulk PZT is more fragile in comparison to organic PVDF, but has demonstrated very good mechanical strength in nanowire form [546].

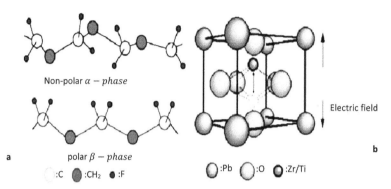

FIGURE 6.72 (a) Schematic diagrams showing crystalline structures of PVDF: (top) nonpolar a-phase, and (bottom) polar b-phase. The dipoles in the nonpolar, a-phase PVDF could be stretched and oriented by an electrical field to become the polar, b-phase structure under electrical poling and mechanical stretching. (b) Schematic diagrams showing crystalline structures of PZT. An electric polarization of PZT can shift up/down of Zr/Ti atom and remain their positions after applying and removing an external electric field for the piezoelectric property.

6.21 THERMAL ENGINEERING DESIGN OF CLOTHING

Clothing plays a very important role in the daily lives of human beings, as it contributes to biological health and psychological happiness in our lives. More and more modern consumers understand the importance of textiles and thus have come to prefer the apparel products with high added values in terms of functional performance [547]. Currently, clothing design focuses not only on the pattern and fashion design, but also pays more attention to the functional performance of the clothing, making the clothing more smart and satisfying human needs in various environments.

From the points of view of biology and physiology, people will be aware of discomfort in terms of warmth or coolness if the temperature of any part of the skin changes by more than 4.5°C, and will be fatal if the core body temperature rises or falls by more than 1.5°C [547].

Actually, there is a thermoregulatory system inside the human body to maintain the body thermal comfort or even being survived in various external environments. When the body temperature drops or rises, the human body must generate or dissipate heat to allow the body temperature to remain in the reasonable range. However, when people are exposed in an extremely hot or cold environment, the thermoregulatory system is not strong enough to maintain the balance between the rates of heat loss and heat generated inside the body. The clothing, as the barrier between the body and environment, needs to be sensitive enough to take the outside environment into consideration and generate a reasonable thermal microenvironment around the body to help it deal with/cope in extreme weather conditions.

Recently, with the successful multidisciplinary teamwork, more and more innovative textile materials and structures are developed for the study of biology and health [547]. Heat generating/storing fiber/fabrics, micro and nano-composite materials, smart phase change material and intelligent coating/membrane are developed and available for clothing functional design. The clothing with superior thermal performance is being known to consumers and regarded as an important concern in their buying decision.

Clothing thermal engineering design is a systematic solution of designing clothing with superior thermal performance for people to deal with various environments, which considers the whole wearing situations, including not only the clothing material, but also the biological behaviors of human body and the wearing environments. Figure 6.72 illustrates all the key issues to be considered in the clothing thermal engineering design. The final product of clothing is supposed to be achieved considering all the physical and physiological phenomena involved during the wearing process. For instance, the design of clothing worn in the hot environment or a certain activity situation should consider the hot sensitive parts of the body which easily accumulate sweat. The design of clothing worn in the cold environment should consider the cold sensitive parts of the body which need more thermal protection. The thermal performance of the designed cloth needs to be suitable for the wearing situations.

The computer plays a critical role in the clothing thermal engineering because it offers a virtual platform for the users to perform their design in the following ways:

(i) the rapid provision of numerical and graphic representation to the traditional qualitative trial-and-error method of clothing design;

(ii) detailed product design, which is manually practiced in design/workshop office in traditional way;

(iii) and modeling the involved behaviors/mechanisms and visualizing product performance.

This computerized engineering method makes a great advance in both the computer applications and functional clothing engineering design. However, it cannot be taken for granted that this method can be realized individually by computer technologies or engineering design or their simple combination. It is an engineering application of multidisciplinary knowledge which makes effective communications and integration between the research studies in different areas.

In order to establish a theoretical understanding of the knowledge behind this computerized thermal engineering system, it is necessary to investigate the physical and physiological behaviors involved in the wearing system and their mathematical representations in the virtual environment. Also it is necessary to devise effective strategies to diffuse the computer technologies into the clothing thermal engineering design.

6.21.1 THERMAL BEHAVIORS IN THE CLOTHING WEARING SYSTEM

Clothing is one of the most intimate objects in people's daily life since it covers most of the human body most of the time. People may keep having subjective psychological feelings of the clothing and consciously judging the warm/cold/comfort sensation during the wearing time. On the basis of wearing experience, people can make a rough evaluation of the thermal function of clothing and choose suitable clothing for their daily activities. However, as projected in the thermoscopic world, the wearing situation of people can be regard as a complex and interactive multicomponent system. Figure 6.73 shows the components of the clothing wearing system. The thermal behaviors involved in the wearing situation may be categorized as [548]:

(i) heat and moisture transfer in the textile materials. This is the physical behavior mainly deciding the thermal performance of clothing. It can be regarded as the following physical process:

- the heat transfer process in the textile material in terms of conduction, convection and radiation;
- the vapor moisture transfer process in the textile material in terms of diffusion and convection;
- the liquid water transfer process in the textile material;
- phase change process in the textile material. It is an approach allowing the heat and moisture transfer in a coupled way, including moisture condensation/evaporation in the fabric air void volume, moisture absorption/desorption of fibers, and microencapsulated phase change materials;
- influence of functional treatments of textiles on the heat and moisture transfer process, such as, waterproof fabric, moisture management treatment, PCM coating and heating fabrics.

(ii) thermoregulatory behaviors of the human body, such as, sweating, shivering and biological metabolism, and heat and moisture exchange of body skin and environment.

(iii) interactions between the inner clothing and body skin.

(iv) the climatic conditions of wearing situation in terms of temperature, relative humidity and wind velocity, which influence the heat and moisture behaviors of textile materials and the human body.

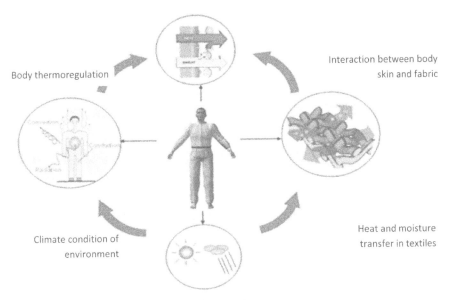

FIGURE 6.73 Components of the clothing wearing system.

6.21.2 THEORETICAL MODELING OF THE CLOTHING WEARING SYSTEM

6.21.2.1 HEAT AND MOISTURE TRANSFER IN THE TEXTILE MATERIALS

Normally, the heat and moisture transfer processes in the textile material occur when there are gradients of temperature and water vapor pressure across textile structures, and these two processes are often coupled accompanying with the appearance of phase change process.

6.21.2.2 HEAT TRANSFER PROCESS

The overall heat transfer in textile materials is the sum of contributions through the fiber and interstitial medium, which may involve multiple transfer mechanisms in terms of conduction, convection and radiation. Theoretically, conduction heat transfer always occurs in the solid fiber material and the medium trapped in the spaces between the fibers as long as a temperature gradient is presented. Convection heat transfer will be obviously observed if the medium is gaseous and if the space is large enough, that is to say, the more porous the textile material is, the more effectively convection takes place. Radiation heat transfer can be ignored when the temperature gradient is small. Consequently, the heat transfer via conduction is the most dominant transfer mechanism.

In the engineering applications, thermal conductivity is usually adopted to express the thermal properties of materials because thermal conduction is better documented and mathematically analyzed [549]. Unlike other porous materials, in textiles the air filled in the space between fibers has substantially bigger proportion than that of the fibers and the thermal conductivity of fiber is much smaller than air. The heat flow by thermal conduction at any position (x) inside the textile structure can be expressed by the Fourier's law:

$$Q_k(x)_c = -k\frac{dT(x)}{dx} \tag{7}$$

$$k = (1 - \epsilon)k_f + k_a \tag{8}$$

The effective thermal conductivity of the textile structure is the combination of the conductivities of the air and the solid fiber.

The radioactive heat transfers in the ways of emitting or absorbing electromagnetic waves when the standard temperature of the textile material is above zero. The intensity of radioactive is depended on the ratio between the radiation penetration depth and the thickness of textile material, besides the temperature difference between the two surfaces of textile material. Farnsworth reported that when the radiation depth is similar to the thickness of fabric, the amount of radiation should be taken into account as it is comparable to the amount of the conduction heat flow, and the radiation heat flux in fabric can be expressed by [550]:

$$\frac{\partial F_R}{\partial x} = -\beta F_R + \beta\sigma T^4 \frac{\partial F_L}{\partial x} = \beta F_L - \beta\sigma T^4 \tag{9}$$

$$\beta = \frac{(1 - \varepsilon)}{r}\varepsilon_r \tag{10}$$

6.21.2.3   MOISTURE TRANSFER PROCESS

Thanks to the porosity of the fabric, the interstices between fibers provide the space for moisture to flow away. There are four ways of moisture (in the phase of water vapor or liquid) transfer occurring in textile materials as summarized by Mecheels [551]:

(i)   diffusion through the space between the fibers,
(ii)   absorption/desorption by the fiber materials,
(iii) transfer of liquid water through capillary interstices in yarns/fibers, and
(iv) migration of liquid water on the fiber surface.

There are many similarities between heat conduction and moisture diffusion in the textile material. When the system scale, material properties and initial and boundary conditions are similar, the governing equations, analysis methods and results would be analogous for these two processes [552]. When there is a difference between the water vapor concentration on the fibers' surface and that of the air in the fiber interstices, there will be a net exchange of moisture. The water vapor diffusion through the textile material can be described by the First Fick's law [552]:

$$Q_w = D_a \frac{\Delta C}{L} \tag{11}$$

Since the fiber has a small proportion of volume in the fabric, the main contribution of moisture flux is from the diffusions process through the air in the fiber interstices. However, it was identified by Wenhner et al. [553] that absorption of moisture by the fiber also importantly affected the response of fabric to the moisture gradient. The water vapor concentration on the fiber surface, theoretically, depends on the amount of absorbed moisture onto the surface and the local temperature of the fiber. The fiber will keep absorbing as much moisture as it can until it reaches a saturated status with respect to the absorption rate. And when the fiber becomes saturated, additional vapor moisture may condense into liquid phase onto the fiber surface. With regard to the physic nature of fibrous materials, condensate water may be held on the surface of the fiber and be relative immobile, or may be transferred across the textile structure by capillary actions.

The moisture absorption capacity of fibers is described by the property of hygroscopicity (also called moisture regain), which means the amount of moisture that the fiber contains when placed in an environment at certain temperature and relative humidity. In 1967, Nordonb land David proposed an exponential relationship to describe the change rate of water content of the fibers, and developed a numerical solution with computer technology at that time. Li and Holcombe, in 1992, devised a new absorption rate equation by analyzing the two-stage sorption kinetics of wool fibers and incorporating it with more realistic boundary condition [554]. They assumed the water vapor uptake rate of fiber is composed by two components associated with the two stages of sorption:

$$\frac{\partial C_f}{\partial t} = (1 - \alpha)R_1 + \alpha R_2 \, 0 \le \alpha \le 1 \tag{12}$$

$$R_1 = \frac{\partial C_f}{\partial t} = \frac{1}{r}\frac{\partial}{\partial r}\frac{(rD_f \partial C_f)}{\partial r} \tag{13}$$

R_1 can be obtained by regarding the sorption/desorption process as Fickian diffusion. R_2 is identified by experimental data and related to the local temperature, humidity and sorption history of the fibers.

6.21.2.4 LIQUID TRANSFER PROCESS

When liquid water transfers across the textile material, it will experience wetting and wicking stages, in which wetting of textile is prerequisite for the wicking process [555]. Both wetting and wicking are determined by surface tensions between the solid-vapor-liquid interfaces. In view of the macroscopic world, these tensions are the energy that must be supplied to increase the surface/interface area by one unit. The liquid water put in touch with fibrous material comes to an equilibrium state with regard to minimization of interfacial free energy on the surface. The force involved in the equilibrium can be expressed with the well-known Young's equation:

$$\gamma_{LV}.\cos\theta = \gamma_{SV} - \gamma_{SL}$$

(14)

The term γ_{LV} is also usually regarded as the surface tension of the liquid. θ is equilibrium contact angle, which is the consequence of wetting instead of the cause of it. The term $\gamma_{LV}.cos\,\theta$ is defined as adhesion tension or specific wettability of textile material. This equation shows that wettability increases with the decreased equilibrium contact angle θ. The equilibrium contact angle is an intrinsic value described by the Young equation for an ideal system. However, precise measurement of surface tension is not commonly possible. The experimentally measured contact angle between the fiber and the liquid can be observed on a macroscopic scale and is an apparent physical property.

When textiles surface is fully wet by the liquid, the wicking process will happen spontaneously, where the liquid water transports into the capillary space formed between fibers and yarns by capillary force. The fibrous textile assembly is a complex nonhomogeneous capillary system due to irregular capillary spaces. These spaces have various dimensions and discontinuous radius distributions. The practical engineering field, an indirectly determined effective capillary radius, is adopted to represent the no-uniform capillary spaces in yarns and fabrics. If the penetration of liquid is limited to the capillary space and the fiber does not absorb the liquid, the wicking process is called capillary penetration, and the penetration is originally driven by the wettability of the fiber, which is decided by the chemical nature and geometry structure of its surface [555]. Ito and Muraoka pointed out that the wicking process will be suppressed with the decreased number of fibers in the textiles [556]. When the number of fibers becomes greater, water moves along the void spaces even between the untwisted fibers, which indicates that sufficient number of continuity of pores is very important to the wicking process.

6.21.3 COUPLED HEAT AND MOISTURE TRANSFER

The research of the heat transfer and moisture diffusion in textile materials was initially regarded as independent under the assumption that the temperature and moisture concentration of the clothing is steady over a period of time. In this steady-state condition, there is no need to address the interactions between heat and moisture transfer process [557].

However, under some transient situations where some phase change processes happen, such as, moisture sorption/desorption and evaporation/condensation, these two processes are coupled and interact significantly.

The moisture absorption/desorption capability of the solid fiber depends on the relative humidity of the enclosed air in the microclimate around the fiber and the type of fiber material. When fibers absorb moisture, heat is generated and released. Consequently, the temperature of fiber will rise and thus results in an increase of dry heat flow and a decrease in latent heat flow across the fabric [558]. The absorbed/desorbed moisture of fibers and the water vapor in the enclosed microclimate in textiles compose the water content of the textile material, which can originate from the wicking process or result from condensation in case of the fully saturated water vapor in fibrous materials [559].

Similar to the phase change process of moisture sorption/desorption, liquid condensation/evaporation pose an impact on the flow of heat and moisture across the textiles by acting as a heat source or be merged into the heat transfer process. Condensation is a physical phenomenon which commonly takes place when the fibrous material is exposed to a large temperature gradient and high humid source, both of which cause the local relative humidity to attain 100% or full saturation. And provided that there is still extra moisture diffusing into the fibrous material, condensation continues. The condition of condensation is different from the transient process of moisture sorption/desorption. When the relative humidity of surrounding microclimate is less than 100%, evaporation occurs.

The first model describing the transient heat and moisture transfer process in porous textile material was developed by Henry [560] in terms of two differential equations respectively for mass and heat governing formulation, as listed in the following. A simple linearity assumption was made in this model to describe the moisture absorption/desorption of fibers to obtain analytical solution. This modeling work established a basic framework of modeling the complicated coupled process of heat and moisture transfer through the textiles material.

$$\epsilon \cdot \frac{\partial C_a}{\partial t} = \frac{D_a \cdot \epsilon}{\tau_a} \cdot \frac{\partial^2 C_a}{\partial x^2} - (1 - \epsilon) \cdot \Gamma_f \tag{15}$$

$$C_v \cdot \frac{\partial T}{\partial t} = K_{fab} \cdot \frac{\partial^2 T}{\partial x^2} + (1 - \epsilon) \cdot \lambda_v \cdot \Gamma_f \tag{16}$$

$$\Gamma_f = \frac{\partial C_f}{\partial t} = const + a_1 C_a + a_2 T \tag{17}$$

Henry assumed the moisture sorption rate can be a linear relationship between temperature and moisture concentration, allowing him to solve the equations analytically.

Ogniewicz and Tien [560] proposed a model considering the heat transport that happened by the ways of conduction, convection and condensation in a pendular state. That model ignored moisture sorption and was lack of a clear definition of the volumetric relationship between the gas phases and the liquid phase. Motakef [561] extended this analysis to describe mobile condenses, in which the moisture condensation was taken into account with simultaneous mass and heat transfer process.

$$\frac{\partial C_a}{\partial t} = D_a \frac{\partial C_a}{\partial x} + \Gamma_{1g} \tag{18}$$

$$C_v \cdot \frac{\partial T}{\partial t} = K \frac{\partial^2 T}{\partial x^2} - \lambda_{1g} \cdot \Gamma_{1g} \tag{19}$$

$$\rho_c \epsilon \frac{\partial \theta}{\partial t} = \rho_c \epsilon \frac{\partial}{\partial t} (D_l(\theta)) \frac{\partial \theta}{\partial x} - \Gamma_{1g} \tag{20}$$

In Motakef's model, a concept of critical liquid content was introduced to address the liquid diffusivity. When the liquid content θ is below the CLC, the liquid is in the pendular state and has no tendency to diffuse. When the liquid content θ is beyond the CLC, a liquid diffusivity D_1 (θ) was introduced to describe the liquid transfer by surface tension force from regions of higher liquid content to the drier regions. D_1 (θ) is a complicated function of the internal geometry and structure of the medium.

Fan and Luo [559] incorporated the new two-stage moisture sorption/desorption model of fibers into the dynamic heat and moisture transfer model for porous clothing assemblies. They considered the radiation heat transfer and the effect of water content of fibers on the thermal conductivity of fiber material. Further, Fan and his co-workers improved the model by introducing moisture bulk flow, which was caused by the vapor-pressure gradients and super-saturation state [559]. This improvement made up for the ignorance of liquid water diffusion in the porous textile material in previous models. The equations of the model are listed as follows:

$$\epsilon \frac{\partial C_a}{\partial t} = -\epsilon \frac{\partial C_a}{\partial x} + \frac{D_a \epsilon}{\tau} \frac{\partial^2 C_a}{\partial x^2} - \Gamma(x,t) \tag{21}$$

$$\rho(1-\varepsilon) \cdot \frac{\partial(w-w_f)}{\partial t} = \rho(1-\varepsilon) \cdot D_l \frac{\partial^2(w-w_f)}{\partial x} + \Gamma(x,t) \tag{22}$$

$$C_v(x,t) \frac{\partial T}{\partial t} = -\varepsilon\mu C_{va}(x,t) \frac{\partial T}{\partial x} + \frac{\partial}{\partial x}\left((x,t) \frac{\partial T}{\partial x} \right) - \frac{\partial F_R}{\partial x} + \frac{\partial F_L}{\partial x} + \lambda(x,t)\Gamma(x,t) \tag{23}$$

Γ (x,t) accounts for moisture change due to absorption/desorption of fibers and the water condensation/evaporation. The diffusion coefficient of free water in the fibrous batting is assumed with a constant value with reference to some previous work. C_v is the effective volumetric heat capacity of fibrous material.

In 2002 year, Li and Zhu reported a new model for simulation of coupled heat and moisture transfer processes, considering the capillary liquid diffusion process in textile [562], which developed the liquid diffusion coefficient as a function of fiber surface energy, contact angle, and fabric pore size distribution. Based on this new model, Wang et al. considered more the radiative heat transfer and moisture sorption and condensation in the porous textile, achieving more accurate simulation for the realistic situation [563]. The governing equations of the model are shown as follows:

$$\epsilon \frac{\partial(C_a \varepsilon_a)}{\partial t} = -\frac{1}{\tau_a}\frac{\partial}{\partial x}\left(D_a \frac{\partial(C_a \varepsilon_a)}{\partial t}\right) - \varepsilon_f \xi_1 \Gamma_f + \Gamma_{1g} \tag{24}$$

$$\frac{\partial(\rho_l \varepsilon_l)}{\partial t} = -\frac{1}{\tau_l}\frac{\partial}{\partial x}\left(D_l \frac{\partial(\rho_l \varepsilon_l)}{\partial t}\right) - \varepsilon_f \xi_2 \Gamma_f + \Gamma_{1g} \tag{25}$$

$$C_v \frac{\partial T}{\partial t} = \frac{\partial}{\partial x}\left(K_{min}(x)\frac{\partial T}{\partial x}\right) + \frac{\partial F_R}{\partial x} - \frac{\partial F_L}{\partial x} + \varepsilon_f \Gamma_f(\xi_1 \lambda_v + \xi_2 \lambda_l) - \lambda_{1g}\Gamma_{1g} \tag{26}$$

The previous research about the heat and moisture transfer in textile material limited their focus on a single layer of porous textiles. Li and Wang extended the model for coupled heat and moisture transfer to multilayer fabric assemblies [563]. They described the geometrical features, layer relationships and blend fibers of the multilayer fabric assemblies by the following definitions:

$$l_{(i-1)} = l_{i1}(2 < i < n) \tag{27}$$

$$\text{Contact}_{in} = 1 l_{ij} = 0 (1 \le i \le n, j = 0,1) \tag{28}$$

$$0 \qquad\qquad\qquad l_{ij} \ne 0$$

$$\bar{P} = \sum_{ti=1}^{tn} f_{ti} P_{ti}(\text{tn} \ge n) \tag{29}$$

Contact is defined to express the contact situation at boundaries between layers, which determines the heat and moisture transfer behavior at the layer boundary. The symbol of is the weighted mean property of all blend fibers in the fabric based on their fractions f_{ti}. In addition, they individually developed boundary conditions for different fabric layers to achieve the numerical solutions.

6.21.4 WATERPROOF FABRIC

Waterproof fabric, which is laminated or coated with microporous or hydrophilic films, is frequently used in the design of functional clothing for the weather of low temperature, wind, rain, and even more extreme situations. With waterproof fabric, the clothing can effectively protect the body from the wind and water; as well as reduce the heat loss from the body to the environment. These functions of waterproof fabric, scientifically, are achieved by significantly affecting the processes of heat and moisture transfer through the textile products.

The term 'water vapor permeability' (WVP, g.m^{-2}.day^{-1}) is commonly used to measure the breathability of the fabric, which indicates the moisture transfer resistance in the heat and mass transfer processes. This property can be obtained by the experimental measurement with the Evaporation and Desiccant methods. The calculation formulation is expressed according to the first Fick's law of diffusion [554]:

$$\text{WVP} = \frac{Q_W}{tA} = \frac{\Delta c}{W_n} \tag{30}$$

where, Q_w is the weight loss/gain in grams over a period time t through an area A. Thus, the simulation of the thermal effect of the waterproof fabric can be realized by specifying the heat and moisture transfer coefficients at the boundaries, as shown by the following formula:

$$H_{mn} = \frac{1}{W_n + \frac{1}{h_{mn}}} \tag{31}$$

$$H_{cn} = \frac{1}{R_n + \frac{1}{h_{nc}}} \tag{32}$$

When the fabric is laminated or coated with microporous or hydrophilic films as being waterproof, the combined mass and heat transfer coefficients H_{mn} and H_{cn} ($n = 0.1$) can be obtained by integrating the water vapor resistance W_n and the thermal resistance R_n of the waterproof fabric.

6.21.5 PHASE CHANGE MATERIAL FABRIC

The PCM which has the ability to change its phase state within a certain temperature range, such as, from solid to liquid or from liquid to solid, is microencapsulated inside the textile fabrics in the functional clothing design in recent years to improve the thermal performance of clothing when subjected to heating or cooling by absorbing or releasing heat during a phase change at their melting and crystallization points.

With PCM technology, the temperature of clothing is able to gain a change delay due to the energy released/absorbed from the PCM when exposed to a very hot/cold environment.

In the textile application, the PCM is enclosed in small plastic spheres with diameters of only a few micrometers. These microscopic spheres containing PCM are called PCM microcapsules, and are either embedded in the fibers or coated on the surface of the fabric.

Research on qualifying the effect of the PCM fabric in clothing on the heat flow from the body was experimentally conducted by Shim. He measured the effect of PCM clothing on heat loss and gain from the manikin which moved from a warm environment to a cold environment and back again. Ghali et al. [564] analyzed the sensitivity of the amount of PCM inside the textile material on the fabric thermal performance. The percentage of PCM in the fabric was found to influence the length of time period during which the phase change occurs.

Also, they drew a conclusion that under steady-state environmental conditions, PCM has no effect on the thermal performance; while when there is a sudden change in the ambient temperature, PCM can delay the transient response and decrease the heat loss from the human body.

In order to investigate the mechanisms of thermal regulation of the PCM on the heat and moisture transfer in textiles, Li and Zhu developed a mathematical model to describe the energy loss rate from the microspheres which is considered to be a sphere consisting of solid and liquid phases [565], as shown in the following equations:

$$\frac{\partial T_{ms}(x,r,t)}{\partial t} = a_{ms} \frac{1}{r^2} \frac{\partial}{\partial r} (r^2 \frac{\partial T_{ms}(x,r,t)}{\partial r}) \text{Spherical core} \tag{33}$$

$$\frac{\partial T_{ms}(x,r,t)}{\partial t} = a_{ms} \frac{1}{r^2} \frac{\partial}{\partial r} (r^2 \frac{\partial T_{ms}(x,r,t)}{\partial r}) \text{Spherical shell} \tag{34}$$

These energy governing equations are developed on the radial coordinate. r denotes the radius of the latest phase interface in micro-PCM. The smaller is the radius of the microspheres, the more significant is the effect of the PCM.

6.21.6 HUMAN BODY THERMOREGULATORY SYSTEM

To sustain life in various environments, the human body must have the ability to keep the temperature of core and skin at a reasonable range under a variety of external conditions, that is, the core body temperature should be maintained at $37.0\pm0.5°C$, and the skin temperature should be managed at approximately 33°C. In the human body, the regulation of the body temperature is implemented by the thermoregulation system, which responds to produce/dissipate heat when the body core temperature drops/rises. The basic mechanism of the body thermoregulation system involves two processes:

(i) when the body feels warm, the blood vessels react with vasodilatation and the glands begin to perform the sweating process;

(ii) when the body feels cold, the blood vessels reduce the blood flow to the skin and increase the heating production by muscle shivering. The schematics of thermoregulatory system of the human body.

The thermal mathematical models for human body can be classified for a single part and/or for the entire body. The models for a single part of the body, which was usually developed by physiologists, most of them are too complicated to simulate the physiological and anatomical details of the specific parts. It seems that these partial models theoretically can be added together to form a complete representation of the thermal exchange of the whole body. However, Fu claimed that such methods were not practical due to the fact that the connection between these models is very difficult, and adding the models together would need a supercomputer even if the connection between models is feasible. The thermal models for the entire body, reviewed by Fu, can be characterized by the following classifications:

(i)   one-node models,

(ii) two-node thermal models,

(iii) multinode models,

(iv) multielement models.

Though most of these models are likely to produce acceptable simulation results under the condition that the temperature is relatively uniform throughout the body, the multinode and multielement models perform better when large temperature gradients exist within the body because of the greater amount of details provided about the body temperature fields.

6.21.7 ONE-NODE THERMAL MODELS

One-node thermal models, in which the human body is represented by one node, are also called empirical models. They usually depend on experiments to determine the thermal response of the human body, and therefore, are not mathematical models in a phenomenological sense. A well-known empirical prediction model for the entire human body was reported by Givoni and Goldman. It was derived by fitting curves to the experimental data obtained from the subjects exposed to various environments.

6.21.8 TWO-NODE THERMAL MODELS

Two-node thermal models tend to divide the entire human body into two concentric shells of an outer skin layer and a central core representing internal organs, bone, muscle and tissue. The temperature of each node is assumed to be uniform. The energy balance equations are usually developed for each node and solved to produce the skin and core temperature and other thermal responses.

The early two-node models were not widely used by people due to the lack of sufficient consideration of the complicated physiological phenomena of the human body. Gagge et al. [566] introduced a more complete two-node model for the entire body, which includes the unsteady-state energy balance equation for the entire human body and two energy balance equations for the skin node and core node, as listed in the following:

$$S = M - E_{res} - C_{res} - E_{sk} - R - C - W \tag{35}$$

$$S_{cr} = M - E_{res} - C_{res} - W - (K_{min} + c_{pbl}V_{bl})(T_{cr} - T_{sk}) \tag{36}$$

$$S_{sk} = (K_{min} + c_{pbl}V_{bl})(T_{cr} - T_{sk}) - E_{sk} - R - C \tag{37}$$

Gagge et al. [566] later improved their two-node model by the development of the thermal control functions for the blood flow rate, the sweat rate, and the shivering metabolic rate.

$$V_{bl} = [6.3 + 200(WARM_{cr})]/[1+0.1(COLD_{cr}] \tag{38}$$

$$RSW = 4.72 \times 10^{-5}.WARM_{bm}.e^{\left(\frac{WARM_{cr}}{10.7}\right)} \tag{39}$$

$$M = 58.2 + 19.4.COLD_{cr}.COLD_{sk} \tag{40}$$

Due to the two-node nature of this model, it is able to be applied easily and simply with the straightforward numerical solution. Smith pointed out that Gagge's model was applicable for situations with moderate levels of activity and uniform environment conditions. However, due to the limitation imposed by only two nodes, Gagge's model can only be applied under uniform environmental circumstances.

6.21.9 THE MULTI-NODE MODELS

Multi-node models divided the entire human body into more than two nodes and developed energy balance equation for each node as well the control functions for blood flow rate, shivering metabolic rate, and so on. Stolwijk et al. [567] presented a more complex multinode mathematical thermal model of the entire human body, in which many efforts are made to the statement of the thermal controller. This model firstly divided the body into six cylindrical parts of head, trunk, arm, hands, legs and feet and a spherical body part comprising the head. Each part is further divided into four concentric shells of core, muscle, fat and skin tissue layers. Specifically, in this model all the blood circulation in the human body is regarded as a node called the central blood pool, which is the only communication connecting each body part.

Therefore, Stolwijk's model is also called a 25-node model, and energy balance equations are developed for each node with the assumption of uniform temperature in each layer, which includes heat accumulation, blood convection, tissue conduction, metabolic generation, respiration and heat transfer to the environment. The descriptions for these equations are shown as follow:

Core layer: $C(i,1)\frac{dT(i,1)}{dt}$=Q(i,1)-B(i,1)-D(i,1)-RES(i,1) $\qquad$ (41)

Muscle layer: $C(i,2)\frac{dT(i,2)}{dt}$=Q(i,2)-B(i,2)+D(i,1)-D(i,2) $\qquad$ (42)

Fat layer: $C(i,3)\frac{dT(i,3)}{dt}$=Q(i,3)-B(i,3)+D(i,2)-D(i,3) $\qquad$ (43)

Skin layer: $C(i,4)\frac{dT(i,4)}{dt}$=Q(i,4)-B(i,4)+D(i,3)-E(i,4)-Q(I,4) $\qquad$ (44)

Similar to Gagge's model [566], Stolwijk et al. [566] also developed the thermal control functions in terms of tissue temperature signals, in which the warm signal and cold signal are corresponding to warm and clod receptors of the skin and are calculated by the error signal. These controller equations produce the signals to drive the regulator, including total efferent sweat command, total efferent shivering command, total efferent skin vasodilation command, and total efferent skin vasoconstriction command.

Stolwijk's model [566] has made much advancement compared to previous multinode models as it is not only capable of calculating the spatial temperature distribution for each node, but also has improved the representation of the human's circulatory system since the blood circulation is the most important function of human body. This model has been validated with the good agreements between the experimental and predicted results of most cases. The limitation of this model is that it cannot be used for the highly nonuniform environmental situations caused from the negligence of spatial tissue temperature gradients.

6.21.10   MULTI-ELEMENT MODELS

The greater difference between the multi-element thermal models and the two-node or multinode models is that it divides the human body into several parts or elements without further division, and the temperature of each part or elements is no longer assumed as uniform. With the lifting of node uniform assumption, the mathematical descriptions of thermal functions, circulation, respiration etc. have also become more detailed to correspond with the detailed temperature filed.

Wissler [557] developed a multi-element model for the entire body by dividing the human body into six elements: head, torso, two arms and two legs, which were connected by the heart and lung where venous streams were mixed. Later on, Wissler improved his model and divided the human body into 15 elements to represent the head, thorax, abdomen, and the proximal, medical, and distal segments of the arms and legs, which were connected by the vascular system composed of arteries, veins and capillaries. Energy balance equations for each element and the arterial and venous pools were developed with the assumption that the blood temperature of arteries and veins in each element were uniform, and the thermal control equations for blood flow rates, the shivering metabolic rate and the sweat rate were built up.

The limitations of this model are that it is not applicable to the situations where a large internal temperature gradient or highly nonuniform environmental conditions exist. Additionally, the effect of the vasodilation and vasoconstriction was not included in the model. Finally, the parameters and constants used in the control equation are not easy to determine.

Smith developed a 3-D, transient, multielement thermal model for the entire human body with detailed control functions for the thermoregulation system. Compared to the previous models, the improvements were that he:

(i)   developed a 3-D temperature description of the human body;
(ii) provided a detailed description of the circulatory system, the respiratory system and the control system;
(iii) employed the finite-element method to get the numerical solutions of the model, which made a 3-D transient model for the entire human body possible.

The model divided the human body in 15 cylindrical body parts: head, neck, torso, upper arms, thighs, forearms, calves, hand and feet. Each body part is connected only by the blood flow and without tissue connection. The simulation results showed this model works well for situations of human thermal response during sedentary conditions in both uniform and nonuniform environments for either hot or cold stress conditions. However, the behaviors of the model during cold or hot exercising conditions were less satisfactory.

Fu summarized the limitations of the previous models and developed a 3-D transient, mathematical thermal model for the clothed human to simulate the clothed human thermal response under different situations. The main improvement of this model is the addition of the subcutaneous fat layer, the accumulation of moisture on the skin, and the blood perfusion and blood pressure to Smith's model. The development of the human model includes the thermal governing equations of the passive and control systems. Fu's 3-D transient model has a good ability of simulating the human body thermoregulatory system in situations where there exists high temperature vibration and even in the extremely atrocious weather conditions.

Though the multi-elements models can predict the thermal status of the human body in more detail, however, it should be noticed that there are many difficulties in applying the multi-element models into clothing engineering design due to the following considerations:

(i)   the multi-element model requires the clothing to be 3-D meshed and modeled, which may cause great complexity in the integration of the models of clothing and human body, and the computation load is very intensive;

(ii) the many parameters involved in the multi-element models have high demanding on the data availability in the engineering application.

6.21.11 INTERFACES BETWEEN THE BODY SKIN AND CLOTHING

In the daily life, the clothing acts as an important barrier for heat and vapor transfer between the skin and the environment, protecting against extreme heat and cold, but meanwhile hampering the loss of superfluous heat during physical effort. This barrier is formed by the clothing materials themselves and by the air they enclose as well as the still air bound to the outer surfaces of the clothing.

Some researchers experimentally observed the phenomena of heat and moisture that exchange actively between the clothing and skin, and found the exchanged amount is considerable compared to the total increase/decrease volume. The maximum heat flow from the skin to clothing depends on the heat conductivity of the inner layer of clothing and covering area of skin. Also, the heat exchange between the human body and clothing is dependent on the external parameters, such as, air temperature, air humidity, and wind speed.

The clothing with the least porous, greatest thickness and lowest permeability will provide the greatest protection to heat and perspiration from the skin to environment provided with least porous and thickest thickness.

Li and Holcombe [568] reported a new model by interfacing the model for a naked body with a heat and moisture transfer model of a fabric. They developed the boundary condition between the body and clothing by quantifying the heat and mass flow.

At the fabric-skin interface

Heat:

$$M_t = h_{ti}(T_{sk} - T_{fi}) \tag{45}$$

Mass:

$$M_d = h_{ci}(C_{sk} - C_{fi}) + L_{sk}\frac{\partial C_{sk}}{\partial t} \tag{46}$$

At the interface between fabric and ambient air
Heat:

$$K\frac{\partial T}{\partial x} = -h_t(T - T_{ab}) \tag{47}$$

Heat:

$$D_a\frac{\partial C_a}{\partial x} = -h_c(C_a - C_{ab}) \tag{48}$$

Since the air spacing between the skin and the fabric continuously varies in time depending on the level of activity and the location, the thermal transfer processes are influenced by the ventilating motion of air through the fabric initialed from the relative motion between the body and surrounding environment, such as, in the walking situations.

Ghali et al. [569] developed a 1D model of the human body. The oscillating trapped air layer gap width and the periodically ventilated fabric predict the effect of walking on exchanges of heat and mass. Murakami et al. [570] presented a numerical simulation of the combined radiation and moisture transport for heat release from a naked body in a house where continuous slight air flow exists by considering the thermal interaction between the human body and the environment and the intrinsic complex air situation in the real world.

Though much attention has been paid to the simulation of the thermal behaviors in the integrated system of human body, clothing and environment, and some numerical algorithms were reported for the simulation, these research studies put their focuses only on the scientific exploration and investigation. Few are developed systemically with a user-oriented purpose and used for clothing functional design.

6.21.12   CLOTHING THERMAL ENGINEERING DESIGN

6.21.12.1   CLOTHING THERMAL FUNCTIONAL DESIGN

The clothing thermal functional design, if following the traditional way of the clothing design and production, begins from the conception design and the prototypes making. As a result, a series of testing configured with experimental protocols will be performed by employing wearing subjects or by thermal manikins, and related thermal data will be measured using various equipment's during the experiments. Based on the analysis of experimental data, designers attempt to find the difference of measured thermal functions of clothing and their design concepts to obtain feedback to improve their design. After the iterative trial and error process, the final products can be put on the market. This traditional design process is very expensive, time-consuming and tedious due to the real prototypes making, experimental testing and burdensome data analysis.

These shortcomings of the traditional design method make it difficult to satisfy the requirements of designers and manufacturers. People come to resort to the powerful capacity of the computer in the design process of thermal functional clothing. Antunano et al. [571] employed a computer model and heat-humidity index to evaluate the heat stress in protective clothing. Schewenzferier et al. [572] optimized thermal protective clothing by using a knowledge bank concept and a learning expert system. Computer has acted a role in the clothing thermal functional design. However, it still cannot directly help the designer to preview the thermal performance of clothing, which is a crucial function for clothing thermal functional design.

James et al. [573] applied the commercial software of computational fluid dynamics in their strategy to simulate the heat and moisture diffusive and convective transport as well as effect of sweating to predict the performance of chemical and steam/fire protective clothing. Kothari et al. [574] simulated the convective heat transfer through textiles with the help of computational fluid dynamics to observe the effects of convection on the total heat transfer of the fabric. The software tools like CFD provide a possible pathway for the user to simulate the heat and fluid distribution in the clothing. However, these tools do not take into account the structural features of the textile materials and the special features of the heat and moisture transfer process in textile materials that are related to the physical properties and chemical compositions of the textile materials. They cannot reflect the practical wearing situation and preview the true complex thermal behaviors in clothing.

In order to obtain scientific simulation of clothing thermal performance, some researchers have made efforts to apply the theoretical models describing the complex heat and moisture behaviors in clothing wearing system to the clothing thermal functional design. Parsonin [575] adopted thermal models for the clothed body including human thermoregulation and clothing to work as tools for evaluating clothing risks and controls. Prasad et al. in 2002 [576] constructed a detail mathematical model to study transient heat and moisture transfer through wet thermal liners and evaluated the thermal performance of fire fighter protective clothing.

Their research has made good exploration in clothing thermal functional design with the computer tools. Design and evaluation models, experts systems, applications of CFD software, and theoretical simulation models have been used to help to carry out clothing thermal functional designs. They either simply focus on evaluating certain thermal properties of clothing or need specialized knowledge to understand. They are very difficult to be applied as engineering tools for the general designers.

6.21.12.2. CAD SYSTEMS FOR CLOTHING THERMAL ENGINEERING DESIGN

The application of CAD technologies for clothing design is a significant sign of revolutionary advancement in the development of computerization and automation in the clothing industry. Clothing designers/engineers are offered a number of flexibilities in their design with CAD systems, such as, the usage of textile material, exploration of functional design and products display.

Currently many CAD packages are available, targeting at pattern design, garment construction, fashion design and physical fitting simulation, which have made many achievements in catering for different requirements of the clothing industry [577, 578]. They are helpful to shorten the design cycle and save time and money on the prototypes preparation as well as improve productivity considerably. Recently, 3D clothing design and visualizations have been developed to simulate and visualize the physical performance of the clothing wear on the body in 3D virtual ways

[577, 579], which enable the designs to be more realistic and make detailed analysis and evaluation of the clothing mechanical performance. However, these pioneer achievements are mainly focused on the mechanical behaviors of clothing.

The newest interest in the CAD system for clothing functional design places the focus on the thermal behaviors of clothing. Clothing thermal engineering design with CAD systems and tools is an effective and economical solution of designing clothing with superior thermal performance for wearing in various environments with a feeling of comfort. The CAD system for clothing thermal engineering design aims to create a virtual platform, which offers designers the ability to conceive their products using the engineering method. Thermal engineering design of clothing is the application of a systematic and quantitative way of designing and engineering of clothing with the interdisciplinary combination of physical, physiological, mathematical, computational and software science, and engineering principles to meet with the thermal biological needs of protection, survival and comfort of human body.

The research in these fields discussed in the above sections lays substantial foundation to achieve this strategy. With this engineering design system, designers can simulate the thermal behaviors of the clothing, human body and environment system in specified scenarios, preview and analyze the thermal performance of the clothing, and iteratively improve their designs for desirable thermal functions of clothing.

In order to help the designers and manufacturers to quickly carry out clothing thermal functional design, Mao et al. [580] have developed two software systems, respectively, for multilayer and multistyle thermal functional design of clothing. With these tools, the designers and manufacturers may have a quick preview of the different thermal performance when using different textile materials to have a comparison and make decision; or they can consider the different style (long, media or short) of the clothing and its thermal performance on the different body parts. The thermal performance of the sportswear can be quickly simulated and validated with the CAD system before the physical pattern making to reduce the design cycle and lower the design cost.

This new application of CAD technologies demonstrates great potentials in the clothing thermal engineering design, because the capacity of simulating and predicting the thermal performance of clothing is indispensable for designing clothing for thermal protection and comfort. While physical fit and a good-looking fashion style are crucial aims of clothing design, the thermal performance of clothing is another critical aspect that relates to the survival, health and comfort of human beings living in various environmental conditions. As more and more consumers want to wear clothing with higher functional and comfort performance, there is an urgent need for a CAD system to design clothing and analyze its thermal performance effectively and efficiently, desirable functions, and then produce real products. The theoretical research of physics, chemistry related to textile and clothing, physiological thermo-regulations of the human body, and the dynamic interactions between the clothing and body lead to thermal engineering design of clothing to achieve desirable thermal functions. The computational simulation can be enabled by related mathematical models, which is the substantial foundation for the engineering design, and the CAD systems provide a friendly tool for users through a series of functionalities to quickly carry out engineering design of clothing for desirable thermal functions. This strategy of thermal engineering design of clothing can hopefully speed up the design cycle and reduce the design cost.

6.22   CONCLUDING REMARKS

With only a few exceptions, most of the open literature on yarns from electrospun fibers focuses on the process of yarn formation, rather than on the yarns obtained. Although the process is certainly an important aspect, researchers should bear in mind that ultimately the intended end-user of their results will be the fibers and textiles manufacturing industry. With this in mind, future research should pay more attention to the properties of the yarns obtained, and report more on these with specific focus on tenacity, elasticity and linear density values.

To the best of our knowledge, there is currently no commercially available continuous nanofiber yarn produced through electrospinning. This is likely to change in the very near future and will lead to rapid worldwide evaluation of the product for numerous potential applications. This will also lead to evaluation of nanofiber yarn properties under 'real world' circumstances and results should indicate where further work is required.

Although electrospinning is more than a hundred years old and processes for producing electrospun nanofiber yarns have existed for more than 60 years, little is known about the mechanical and other properties of nanofibers, and especially their twisted yarns. Some recent work has focused on some of the properties of twisted yarns of specific polymers, as discussed in Sections 3.4 and 3.5, but many of the unknowns still need to be investigated.

There are certain drawbacks of the electrospinning process, such as, low production rates, the requirement for proportionately large quantities of industrial solvents, and the necessity for electrospinning through small, needlelike capillaries that tend to get blocked over time-through precipitation of small amounts of the spinning polymer at the capillary mouth. There are also certain challenges related to repeatedly electrospinning core-shell and hollow core fibers and obtaining sub-100 nm fiber diameters of all spinnable polymers without the formation of bead defects. Future work will undoubtedly be aimed at eliminating these drawbacks. Possible solutions to these problems could include multiple-jet, needleless spinning to overcome low production rates and needle blockage. The amounts of industrial solvents used in electrospinning could possibly be reduced or eliminated through application of supercritical fluid techniques or water-based emulsion chemistry.

In the field of tissue engineering, more information on and a better understanding of the wettability and permeability of nanofiber yarns, as well as their structural properties as a function of biodegradation, should lead to the development of highly functional tissue scaffolds and wound dressings.

Successful electrospinning of other materials, such as, metals and nonoxide ceramics, and better control over the crystallinity of electrospun polymer fibers will lead to significant advances in nanofiber reinforced composite materials.

Quality control methods for nanofiber yarn production processes will need to be developed for commercialization purposes, and other new problems, which will also arise once nanofiber yarns are spun on an industrial scale, will have to be dealt with. On a purely esthetic level, once nanofibers are incorporated into wearable textiles, the question of coloration will arise. This might pose some problems, since fibers with diameters smaller than the optically visible wavelength range are seen through diffraction of light, not reflection, and therefore they usually appear white under normal circumstances.

KEYWORDS

- **electrospun nanofibers**
- **modern applications**
- **nanoengineered materials**
- **nanotechnology**
- **textile industry**

REFERENCES

1. Selin, C., *Expectations and the Emergence of Nanotechnology.* Science, Technology & Human Values, 2007, 32(2), 196–220.
2. Mansoori, G. A., *Principles of Nanotechnology: Molecular-Based Study of Condensed Matter in Small Systems.* 2005, World Scientific.
3. Peterson, C. L., *Nanotechnology: From Feynman to the Grand Challenge of Molecular Manufacturing.* Technology and Society Magazine, IEEE, 2004, 23(4), 9–15.
4. Gleiter, H., *Nanostructured Materials: Basic Concepts and Microstructure.* Acta Materialia, 2000, 48(1), 1–29.
5. Peterson, C. L., *Nanotechnology-Evolution of the Concept.* Journal of the British Interplanetary Society, 1992, 45, 395–400.
6. Yadugiri, V. T., R. Malhotra, *'Plenty of Room'-Fifty Years after the Feynman Lecture.* Current Science, 2010, 99(7), 900–907.
7. Wilson, K., *History of Textiles.* 1979, Westview Press.
8. Harris, J., *Textiles, 5,000 Years: An International History and Illustrated Survey.* 1995, New York City, United States: Harry N. Abrams, Inc.

9. Sawhney, A. P. S., et al., *Modern Applications of Nanotechnology in Textiles.* Textile Research Journal, 2008, 78(8), 731–739.

10. Tao, X., *Smart Fibers, Fabrics and Clothing.* Vol. 20, 2001, Woodhead publishing.

11. Tyrer, R. B., *The Demographic and Economic History of the Audiencia of Quito: Indian Population and the Textile Industry, 1600–1800,* 1976, University of California, Berkeley.

12. Chapman, S. D., S. Chassagne, *European Textile Printers in the Eighteenth Century: A Study of Peel and Oberkampf.* 1981, Heinemann Educational Books, Pasold Fund.

13. Jeremy, D. J., *Transatlantic Industrial Revolution: The Diffusion of Textile Technologies between Britain and America, 1790–1830s.* 1981, MIT Press.

14. Hekman, J. S., *The Product Cycle and New England Textiles.* The Quarterly Journal of Economics, 1980, 94(4), 697–717.

15. Mantoux, P., *The Industrial Revolution in the Eighteenth Century: An Outline of the Beginnings of the Modern Factory System in England.* 2013, Routledge.

16. French, G. J., *Life and Times of Samuel Crompton of Hall-in-the-Wood: Inventor of the Spinning Machine Called the Mule.* 1862, Charles Simms and Co.

17. O'brien, P., *The Micro Foundations of Macro Invention: The Case of the Reverend Edmund Cartwright.* Textile History, 1997, 28(2), 201–233.

18. Aspin, C., *The Cotton Industry.* 1981, Osprey Publishing.

19. Smout, T. C., *The Development and Enterprise of Glasgow 1556-1707.* Scottish Journal of Political Economy, 1959, 6(3), 194–212.

20. Sale, K., *Rebels against the Future: The Luddites and their War on the Industrial Revolution: Lessons for the Computer Age.* 1995, Basic Books.

21. Aspin, C., *The Woollen Industry.* 1982, Osprey Publishing.

22. Beckert, S., *Emancipation and Empire: Reconstructing the Worldwide Web of Cotton Production in the Age of the American Civil War.* The American Historical Review, 2004, 109(5), 1405–1438.

23. Heberlein, G., *Georges Heberlein,* 1935, Google Patents.

24. Townswnd, B. A., *Full Circle in Cellulose.* 1993, New York: Elsevier Applied Science.

25. Hicks, E. M., et al., *The Production of Synthetic-Polymer Fibres.* Textile Progress, 1971, 3(1), 1–108.

26. Cook, J. G., *Handbook of Textile Fibres: Man-Made Fibres.* Vol. 2, 1984, Elsevier.

27. Conrad, D. J., et al., *Ink-Printed, Low Basis Weight Nonwoven Fibrous Webs and Method,* 1995, Google Patents.

28. Sawhney, A. P. S., et al., *Modern Applications of Nanotechnology in Textiles.* Textile Research Journal, 2008, 78(8), 731–739.

29. Brown, P., K. Stevens, *Nanofibers and Nanotechnology in Textiles.* 2007, Elsevier.

30. Valko, E. I., *Textile Material,* 1966, Google Patents.

31. Higgins, L., M. E. Anand, *Textiles Materials and Products for Activewear and Sportswear.* Technical Textile Market, 2003, 52, 9–40.

32. Hunter, L., *Mohair, Cashmere and other Animal Hair Fibres,* in *Handbook of Natural Fibres,* R. M. Kozłowski, Editor. 2012, Woodhead Publishing. 196–290.

33. Kuffner, H., C. Popescu, *Wool Fibres,* in *Handbook of Natural Fibres,* R. M. Kozłowski, Editor. 2012, Woodhead Publishing. 171–195.

34. Roff, W. J., J. R. Scott, *Silk and Wool,* in *Fibres, Films, Plastics and Rubbers,* W. J. Roff and J. R. Scott, Editors. 1971, Butterworth-Heinemann. 188–196.

35. Gordon, S., *Identifying Plant Fibres in Textiles: The Case of Cotton,* in *Identification of Textile Fibers,* M. M. Houck, Editor. 2009, Woodhead Publishing. 239–258.

36. Shahid ul, I., M. Shahid, F. Mohammad, *Perspectives for Natural Product Based Agents Derived from Industrial Plants in Textile Applications – A Review.* Journal of Cleaner Production, 2013, 57(0), 2–18.

37. Ansell, M. P., L. Y. Mwaikambo, *The Structure of Cotton and other Plant Fibres*, in *Handbook of Textile Fibre Structure*, S. J. Eichhorn, et al., Editors. 2009, Woodhead Publishing. 62–94.

38. Brown, R. C., et al., *Pathogenetic Mechanisms of Asbestos and other Mineral Fibres*. Molecular Aspects of Medicine, 1990, 11(5), 325–349.

39. Porter, R. M., *Glass Fiber Compositions*, 1991, Google Patents.

40. Milašius, R., V. Jonaitienė, *Synthetic Fibres for Interior Textiles*, in *Interior Textiles*, T. Rowe, Editor. 2009, Woodhead Publishing. 39–46.

41. Rugeley, E. W., T. A. Feild Jr, J. L. Petrokubi, *Synthetic Textile Articles*, 1947, US Patent

42. Woolman, M. S., E. B. McGowan, *Textiles: A Handbook for the Student and the Consumer*. 1915, Macmillan.

43. Todd, M. P., *Hand-loom Weaving: A Manual for School and Home*. 1902, Rand, McNally.

44. Horne, P., S. Bowden, *Knitting and Crochet*. Design, 1974, 75(3), 38–38.

45. Burt, E. C., *Bark-Cloth in East Africa*. Textile History, 1995, 26(1), 75–88.

46. Halls, Z., *Machine-Made Lace 1780–1820 (Pillow and Bobbin to Bobbin and Carriage)*. Costume, 1970, 4(Supplement-1), 46–50.

47. Smith, C. W., J. T. Cothren, *Cotton: Origin, History, Technology, and Production*. Vol. 4, 1999, John Wiley & Sons.

48. Kriger, C. E., *Guinea Cloth': Production and Consumption of Cotton Textiles in West Africa before and during the Atlantic Slave Trade*. The Spinning World. A Global History of Cotton Textiles, 2009, 1200–1850.

49. Betrabet, S. M., K. P. R. Pillay, R. L. N. Iyengar, *Structural Properties of Cotton Fibers: Part II: Birefringence and Structural Reversals in Relation to Mechanical Properties*. Textile Research Journal, 1963, 33(9), 720–727.

50. Alston, J. M., J. D. Mullen, *Economic Effects of Research into Traded Goods: The Case of Australian Wool*. Journal of Agricultural Economics, 1992, 43(2), 268–278.

51. Company, I. T., I. C. Schools, *Wool, Wool Scouring, Wool Drying, Burr Picking, Carbonizing, Wool Mixing, Wool Oiling: Woolen Carding, Woolen Spinning, Woolen and Worsted Warp Preparation*. Vol. 79, 1905, International Textbook Company.

52. Parkes, C., *The Knitter's Book of Wool: The Ultimate Guide to Understanding, Using, and Loving this Most Fabulous Fiber*. 2011, Random House LLC.

53. D. Arden, L. D., A. Pfister, *Woolen and Worsted Yarn for an Elizabethan Knitted Suite*. 2003.

54. Knapton, J. J. F., et al., *The Dimensional Properties of Knitted Wool Fabrics Part I: The Plain-Knitted Structure*. Textile Research Journal, 1968, 38(10), 999–1012.

55. Oikonomides, N., *Silk Trade and Production in Byzantium from the Sixth to the Ninth Century: The Seals of Kommerkiarioi*. Dumbarton Oaks Papers, 1986, 40, 33–53.

56. Lock, R. L., *Process for Making Silk Fibroin Fibers*, 1993, Google Patents.

57. Kaplan, D. L. *Silk Polymers*. in *Workshop on Silks: Biology, Structure, Properties, Genetics (1993, Charlottesville, Va.)*. 1994. American Chemical Society.

58. Bolton, E. K., *Chemical Industry Medal. Development of Nylon*. Industrial & Engineering Chemistry, 1942, 34(1), 53–58.

59. Li, L., et al., *Formation and Properties of Nylon-6 and Nylon-6/Montmorillonite Composite Nanofibers*. Polymer, 2006, 47(17), 6208–6217.

60. Harazoe, H., M. Matsuno, S. Noda, *Method of Manufacturing Polyesters*, 1996, Google Patents.

61. Mijatovic, D., J. C. T. Eijkel, A. V. D. Berg, *Technologies for Nanofluidic Systems: Top-Down vs. Bottom-Up—A Review*. Lab on a Chip, 2005, 5(5), 492–500.

62. Patra, J. K., S. Gouda, *Application of Nanotechnology in Textile Engineering: An Overview*. Journal of Engineering and Technology Research, 2013, 5(5), 104–111.

63. Mohanraj, V. J., Y. Chen, *Nanoparticles-A Review*. Tropical Journal of Pharmaceutical Research, 2007, 5(1), 561–573.

64. Mehnert, W., K. Mäder, *Solid Lipid Nanoparticles: Production, Characterization and Applications*. Advanced Drug Delivery Reviews, 2001, 47(2), 165–196.

65. Roney, C., et al., *Targeted Nanoparticles for Drug Delivery through the Blood–Brain Barrier for Alzheimer's Disease*. Journal of Controlled Release, 2005, 108(2), 193–214.

66. Joshi, M., A. Bhattacharyya, *Nanotechnology–A New Route to High-Performance Functional Textiles*. Textile Progress, 2011, 43(3), 155–233.

67. Kaounides, L., H. Yu, T. Harper, *Nanotechnology Innovation and Applications in Textiles Industry: Current Markets and Future Growth Trends*. Materials Science and Technology, 2007, 22(4), 209–237.

68. Kathiervelu, S. S., *Applications of Nanotechnology in Fibre Finishing*. Synthetic Fibres, 2003, 32(4), 20–22.

69. Wang, C. C., C. C. Chen, *Physical Properties of the Crosslinked Cellulose Catalyzed with Nanotitanium Dioxide under UV Irradiation and Electronic Field*. Applied Catalysis A: General, 2005, 293, 171–179.

70. Price, D., et al., *Burning Behaviour of Foam/Cotton Fabric Combinations in the Cone Calorimeter*. Polymer Degradation and Stability, 2002, 77(2), 213–220.

71. Lamba, N. M. K., K. A. Woodhouse, S. L. Cooper, *Polyurethanes in Biomedical Applications*. 1997, CRC press.

72. Qian, L., J. P. Hinestroza, *Application of Nanotechnology for High Performance Textiles*. Journal of Textile and Apparel, Technology and Management, 2004, 4(1), 1–7.

73. Singh, V. K., et al. *Applications and Future of Nanotechnology in Textiles*. in *National Cotton Council Beltwide Cotton Conference*. 2006.

74. Gibson, P., H. S. Gibson, C. Pentheny, *Electrospinning Technology: Direct Application of Tailorable Ultrathin Membranes*. Journal of Industrial Textiles, 1998, 28(1), 63–72.

75. Snyder, R. G., *The Bionic Tailor: TC2's 3-D Body Scanner*, 2001, IEEE-INST Electrical Electronics Engineers INC 345 E 47TH ST NY 10017–2394 USA.

76. Okabe, H., et al. *Three Dimensional Apparel CAD System*. in *ACM SIGGRAPH Computer Graphics*. 1992. ACM.

77. Yan, H., S. S. Fiorito, *Communication: CAD/CAM Adoption in US Textile and Apparel Industries*. International Journal of Clothing Science and Technology, 2002, 14(2), 132–140.

78. Xing, X., Y. Wang, B. Li, *Nanofibers Drawing and Nanodevices Assembly in Poly (Trimethylene Terephthalate)*. Optics Express, 2008, 16(14), 10815–10822.

79. Ikegame, M., K. Tajima, T. Aida, *Template Synthesis of Polypyrrole Nanofibers Insulated within One-Dimensional Silicate Channels: Hexagonal versus Lamellar for Recombination of Polarons into Bipolarons*. Angewandte Chemie International Edition, 2003, 42(19), 2154–2157.

80. He, L., et al., *Fabrication and Characterization of Poly (L-Lactic Acid) 3D Nanofibrous Scaffolds with Controlled Architecture by Liquid–Liquid Phase Separation from a Ternary Polymer–Solvent System*. Polymer, 2009, 50(16), 4128–4138.

81. Hartgerink, J. D., E. Beniash, S. I. Stupp, *Self-Assembly and Mineralization of Peptide-Amphiphile Nanofibers*. Science, 2001, 294(5547), 1684–1688.

82. Li, D., Y. Xia, *Fabrication of Titania Nanofibers by Electrospinning*. Nano Letters, 2003, 3(4), 555–560.

83. Ramakrishna, S., et al., *Electrospun Nanofibers: Solving Global Issues*. Materials Today, 2006, 9(3), 40–50.

84. Reneker, D. H., I. Chun, *Nanometer Diameter Fibres of Polymer, Produced by Electrospinning*. Nanotechnology, 1996, 7(3), 216–233.

85. Frenot, A., I. S. Chronakis, *Polymer Nanofibers Assembled by Electrospinning.* Current Opinion in Colloid & Interface Science, 2003, 8(1), 64–75.

86. Dersch, R., et al., *Electrospun Nanofibers: Internal Structure and Intrinsic Orientation.* Journal of Polymer Science Part A: Polymer Chemistry, 2003, 41(4), 545–553.

87. Yang, F., et al., *Electrospinning of Nano/Micro Scale Poly (L-Lactic Acid) Aligned Fibers and their Potential in Neural Tissue Engineering.* Biomaterials, 2005, 26(15), 2603–2610.

88. Moroni, L., et al., *Fiber Diameter and Texture of Electrospun PEOT/PBT Scaffolds Influence Human Mesenchymal Stem Cell Proliferation and Morphology, and the Release of Incorporated Compounds.* Biomaterials, 2006, 27(28), 4911–4922.

89. Jayakumar, R., et al., *Novel Chitin and Chitosan Nanofibers in Biomedical Applications.* Biotechnology Advances, 2010, 28(1), 142–150.

90. Zhang, Y., et al., *Recent Development of Polymer Nanofibers for Biomedical and Biotechnological Applications.* Journal of Materials Science: Materials in Medicine, 2005, 16(10), 933–946.

91. Nazarov, R., H.-J. Jin, D. L. Kaplan, *Porous 3-D scaffolds from regenerated silk fibroin.* Biomacromolecules, 2004, 5(3), 718–726.

92. Smit, E. A., *Studies towards High-throughput Production of Nanofiber Yarns,* 2008, Stellenbosch University: Stellenbosch.

93. Subbiah, T., et al., *Electrospinning of Nanofibers.* Journal of Applied Polymer Science, 2005, 96(2), 557–569.

94. Murugan, R., S. Ramakrishna, *Design Strategies of Tissue Engineering Scaffolds with Controlled Fiber Orientation.* Tissue Engineering, 2007, 13(8), 1845–1866.

95. Badrossamay, M. R., et al., *Nanofiber Assembly by Rotary Jet-Spinning.* Nano letters, 2010, 10(6), 2257–2261.

96. Srary, J., *Method of and Apparatus for Ringless Spinning of Fibers,* 1970, Google Patents.

97. Katta, P., et al., *Continuous Electrospinning of Aligned Polymer Nanofibers onto a Wire Drum Collector.* Nano Letters, 2004, 4(11), 2215–2218.

98. M. Creight, D. J., et al., *Short Staple Yarn Manufacturing.* Vol. 700, 1997, NC, USA: Carolina Academic Press Durham,

99. Pan, H., et al., *Continuous Aligned Polymer Fibers Produced by a Modified Electrospinning Method.* Polymer, 2006, 47(14), 4901–4904.

100. Dzenis, Y. A., *Spinning Continuous Fibers for Nanotechnology.* Science 2004, 304 1917–1919.

101. Jacobs, V., R. D. Anandjiwala, M. Maaza, *The Influence of Electrospinning Parameters on the Structural Morphology and Diameter of Electrospun Nanofibers.* Journal of Applied Polymer Science, 2010, 115(5), 3130–3136.

102. Sarkar, K., et al., *Electrospinning to Forcespinning.* Materials Today, 2010, 13(11), 12–14.

103. Smit, A. E., U. Buttner, R. D. Sanderson, *Continuous Yarns from Electrospun Nanofibers.* Nanofibers and Nanotechnology in Textiles. 2007, 45.

104. Hong, Y. T., et al., *Filament Bundle Type Nano Fiber and Manufacturing Method Thereof,* 2010, Google Patents.

105. Zhou, F. L., R. H. Gong, *Manufacturing Technologies of Polymeric Nanofibers and Nanofiber Yarns.* Polymer International, 2008, 57(6), 837–845.

106. Teo, W. E., S. Ramakrishna, *A Review on Electrospinning Design and Nanofiber Assemblies.* Nanotechnology, 2006, 17(14), p. R89-R106.

107. Kleinmeyer, J., J. Deitzel, J. Hirvonen, *Electro Spinning of Submicron Diameter Polymer Filaments,* 2003, Google Patents.

108. Childs, H. R., *Process of Electrostatic Spinning,* 1944, US Patent

109. Ko, F. K., *Nanofiber Technology: Bridging the Gap between Nano and Macro World,* in *Nanoengineered Nanofibrous Materials.* 2004. 1–18.

110. Murray, M. P., *Cone Collecting Techniques for Whitebark Pine.* Western Journal of Applied Forestry, 2007, 22(3), 153–155.

111. Hermes, J., *Apparatus for Winding Yarn,* 1986, Google Patents.

112. Yadav, A., et al., *Functional Finishing in Cotton Fabrics Using Zinc Oxide Nanoparticles.* Bulletin of Materials Science, 2006, 29(6), 641–645.

113. Perelshtein, I., et al., *A One-Step Process for the Antimicrobial Finishing of Textiles with Crystalline TiO$_2$ Nanoparticles.* Chemistry-A European Journal, 2012, 18(15), 4575–4582.

114. Jiu, J. T., et al., *The Preparation of MgO Nanoparticles Protected by Polymer.* Chinese Journal of Inorganic Chemistry, 2001, 17(3), 361–365.

115. Soane, D. S., et al., *Nanoparticle-Based Permanent Treatments for Textiles,* 2003, Google Patents.

116. Song, X. Q., et al., *The Effect of Nano-Particle Concentration and Heating Time in the Anti-Crinkle Treatment of Silk.* Journal of Jilin Institute of Technology, 2001, 22, 24–27.

117. Russell, E., *Nanotechnologies and the Shrinking World of Textiles.* Textile Horizons, 2002, 9(10), 7–9.

118. Wong, Y. W. H., et al., *Selected Applications of Nanotechnology in Textiles.* AUTEX Research Journal, 2006, 6(1), 1–8.

119. Dastjerdi, R., M. Montazer, *A Review on the Application of Inorganic Nano-Structured Materials in the Modification of Textiles: Focus on Anti-Microbial Properties.* Colloids and Surfaces B: Biointerfaces, 2010, 79(1), 5–18.

120. Dastjerdi, R., M. Montazer, S. Shahsavan, *A New Method to Stabilize Nanoparticles on Textile Surfaces.* Colloids and Surfaces A: Physicochemical and Engineering Aspects, 2009, 345(1–3), 202–210.

121. Nazari, A., M. Montazer, M. B. Moghadam, *Introducing Covalent and Ionic Cross-linking into Cotton through Polycarboxylic Acids and Nano TiO2.* Journal of The Textile Institute, 2012, 103(9), 985–996.

122. Gorjanc, M., et al., *The Influence of Water Vapor Plasma Treatment on Specific Properties of Bleached and Mercerized Cotton Fabric.* Textile Research Journal, 2010, 80(6), 557–567.

123. Mihailović, D., et al., *Functionalization of Cotton Fabrics with Corona/Air RF Plasma and Colloidal TiO2 Nanoparticles.* Cellulose, 2011, 18(3), 811–825.

124. Karimi, L., et al., *Effect of Nano TiO2 on Self-cleaning Property of Cross-linking Cotton Fabric with Succinic Acid Under UV Irradiation.* Photochemistry and photobiology, 2010, 86(5), 1030–1037.

125. Yuranova, T., et al., *Antibacterial Textiles Prepared by RF-Plasma and Vacuum-UV Mediated Deposition of Silver.* Journal of Photochemistry and Photobiology A: Chemistry, 2003, 161(1), 27–34.

126. M. Quaid, M., P. Beesley, *Extreme Textiles: Designing for High Performance.* 2005, Princeton Architectural Press.

127. Mahltig, B., H. Haufe, H. Böttcher, *Functionalization of Textiles by Inorganic Sol–Gel Coatings.* Journal of Materials Chemistry, 2005, 15(41), 4385–4398.

128. Kang, J. Y., M. Sarmadi, *Textile Plasma Treatment Review–Natural Polymer-Based Textiles.* American Association of Textile Chemists and Colorists Review, 2004, 4(10), 28–32.

129. Rahman, M., et al., *Tool-Based Nanofinishing and Micromachining.* Journal of Materials Processing Technology, 2007, 185(1), 2–16.

130. Wang, R. H., J. H. Xin, X. M. Tao, *UV-Blocking Property of Dumbbell-Shaped ZnO Crystallites on Cotton Fabrics.* Inorganic Chemistry, 2005, 44(11), 3926–3930.

131. Lee, H. J., S. Y. Yeo, S. H. Jeong, *Antibacterial Effect of Nanosized Silver Colloidal Solution on Textile Fabrics.* Journal of Materials Science, 2003, 38(10), 2199–2204.

132. Sennett, M., et al., *Dispersion and Alignment of Carbon Nanotubes in Polycarbonate.* Applied Physics A, 2003, 76(1), 111–113.

133. Weiguo, D., *Research on Properties of Nano Polypropylene/TiO₂ Composite Fiber.* Journal of Textile Research, 2002, 23(1), 22–23.

134. Meier, U., *Carbon Fiber-Reinforced Polymers: Modern Materials in Bridge Engineering.* Structural Engineering International, 1992, 2(1), 7–12.

135. Huang, Z. M., et al., *A Review on Polymer Nanofibers by Electrospinning and their Applications in Nanocomposites.* Composites Science and Technology, 2003, 63(15), 2223–2253.

136. Daoud, W. A., J. H. Xin, *Low Temperature Sol-Gel Processed Photocatalytic Titania Coating.* Journal of Sol-Gel Science and Technology, 2004, 29(1), 25–29.

137. Hartley, S. M., et al. *The next Generation of Chemical and Biological Protective Materials Utilizing Reactive Nanoparticles.* in *24th Army Science Conference.* 2004. Orlando, Florida, USA.

138. Wang, R. H., et al., *ZnO Nanorods Grown on Cotton Fabrics at Low Temperature.* Chemical Physics Letters, 2004, 398(1), 250–255.

139. Zhang, J., et al., *Hydrophobic Cotton Fabric Coated by a Thin Nanoparticulate Plasma Film.* Journal of Applied Polymer Science, 2003, 88(6), 1473–1481.

140. Yang, H., S. Zhu, N. Pan, *Studying the Mechanisms of Titanium Dioxide as Ultraviolet-Blocking Additive for Films and Fabrics by an Improved Scheme.* Journal of Applied Polymer Science, 2004, 92(5), 3201–3210.

141. El-Molla, M. M., et al., *Nanotechnology to Improve Coloration and Antimicrobial Properties of Silk Fabrics.* Indian Journal of Fibre and Textile Research, 2011, 36(3), 266–271.

142. Mondal, S., *Phase Change Materials for Smart Textiles–An Overview.* Applied Thermal Engineering, 2008, 28(11), 1536–1550.

143. Hu, J., *Advances in Shape Memory Polymers.* 2013, Elsevier.

144. Spencer, B. F., M. E. R. Sandoval, N. Kurata, *Smart Sensing Technology: Opportunities and Challenges.* Structural Control and Health Monitoring, 2004, 11(4), 349–368.

145. Diamond, D., et al., *Wireless Sensor Networks and Chemo-/Biosensing.* Chemical Reviews, 2008, 108(2), 652–679.

146. Tao, X. M., *Wearable Electronics and Photonics.* 2005, Elsevier.

147. Koncar, V., *Optical Fiber Fabric Displays.* Optics and Photonics News, 2005, 16(4), 40–44.

148. Sharma, A., et al., *Review on Thermal Energy Storage with Phase Change Materials and Applications.* Renewable and Sustainable Energy Reviews, 2009, 13(2), 318–345.

149. Zalba, B., et al., *Review on Thermal Energy Storage with Phase Change: Materials, Heat Transfer Analysis and Applications.* Applied Thermal Engineering, 2003, 23(3), 251–283.

150. Shin, Y., D. I. Yoo, K. Son, *Development of Thermoregulating Textile Materials with Microencapsulated Phase Change Materials (PCM). IV. Performance Properties and Hand of Fabrics Treated with PCM Microcapsules.* Journal of Applied Polymer Science, 2005, 97(3), 910–915.

151. B. García, L., et al., *Phase Change Materials (PCM) Microcapsules with Different Shell Compositions: Preparation, Characterization and Thermal Stability.* Solar Energy Materials and Solar Cells, 2010, 94(7), 1235–1240.

152. Tyagi, V. V., et al., *Development of Phase Change Materials Based Microencapsulated Technology for Buildings: A Review.* Renewable and Sustainable Energy Reviews, 2011, 15(2), 1373–1391.

153. Nelson, G., *Application of Microencapsulation in Textiles.* International Journal of Pharmaceutics, 2002, 242(1), 55–62.

154. Shim, H., E. A. M. Cullough, B. W. Jones, *Using Phase Change Materials in Clothing.* Textile Research Journal, 2001, 71(6), 495–502.

155. Gao, C., K. Kuklane, I. Holmer, *Cooling Vests with Phase Change Material Packs: The Effects of Temperature Gradient, Mass and Covering Area.* Ergonomics, 2010, 53(5), 716–723.

156. Tang, S. L. P., G. K. Stylios, *An Overview of Smart Technologies for Clothing Design and Engineering.* International Journal of Clothing Science and Technology, 2006, 18(2), 108–128.

157. Buckley, T. M., *Flexible Composite Material with Phase Change Thermal Storage*, 1999, Google Patents.

158. Soares, N., et al., *Review of Passive PCM Latent Heat Thermal Energy Storage Systems towards Buildings' Energy Efficiency.* Energy and Buildings, 2013, 59, 82–103.

159. Park, J. Y., et al., *Growth kinetics of nanograins in SnO₂ fibers and size dependent sensing properties.* Sensors and Actuators B: Chemical, 2011, 152(2), 254–260.

160. Cho, C. G., *Shape Memory Material*, in *Smart Clothing*. 2010. 189–221.

161. Wang, M., X. Luo, D. Ma, *Dynamic Mechanical Behavior in the Ethylene Terephthalate-Ethylene Oxide Copolymer with Long Soft Segment as a Shape Memory Material.* European Polymer Journal, 1998, 34(1), 1–5.

162. Mother, P. T., H. G. Jeon, T. S. Haddad, *Strain Recovery in POSS Hybrid Thermoplastics.* Polymer Preprints (USA), 2000, 41(1), 528–529.

163. Otsuka, K., C. M. Wayman, *Shape Memory Materials.* 1999, Cambridge University Press.

164. B. Laurent, H., H. Dürr, *Organic Photochromism (IUPAC Technical Report).* Pure and Applied Chemistry, 2001, 73(4), 639–665.

165. Mattila, H. R., *Intelligent Textiles and Clothing.* Vol. 3, 2006, CRC press England.

166. Kiekens, P., S. Jayaraman, *Intelligent Textiles and Clothing for Ballistic and NBC Protection.* 2011, Springer.

167. Rantanen, J., et al., *Smart Clothing Prototype for the Arctic Environment.* Personal and Ubiquitous Computing, 2002, 6(1), 3–16.

168. Lind, E. J., et al. *A Sensate Liner for Personnel Monitoring Applications.* in *First International Symposium on Wearable Computers*. 1997. IEEE.

169. Lee, K., D. S. Kwon. *Wearable Master Device Using Optical Fiber Curvature Sensors for the Dizabled.* in *IEEE International Conference on Robotics and Automation*. 2001. IEEE.

170. Thomas, H. L., *Multi-Structure Ballistic Material*, 1998, Google Patents.

171. Rao, Y. J., *Fiber Bragg Grating Sensors: Principles and Applications*, in *Optical Fiber Sensor Technology*. 1998, Springer. 355–379.

172. Deflin, E., V. Koncar. *For Communicating clothing: The Flexible Display of Glass fiber Fabrics is Reality.* in *2nd International Avantex Symposium*. 2002.

173. Rothmaier, M., M. Luong, F. Clemens, *Textile Pressure Sensor Made of Flexible Plastic Optical Fibers.* Sensors, 2008, 8(7), 4318–4329.

174. Marchini, F., *Advanced Applications of Metallized Fibres for Electrostatic Discharge and Radiation Shielding.* Journal of Industrial Textiles, 1991, 20(3), 153–166.

175. Wendorff, J. H., S. Agarwal, A. Greiner, *Electrospinning: Materials, Processing, and Applications.* 2012, John Wiley & Sons.

176. Brotin, T., et al., *[n]-Polyenovanillins (n= 1–6) as New Push-Pull Polyenes for Nonlinear Optics: Synthesis, Structural Studies, and Experimental and Theoretical Investigation of Their Spectroscopic Properties, Electronic Structures, and Quadratic Hyperpolarizabilities.* Chemistry of materials, 1996, 8(4), 890–906.

177. C. Jr, P. H., *Nonwoven Thermal Insulating Stretch Fabric and Method for Producing Same*, 1985, Google Patents.

178. Omastová, M., et al., *Synthesis, Electrical Properties and Stability of Polypyrrole-Containing Conducting Polymer Composites.* Polymer International, 1997, 43(2), 109–116.

179. Thieblemont, J. C., et al., *Thermal Analysis of Polypyrrole Oxidation in Air.* Polymer, 1995, 36(8), 1605–1610.

180. Bashir, T., *Conjugated Polymer-based Conductive Fibers for Smart Textile Applications.* 2013, Chalmers University of Technology.

181. Ruckenstein, E., J. H. Chen, *Polypyrrole Conductive Composites Prepared by Coprecipitation.* Polymer, 1991, 32(7), 1230–1235.

182. Chen, Y., et al., *Morphological and Mechanical Behaviour of an in Situ Polymerized Polypyrrole/Nylon 66 Composite Film.* Polymer Communications, 1991, 32(6), 189–192.

183. Gregory, R. V., W. C. Kimbrell, H. H. Kuhn, *Electrically Conductive Non-Metallic Textile Coatings.* Journal of Industrial Textiles, 1991, 20(3), 167–175.

184. Batchelder, D. N., *Colour and Chromism of Conjugated Polymers.* Contemporary Physics, 1988, 29(1), 3–31.

185. Kazani, I., et al., *Electrical Conductive Textiles Obtained by Screen Printing.* Fibres & Textiles in Eastern Europe, 2012, 20(1), 57–63.

186. Pacelli, M., et al., *Sensing Threads and Fabrics for Monitoring Body Kinematic and Vital Signs*, in *Fibers and Textiles for the Future* 2001. 55–63.

187. Post, E. R., et al., *Electrically Active Textiles and Articles Made Therefrom*, 2001, Google Patents.

188. Jachimowicz, K. E., M. S. Lebby, *Textile Fabric with Integrated Electrically Conductive Fibers and Clothing Fabricated Thereof*, 1999, Google Patents.

189. Marculescu, D., et al., *Electronic Textiles: A Platform for Pervasive Computing.* Proceedings of the IEEE, 2003, 91(12), 1995–2018.

190. Child, A. D., A. R. D. Angelis, *Patterned Conductive Textiles*, 1999, Google Patents.

191. K. Jr, W. C., H. H. Kuhn, *Electrically Conductive Textile Materials and Method for Making Same*, 1990, Google Patents.

192. Davies, C. N., *Air Filtration.* 1973, New York: Academic Press.

193. Kruiēska, I., E. Klata, M. Chrzanowski, *New Textile Materials for Environmental Protection*, in *Intelligent Textiles for Personal Protection and Safety.* 2006. 41–53.

194. Albrecht, F., *Theoretische Untersuchungen über die Ablagerung von Staub aus strömender Luft und ihre Anwendung auf die Theorie der Staubfilter.* Physikalische Zeitschrift, 1931, 23, 48–56.

195. Walkenhorst, W., *Physikalische Eigenschaften von Stäuben sowie Grundlagen der Staubmessung und Staubbekämpfung*, in *Pneumokoniosen.* 1976, Springer. 11–70.

196. Langmuir, I., *Report on Smokes and Filters*, in *Section I* 1942, US Office of Scientific Research and Development.

197. Wente, V. A., *Superfine Thermoplastic Fibers.* Industrial & Engineering Chemistry, 1956, 48(8), 1342–1346.

198. Buntin, R. R., D. T. Lohkamp, *Melt Blowing-One-Step WEB Process for New Nonwoven Products.* Tappi, 1973, 56(4), 74–77.

199. Angadjivand, S., R. Kinderman, T. Wu, *High Efficiency Synthetic Filter Medium*, 2000, Google Patents.

200. Zeleny, J., *The electrical discharge from Liquid Points, and a Hydrostatic Method of measuring the electric intensity at their surface.* Physical Review, 1914, 3(2), 69–91.

201. Formhals, A., *Process and Apparatus Fob Pbepabing*, 1934, Google Patents.

202. Taylor, G., *Disintegration of Water Drops in an Electric Field.* Proceedings of the Royal Society of London. Series A. Mathematical and Physical Sciences, 1964, 280(1382), 383–397.

203. Gibson, P., H. S. Gibson, D. Rivin, *Transport Properties of Porous Membranes Based on Electrospun Nanofibers.* Colloids and Surfaces A: Physicochemical and Engineering Aspects, 2001, 187, 469–481.

204. Patarin, J., B. Lebeau, R. Zana, *Recent Advances in the Formation Mechanisms of Organized Mesoporous Materials.* Current Opinion in Colloid & Interface Science, 2002, 7(1), 107–115.

205. Weghmann, A. *Production of Electrostatic Spun Synthetic Microfiber Nonwovens and Applications in Filtration.* in *Proceedings of the 3rd World Filtration Congress, Filtration Society.* 1982. London.

206. Yarin, A. L., E. Zussman, *Upward Needleless Electrospinning of Multiple Nanofibers.* Polymer, 2004, 45(9), 2977–2980.

207. Majeed, S., et al., *Multi-Walled Carbon Nanotubes (MWCNTs) Mixed Polyacrylonitrile (PAN) Ultrafiltration Membranes.* Journal of Membrane Science, 2012, 403, 101–109.

208. Macedonio, F., E. Drioli, *Pressure-Driven Membrane Operations and Membrane Distillation Technology Integration for Water Purification.* Desalination, 2008, 223(1), 396–409.

209. Merdaw, A. A., A. O. Sharif, G. A. W. Derwish, *Mass Transfer in Pressure-Driven Membrane Separation Processes, Part II.* Chemical Engineering Journal, 2011, 168(1), 229–240.

210. Van Der Bruggen, B., et al., *A Review of Pressure-Driven Membrane Processes in Wastewater Treatment and Drinking Water Production.* Environmental Progress, 2003, 22(1), 46–56.

211. Cui, Z. F., H. S. Muralidhara, *Membrane Technology: A Practical Guide to Membrane Technology and Applications in Food and Bioprocessing.* 2010, Elsevier. 288.

212. Shirazi, S., C. J. Lin, D. Chen, *Inorganic Fouling of Pressure-Driven Membrane Processes — A Critical Review.* Desalination, 2010, 250(1), 236–248.

213. Pendergast, M. M., E. M. V. Hoek, *A Review of Water Treatment Membrane Nanotechnologies.* Energy & Environmental Science, 2011, 4(6), 1946–1971.

214. Hilal, N., et al., *A comprehensive review of nanofiltration membranes: Treatment, pretreatment, modeling, and atomic force microscopy.* Desalination, 2004, 170(3), 281–308.

215. Srivastava, A., S. Srivastava, K. Kalaga, *Carbon Nanotube Membrane Filters*, in *Springer Handbook of Nanomaterials.* 2013, Springer. 1099–1116.

216. Colombo, L., A. L. Fasolino, *Computer-Based Modeling of Novel Carbon Systems and Their Properties: Beyond Nanotubes.* Vol. 3, 2010, Springer. 258.

217. Polarz, S., B. Smarsly, *Nanoporous Materials.* Journal of Nanoscience and Nanotechnology, 2002, 2(6), 581–612.

218. Gray-Weale, A. A., et al., *Transition-state theory model for the diffusion coefficients of small penetrants in glassy polymers.* Macromolecules, 1997, 30(23), 7296–7306.

219. Rigby, D., R. Roe, *Molecular Dynamics Simulation of Polymer Liquid and Glass. I. Glass Transition.* The Journal of chemical physics, 1987, 87, 7285.

220. Freeman, B. D., Y. P. Yampolskii, I. Pinnau, *Materials Science of Membranes for Gas and Vapor Separation.* 2006, Wiley. com. 466.

221. Hofmann, D., et al., *Molecular Modeling Investigation of Free Volume Distributions in Stiff Chain Polymers with Conventional and Ultrahigh Free Volume: Comparison Between Molecular Modeling and Positron Lifetime Studies.* Macromolecules, 2003, 36(22), 8528–8538.

222. Greenfield, M. L., D. N. Theodorou, *Geometric Analysis of Diffusion Pathways in Glassy and Melt Atactic Polypropylene.* Macromolecules, 1993, 26(20), 5461–5472.

223. Baker, R. W., *Membrane Technology and Applications.* 2012, John Wiley & Sons. 592

224. Strathmann, H., L. Giorno, E. Drioli, *Introduction to Membrane Science and Technology.* 2011, Wiley-VCH Verlag & Company. 544.

225. Chen, J. P., et al., *Membrane Separation: Basics and Applications*, in *Membrane and Desalination Technologies*, L. K. Wang, et al., Editors. 2008, Humana Press. 271–332.

226. Mortazavi, S., *Application of Membrane Separation Technology to Mitigation of Mine Effluent and Acidic Drainage.* 2008, Natural Resources Canada. 194.

227. Porter, M. C., *Handbook of Industrial Membrane Technology.* 1990, Noyes Publications. 604.

228. Naylor, T. V., *Polymer Membranes: Materials, Structures and Separation Performance.* 1996, Rapra Technology Limited. 136.

229. Freeman, B. D., *Introduction to Membrane Science and Technology. By Heinrich Strathmann.* Angewandte Chemie International Edition, 2012, 51(38), 9485–9485.
230. Kim, I., H. Yoon, K. M. Lee, *Formation of Integrally Skinned Asymmetric Polyetherimide Nanofiltration Membranes by Phase Inversion Process.* Journal of applied polymer science, 2002, 84(6), 1300–1307.
231. Khulbe, K. C., C. Y. Feng, T. Matsuura, *Synthetic Polymeric Membranes: Characterization by Atomic Force Microscopy.* 2007, Springer. 198.
232. Loeb, L. B., *The Kinetic Theory of Gases.* 2004, Courier Dover Publications. 678.
233. Koros, W. J., G. K. Fleming, *Membrane-Based Gas Separation.* Journal of Membrane Science, 1993, 83(1), 1–80.
234. Perry, J. D., K. Nagai, W. J. Koros, *Polymer membranes for hydrogen separations.* MRS bulletin, 2006, 31(10), 745–749.
235. Hiemenz, P. C., R. Rajagopalan, *Principles of Colloid and Surface Chemistry, revised and expanded.* Vol. 14, 1997, CRC Press.
236. McDowell-Boyer, L. M., J. R. Hunt, N. Sitar, *Particle transport through porous media.* Water Resources Research, 1986, 22(13), 1901–1921.
237. Auset, M., A. A. Keller, *Pore-scale processes that control dispersion of colloids in saturated porous media.* Water Resources Research, 2004, 40(3).
238. Bhave, R. R., *Inorganic membranes synthesis, characteristics, and applications.* Vol. 312, 1991, Springer.
239. Lin, V. S.-Y., et al., *A porous silicon-based optical interferometric biosensor.* Science, 1997, 278(5339), 840–843.
240. Hedrick, J., et al. *Templating nanoporosity in organosilicates using well-defined branched macromolecules.* in *MATERIALS RESEARCH SOCIETY SYMPOSIUM PROCEEDINGS.* 1998. Cambridge University Press.
241. Hubbell, J. A., R. Langer, *Tissue engineering* Chem. Eng. News 1995, 13, 42–45.
242. Schaefer, D. W., *Engineered porous materials* MRS Bulletin 1994, 19, 14–17.
243. Hentze, H. P., M. Antonietti, *Porous Polymers and Resins.* Handbook of Porous Solids: 1964–2013.
244. Endo, A., et al., *Synthesis of ordered microporous silica by the solvent evaporation method.* Journal of materials science, 2004, 39(3), 1117–1119.
245. Sing, K., et al., *Physical and biophysical chemistry division commission on colloid and surface chemistry including catalysis.* Pure and Applied Chemistry, 1985, 57(4), 603–619.
246. Kresge, C., et al., *Ordered mesoporous molecular sieves synthesized by a liquid-crystal template mechanism.* nature, 1992, 359(6397), 710–712.
247. Yang, P., et al., *Generalized syntheses of large-pore mesoporous metal oxides with semicrystalline frameworks.* nature, 1998, 396(6707), 152–155.
248. Jiao, F., K. M. Shaju, P. G. Bruce, *Synthesis of Nanowire and Mesoporous Low-Temperature LiCoO2 by a Post-Templating Reaction.* Angewandte Chemie International Edition, 2005, 44(40), 6550–6553.
249. Ryoo, R., et al., *Ordered mesoporous carbons.* Advanced Materials, 2001, 13(9), 677–681.
250. Beck, J., et al., *Chu, DH Olson, EW Sheppard, SB McCullen, JB Higgins and JL Schlenker.* J. Am. Chem. Soc, 1992, 114(10), 834.
251. Zhao, D., et al., *Triblock copolymer syntheses of mesoporous silica with periodic 50 to 300 angstrom pores.* Science, 1998, 279(5350), 548–552.
252. Joo, S. H., et al., *Ordered nanoporous arrays of carbon supporting high dispersions of platinum nanoparticles.* nature, 2001, 412(6843), 169–172.
253. Kruk, M., et al., *Synthesis and characterization of hexagonally ordered carbon nanopipes.* Chemistry of materials, 2003, 15(14), 2815–2823.

254. Rouquerol, J., et al., *Recommendations for the characterization of porous solids (Technical Report)*. Pure and Applied Chemistry, 1994, 66(8), 1739–1758.

255. Schüth, F., K. S. W. Sing, J. Weitkamp, *Handbook of porous solids*. 2002, Wiley-Vch.

256. Maly, K. E., *Assembly of nanoporous organic materials from molecular building blocks*. Journal of Materials Chemistry, 2009, 19(13), 1781–1787.

257. Davis, M. E., *Ordered porous materials for emerging applications*. nature, 2002, 417(6891), 813–821.

258. Morris, R. E., P. S. Wheatley, *Gas storage in nanoporous materials*. Angewandte Chemie International Edition, 2008, 47(27), 4966–4981.

259. Cote, A. P., et al., *Porous, crystalline, covalent organic frameworks*. Science, 2005, 310(5751), 1166–1170.

260. El-Kaderi, H. M., et al., *Designed synthesis of 3D covalent organic frameworks*. Science, 2007, 316(5822), 268–272.

261. Jiang, J. X., et al., *Conjugated microporous poly (aryleneethynylene) networks*. Angewandte Chemie International Edition, 2007, 46(45), 8574–8578.

262. Ben, T., et al., *Targeted synthesis of a porous aromatic framework with high stability and exceptionally high surface area*. Angewandte Chemie, 2009, 121(50), 9621–9624.

263. Eddaoudi, M., et al., *Modular chemistry: secondary building units as a basis for the design of highly porous and robust metal-organic carboxylate frameworks*. Accounts of Chemical Research, 2001, 34(4), 319–330.

264. Lu, G. Q., X. S. Zhao, T. K. Wei, *Nanoporous materials: science and engineering*. Vol. 4, 2004, Imperial College Press.

265. Holister, P., C. R. Vas, T. Harper, *Nanocrystalline materials*. Technologie White Papers, 2003(4).

266. Smith, B. L., et al., *Molecular mechanistic origin of the toughness of natural adhesives, fibers and composites*. nature, 1999, 399(6738), 761–763.

267. Mann, S., G. A. Ozin, *Synthesis of inorganic materials with complex form*. nature, 1996, 382(6589), 313–318.

268. Mann, S., *Molecular tectonics in biomineralization and biomimetic materials chemistry*. nature, 1993, 365(6446), 499–505.

269. Busch, K., S. John, *Photonic band gap formation in certain self-organizing systems*. Physical Review E, 1998, 58(3), 3896.

270. Argyros, A., et al., *Electron tomography and computer visualization of a three-dimensional 'photonic' crystal in a butterfly wing-scale*. Micron, 2002, 33(5), 483–487.

271. Sailor, M. J., K. L. Kavanagh, *Porous silicon–what is responsible for the visible luminescence?* Advanced Materials, 1992, 4(6), 432–434.

272. Koshida, N., B. Gelloz, *Wet and dry porous silicon*. Current opinion in colloid & interface science, 1999, 4(4), 309–313.

273. Hentze, H.-P. and M. Antonietti, *Template synthesis of porous organic polymers*. Current Opinion in Solid State and Materials Science, 2001, 5(4), 343–353.

274. Nakao, S.-i., *Determination of pore size and pore size distribution: 3. Filtration membranes*. Journal of Membrane Science, 1994, 96(1), 131–165.

275. Barrer, R. M., *Zeolites and their synthesis*. Zeolites, 1981, 1(3), 130–140.

276. Iijima, S., *Helical microtubules of graphitic carbon*. nature, 1991, 354(6348), 56–58.

277. Kaneko, K., K. Inouye, *Adsorption of water on FeOOH as studied by electrical conductivity measurements*. Bulletin of the Chemical Society of Japan, 1979, 52(2), 315–320.

278. Maeda, K., et al., *Control with polyethers of pore distribution of alumina by the sol-gel method*. Chem.& Ind., 1989(23), 807.

279. Matsuzaki, S., M. Taniguchi, M. Sano, *Polymerization of benzene occluded in graphite-alkali metal intercalation compounds*. Synthetic metals, 1986, 16(3), 343–348.

280. Enoki, T., H. Inokuchi, M. Sano, *ESR study of the hydrogen-potassium-graphite ternary intercalation compounds.* Physical Review B, 1988, 37(16), 9163.
281. Pinnavaia, T. J., *Intercalated clay catalysts.* Science, 1983, 220(4595), 365–371.
282. Yamanaka, S., et al., *High surface area solids obtained by intercalation of iron oxide pillars in montmorillonite.* Materials research bulletin, 1984, 19(2), 161–168.
283. Inagaki, S., Y. Fukushima, K. Kuroda, *Synthesis of highly ordered mesoporous materials from a layered polysilicate.* J. Chem. Soc., Chem. Commun., 1993(8), 680–682.
284. Vallano, P. T., V. T. Remcho, *Modeling interparticle and intraparticle (perfusive) electroosmotic flow in capillary electrochromatography.* Analytical chemistry, 2000, 72(18), 4255–4265.
285. Levenspiel, O., *Chemical reaction engineering.* Vol. 2, 1972, Wiley New York etc.
286. Li, Q., et al., *Interparticle and intraparticle mass transfer in chromatographic separation.* Bioseparation, 1995, 5(4), 189–202.
287. Setoyama, N., et al., *Surface characterization of microporous solids with helium adsorption and small angle x-ray scattering.* Langmuir, 1993, 9(10), 2612–2617.
288. Marsh, H., *Introduction to carbon science.* 1989.
289. Kaneko, K., *Determination of pore size and pore size distribution: 1. Adsorbents and catalysts.* Journal of Membrane Science, 1994, 96(1), 59–89.
290. Seaton, N., J. Walton, *A new analysis method for the determination of the pore size distribution of porous carbons from nitrogen adsorption measurements.* Carbon, 1989, 27(6), 853–861.
291. Dullien, F. A., *Porous media: fluid transport and pore structure.* 1991, Academic press.
292. Yang, W., et al., *Carbon Nanotubes for Biological and Biomedical Applications.* Nanotechnology, 2007, 18(41), 412001.
293. Bianco, A., et al., *Biomedical Applications of Functionalised Carbon Nanotubes.* Chemical Communications, 2005(5), 571–577.
294. Salvetat, J., et al., *Mechanical Properties of Carbon Nanotubes.* Applied Physics A, 1999, 69(3), 255–260.
295. Zhang, X., et al., *Ultrastrong, Stiff, and Lightweight Carbon-Nanotube Fibers.* Advanced Materials, 2007, 19(23), 4198–4201.
296. Arroyo, M., T. Belytschko, *Finite Crystal Elasticity of Carbon Nanotubes Based on the Exponential Cauchy-Born Rule.* Physical Review B, 2004, 69(11), 115415.
297. Wang, J., et al., *Energy and Mechanical Properties of Single-Walled Carbon Nanotubes Predicted Using the Higher Order Cauchy-Born rule.* Physical Review B, 2006, 73(11), 115428.
298. Zhang, Y., *Single-walled carbon nanotube modeling based on one-and two-dimensional Cosserat continua,* 2011, University of Nottingham.
299. Wang, S., *Functionalization of Carbon Nanotubes: Characterization, Modeling and Composite Applications.* 2006, Florida State University. 193.
300. Lau, K.-t., C. Gu, D. Hui, *A critical review on nanotube and nanotube/nanoclay related polymer composite materials.* Composites Part B: Engineering, 2006, 37(6), 425–436.
301. Choi, W., et al., *Carbon Nanotube-Guided Thermopower Waves.* Materials Today, 2010, 13(10), 22–33.
302. Sholl, D. S., J. Johnson, *Making High-Flux Membranes with Carbon Nanotubes.* Science, 2006, 312(5776), 1003–1004.
303. Zang, J., et al., *Self-Diffusion of Water and Simple Alcohols in Single-Walled Aluminosilicate Nanotubes.* ACS nano, 2009, 3(6), 1548–1556.
304. Talapatra, S., V. Krunglevicuite, A. D. Migone, *Higher Coverage Gas Adsorption on the Surface of Carbon Nanotubes: Evidence for a Possible New Phase in the Second Layer.* Physical Review Letters, 2002, 89(24), 246106.

305. Pujari, S., et al., *Orientation Dynamics in Multiwalled Carbon Nanotube Dispersions Under Shear Flow.* The Journal of chemical physics, 2009, 130, 214903.
306. Singh, S., P. Kruse, *Carbon Nanotube Surface Science.* International Journal of Nanotechnology, 2008, 5(9), 900–929.
307. Baker, R. W., *Future Directions of Membrane Gas Separation Technology.* Industrial & Engineering Chemistry Research, 2002, 41(6), 1393–1411.
308. Erucar, I., S. Keskin, *Screening Metal–Organic Framework-Based Mixed-Matrix Membranes for $CO_2/CH4$ Separations.* Industrial & Engineering Chemistry Research, 2011, 50(22), 12606–12616.
309. Bethune, D. S., et al., *Cobalt-Catalyzed Growth of Carbon Nanotubes with Single-Atomic-Layer Walls.* Nature 1993, 363, 605–607.
310. Iijima, S., T. Ichihashi, *Single-Shell Carbon Nanotubes of 1-nm Diameter.* Nature, 1993, 363, 603–605.
311. Treacy, M., T. Ebbesen, J. Gibson, *Exceptionally high Young's modulus observed for individual carbon nanotubes.* 1996.
312. Wong, E. W., P. E. Sheehan, C. Lieber, *Nanobeam Mechanics: Elasticity, Strength, and Toughness of Nanorods and Nanotubes.* Science, 1997, 277(5334), 1971–1975.
313. Thostenson, E. T., C. Li, T. W. Chou, *Nanocomposites in Context.* Composites Science and Technology, 2005, 65(3), 491–516.
314. Barski, M., P. Kędziora, M. Chwał, *Carbon Nanotube/Polymer Nanocomposites: A Brief Modeling Overview.* Key Engineering Materials, 2013, 542, 29–42.
315. Dresselhaus, M. S., G. Dresselhaus, P. C. Eklund, *Science of Fullerenes and Carbon nanotubes:Ttheir Properties and Applications.* 1996, Academic Press. 965.
316. Yakobson, B., R. E. Smalley, *Some Unusual New Molecules—Long, Hollow Fibers with Tantalizing Electronic and Mechanical Properties—have Joined Diamonds and Graphite in the Carbon Family.* Am Scientist, 1997, 85, 324–337.
317. Guo, Y., W. Guo, *Mechanical and Electrostatic Properties of Carbon Nanotubes under Tensile Loading and Electric Field.* Journal of Physics D: Applied Physics, 2003, 36(7), 805.
318. Berger, C., et al., *Electronic Confinement and Coherence in Patterned Epitaxial Graphene.* Science, 2006, 312(5777), 1191–1196.
319. Song, K., et al., *Structural Polymer-Based Carbon Nanotube Composite Fibers: Understanding the Processing–Structure–Performance Relationship.* Materials, 2013, 6(6), 2543–2577.
320. Park, O. K., et al., *Effect of Surface Treatment with Potassium Persulfate on Dispersion Stability of Multi-Walled Carbon Nanotubes.* Materials Letters, 2010, 64(6), 718–721.
321. Banerjee, S., T. Hemraj-Benny, S. S. Wong, *Covalent Surface Chemistry of Single-Walled Carbon Nanotubes.* Advanced Materials, 2005, 17(1), 17–29.
322. Balasubramanian, K., M. Burghard, *Chemically Functionalized Carbon Nanotubes.* Small, 2005, 1(2), 180–192.
323. Xu, Z. L., F. Alsalhy Qusay, *Polyethersulfone (PES) Hollow Fiber Ultrafiltration Membranes Prepared by PES/nonSolvent/NMP Solution.* Journal of Membrane Science, 2004, 233(1–2), 101–111.
324. Chung, T. S., J. J. Qin, J. Gu, *Effect of Shear Rate Within the Spinneret on Morphology, Separation Performance and Mechanical Properties of Ultrafiltration Polyethersulfone Hollow Fiber Membranes.* Chemical Engineering Science, 2000, 55(6), 1077–1091.
325. Choi, J. H., J. Jegal, W. N. Kim, *Modification of Performances of Various Membranes Using MWNTs as a Modifier.* Macromolecular Symposia, 2007, 249–250(1), 610–617.
326. Wang, Z., J. Ma, *The Role of Nonsolvent in-Diffusion Velocity in Determining Polymeric Membrane Morphology.* Desalination, 2012, 286(0), 69–79.
327. Vilatela, J. J., R. Khare, A. H. Windle, *The Hierarchical Structure and Properties of Multifunctional Carbon Nanotube Fibre Composites.* Carbon, 2012, 50(3), 1227–1234.

328. Benavides, R. E., S. C. Jana, D. H. Reneker, *Nanofibers from Scalable Gas Jet Process.* ACS Macro Letters, 2012, 1(8), 1032–1036.
329. Gupta, V. B., V. K. Kothari, *Manufactured Fiber Technology.* 1997, Springer. 661.
330. Wang, T., S. Kumar, *Electrospinning of Polyacrylonitrile Nanofibers.* Journal of applied polymer science, 2006, 102(2), 1023–1029.
331. Song, K., et al., *Lubrication of Poly (vinyl alcohol) Chain Orientation by Carbon nano-chips in Composite Tapes.* Journal of applied polymer science, 2013, 127(4), 2977–2982.
332. Filatov, Y., A. Budyka, V. Kirichenko, *Electrospinning of microand nanofibers: fundamentals in separation and filtration processes.* 2007, Begell House Inc., Redding, CT.
333. Brown, R. C., *Air filtration: an integrated approach to the theory and applications of fibrous filters.* Vol. 650, 1993, Pergamon press Oxford.
334. Hinds, W. C., *Aerosol technology: properties, behavior, and measurement of airborne particles.* 1982.
335. Greiner, A., J. Wendorff, *Functional self-assembled nanofibers by electrospinning,* in *Self-Assembled Nanomaterials I.* 2008, Springer. 107–171.
336. Jeong, E. H., J. Yang, J. H. Youk, *Preparation of polyurethane cationomer nanofiber mats for use in antimicrobial nanofilter applications.* Materials Letters, 2007, 61(18), 3991–3994.
337. Lala, N. L., et al., *Fabrication of nanofibers with antimicrobial functionality used as filters: protection against bacterial contaminants.* Biotechnology and Bioengineering, 2007, 97(6), 1357–1365.
338. Maze, B., et al., *A simulation of unsteady-state filtration via nanofiber media at reduced operating pressures.* Journal of Aerosol Science, 2007, 38(5), 550–571.
339. Payet, S., et al., *Penetration and pressure drop of a HEPA filter during loading with submicron liquid particles.* Journal of Aerosol Science, 1992, 23(7), 723–735.
340. Fang, J., X. Wang, T. Lin, *Functional applications of electrospun nanofibers.* Nanofibers-production, properties and functional applications, 2011, 287–326.
341. Su, L., et al., *Free-standing palladium/polyamide 6 nanofibers for electrooxidation of alcohols in alkaline medium.* The Journal of Physical Chemistry C, 2009, 113(36), 16174–16180.
342. Kim, H. J., et al., *Pt and PtRh nanowire electrocatalysts for cyclohexane-fueled polymer electrolyte membrane fuel cell.* Electrochemistry Communications, 2009, 11(2), 446–449.
343. Stasiak, M., et al., *Design of polymer nanofiber systems for the immobilization of homogeneous catalysts–Preparation and leaching studies.* Polymer, 2007, 48(18), 5208–5218.
344. Wang, Y., Y. L. Hsieh, *Enzyme immobilization to ultra-fine cellulose fibers via amphiphilic polyethylene glycol spacers.* Journal of Polymer Science Part A: Polymer Chemistry, 2004, 42(17), 4289–4299.
345. Stoilova, O., et al., *Functionalized electrospun mats from styrene–maleic anhydride copolymers for immobilization of acetylcholinesterase.* European Polymer Journal, 2010, 46(10), 1966–1974.
346. Jia, H., et al., *Enzyme-carrying polymeric nanofibers prepared via electrospinning for use as unique biocatalysts.* Biotechnology progress, 2002, 18(5), 1027–1032.
347. Kim, B. C., et al., *Preparation of biocatalytic nanofibers with high activity and stability via enzyme aggregate coating on polymer nanofibers.* Nanotechnology, 2005, 16(7), p. S382.
348. Li, S.-F., J.-P. Chen, W.-T. Wu, *Electrospun polyacrylonitrile nanofibrous membranes for lipase immobilization.* Journal of Molecular Catalysis B: Enzymatic, 2007, 47(3), 117–124.
349. Huang, X.-J., D. Ge, Z.-K. Xu, *Preparation and characterization of stable chitosan nanofibrous membrane for lipase immobilization.* European Polymer Journal, 2007, 43(9), 3710–3718.
350. Xie, J., Y.-L. Hsieh, *Ultra-high surface fibrous membranes from electrospinning of natural proteins: casein and lipase enzyme.* Journal of Materials Science, 2003, 38(10), 2125–2133.

351. Herricks, T. E., et al., *Direct fabrication of enzyme-carrying polymer nanofibers by electrospinning*. Journal of Materials Chemistry, 2005, 15(31), 3241–3245.
352. Wang, G., et al., *Fabrication and characterization of polycrystalline WO3 nanofibers and their application for ammonia sensing*. The Journal of Physical Chemistry B, 2006, 110(47), 23777–23782.
353. Teoh, W. Y., J. A. Scott, R. Amal, *Progress in heterogeneous photocatalysis: from classical radical chemistry to engineering nanomaterials and solar reactors*. The Journal of Physical Chemistry Letters, 2012, 3(5), 629–639.
354. Bhatkhande, D. S., V. G. Pangarkar, A. A. Beenackers, *Photocatalytic degradation for environmental applications–a review*. Journal of Chemical Technology and Biotechnology, 2002, 77(1), 102–116.
355. Han, F., et al., *Tailored titanium dioxide photocatalysts for the degradation of organic dyes in wastewater treatment: a review*. Applied Catalysis A: General, 2009, 359(1), 25–40.
356. Alves, A., et al., *Photocatalytic activity of titania fibers obtained by electrospinning*. Materials Research Bulletin, 2009, 44(2), 312–317.
357. Szilágyi, I. M., et al., *Photocatalytic Properties of WO3/TiO2 Core/Shell Nanofibers prepared by Electrospinning and Atomic Layer Deposition*. Chemical Vapor Deposition, 2013, 19(4-6), 149–155.
358. Shengyuan, Y., et al., *Rice grain-shaped TiO2 mesostructures—synthesis, characterization and applications in dye-sensitized solar cells and photocatalysis*. Journal of Materials Chemistry, 2011, 21(18), 6541–6548.
359. Meng, X., et al., *Growth of hierarchical TiO2 nanostructures on anatase nanofibers and their application in photocatalytic activity*. CrystEngComm, 2011, 13(8), 3021–3029.
360. Zhang, X., et al., *Novel hollow mesoporous 1D TiO2 nanofibers as photovoltaic and photocatalytic materials*. Nanoscale, 2012, 4(5), 1707–1716.
361. Singh, P., K. Mondal, A. Sharma, *Reusable electrospun mesoporous ZnO nanofiber mats for photocatalytic degradation of polycyclic aromatic hydrocarbon dyes in wastewater*. Journal of colloid and interface science, 2013, 394, 208–215.
362. Ganesh, V. A., et al., *Photocatalytic superhydrophilic TiO2 coating on glass by electrospinning*. RSC Advances, 2012, 2(5), 2067–2072.
363. Bedford, N., A. Steckl, *Photocatalytic self cleaning textile fibers by coaxial electrospinning*. ACS applied materials & interfaces, 2010, 2(8), 2448–2455.
364. Bao, N., et al., *Adsorption of dyes on hierarchical mesoporous TiO2 fibers and its enhanced photocatalytic properties*. The Journal of Physical Chemistry C, 2011, 115(13), 5708–5719.
365. Pant, B., et al., *Carbon nanofibers decorated with binary semiconductor (TiO_2/ZnO) nanocomposites for the effective removal of organic pollutants and the enhancement of antibacterial activities*. Ceramics International, 2013, 39(6), 7029–7035.
366. Su, Q., et al., *Effect of the morphology of V_2O_5/TiO_2 nanoheterostructures on the visible light photocatalytic activity*. Journal of Physics and Chemistry of Solids, 2013, 74(10), 1475–1481.
367. Temenoff, J. S., A. G. Mikos, *Review: Tissue Engineering for Regeneration of Articular Cartilage*. Biomaterials, 2000, 21(5), 431–440.
368. Bianco, P., P. G. Robey, *Stem Cells in Tissue Engineering*. Nature, 2001, 414(6859), 118–121.
369. Griffith, L. G., G. Naughton, *Tissue Engineering–current Challenges and Expanding Opportunities*. Science, 2002, 295(5557), 1009–1014.
370. Raghunath, J., et al., *Biomaterials and Scaffold Design: Key to Tissue-Engineering Cartilage*. Biotechnology and Applied Biochemistry, 2007, 46(2), 73–84.
371. Stocum, D. L., *Stem Cells in CNS and Cardiac Regeneration*, in *Regenerative Medicine I*. 2005, Springer. 135–159.

372. Ratcliffe, A., L. E. Niklason, *Bioreactors and Bioprocessing for Tissue Engineering.* Annals of the New York Academy of Sciences, 2002, 961(1), 210–215.

373. Hutmacher, D. W., *Scaffold Design and Fabrication Technologies for Engineering Tissues—State of the Art and Future Perspectives.* Journal of Biomaterials Science, Polymer Edition, 2001, 12(1), 107–124.

374. Lannutti, J., et al., *Electrospinning for Tissue Engineering Scaffolds.* Materials Science and Engineering: C, 2007, 27(3), 504–509.

375. Smith, L. A., P. X. Ma, *Nano-Fibrous Scaffolds for Tissue Engineering.* Colloids and Surfaces B: Biointerfaces, 2004, 39(3), 125–131.

376. Ma, Z., et al., *Potential of Nanofiber Matrix as Tissue-Engineering Scaffolds.* Tissue Engineering, 2005, 11(1–2), 101–109.

377. Noh, H. K., et al., *Electrospinning of Chitin Nanofibers: Degradation Behavior and Cellular Response to Normal Human Keratinocytes and Fibroblasts.* Biomaterials, 2006, 27(21), 3934–3944.

378. Kwon, k., S. Kidoaki, T. Matsuda, *Electrospun Nano-to Microfiber Fabrics Made of Biodegradable Copolyesters: Structural Characteristics, Mechanical Properties and Cell Adhesion Potential.* Biomaterials, 2005, 26(18), 3929–3939.

379. Riboldi, S. A., et al., *Electrospun Degradable Polyesterurethane Membranes: Potential Scaffolds for Skeletal Muscle Tissue Engineering.* Biomaterials, 2005, 26(22), 4606–4615.

380. Hwang, N. S., S. Varghese, J. Elisseeff, *Cartilage Tissue Engineering*, in *Stem Cell Assays.* 2007, Springer. 351–373.

381. Tuzlakoglu, K., et al., *Nano-and Micro-fiber Combined Scaffolds: A New Architecture for Bone Tissue Engineering.* Journal of Materials Science: Materials in Medicine, 2005, 16(12), 1099–1104.

382. Santos, M. I., et al., *Endothelial Cell Colonization and Angiogenic Potential of Combined Nano-and Micro-fibrous Scaffolds for Bone Tissue engineering.* Biomaterials, 2008, 29(32), 4306–4313.

383. Holzwarth, J. M., P. X. Ma, *Biomimetic Nanofibrous Scaffolds for Bone Tissue Engineering.* Biomaterials, 2011, 32(36), 9622–9629.

384. Sill, T. J., H. A. Recum, *Electrospinning: Applications in Drug Delivery and Tissue Engineering.* Biomaterials, 2008, 29(13), 1989–2006.

385. Gautam, S., A. K. Dinda, N. C. Mishra, *Fabrication and Characterization of PCL/gelatin Composite Nanofibrous Scaffold for Tissue Engineering Applications by Electrospinning Method.* Materials Science and Engineering: C, 2013, 33, 1228–1235.

386. Mouthuy, P. A., H. Ye, *Biomaterials: Electrospinning* in *Comprehensive Biotechnology.* 2011, Elsevier: Oxford. 23–36.

387. Shen, Q., et al., *Progress on Materials and Scaffold Fabrications Applied to Esophageal Tissue Engineering.* Materials Science and Engineering: C, 2013, 33, 1860–1866.

388. Jang, J. H., O. Castano, H. W. Kim, *Electrospun Materials as Potential Platforms for Bone Tissue Engineering.* Advanced Drug Delivery Reviews, 2009, 61(12), 1065–1083.

389. Schneider, O. D., et al., *In vivo and* in vitro *Evaluation of Flexible, Cotton wool-like Nanocomposites as Bone Substitute Material for Complex Defects.* Acta Biomaterialia, 2009, 5(5), 1775–1784.

390. Meng, Z. X., et al., *Electrospinning of PLGA/gelatin Randomly oriented and Aligned Nanofibers as Potential Scaffold in Tissue Engineering.* Materials Science and Engineering: C, 2010, 30(8), 1204–1210.

391. Chahal, S., F. S. J. Hussain, M. B. M. Yusoff, *Characterization of Modified Cellulose (MC)/ Poly (Vinyl Alcohol) Electrospun Nanofibers for Bone Tissue Engineering.* Procedia Engineering, 2013, 53, 683–688.

392. Kramschuster, A., L. S. Turng, *Fabrication of Tissue Engineering Scaffolds*. Handbook of Biopolymers and Biodegradable Plastics: Properties, Processing and Applications. Vol. 17, 2013, Elsevier.

393. Shoichet, M. S., *Polymer Scaffolds for Biomaterials Applications*. Macromolecules, 2009, 43(2), 581–591.

394. Orlando, G., et al., *Regenerative Medicine as Applied to General Surgery*. Annals of Surgery, 2012, 255(5), 867–880.

395. Lakshmipathy, U., C. Verfaillie, *Stem Cell Plasticity*. Blood Reviews, 2005, 19(1), 29–38.

396. Yoshimoto, H., et al., *A Biodegradable Nanofiber Scaffold by Electrospinning and its Potential for Bone Tissue Engineering*. Biomaterials, 2003, 24(12), 2077–2082.

397. Barnes, C. P., et al., *Nanofiber Technology: Designing the next Generation of Tissue Engineering Scaffolds*. Advanced Drug Delivery Reviews, 2007, 59(14), 1413–1433.

398. Janković, B., et al., *The Design Trend in Tissue-engineering Scaffolds Based on Nanomechanical Properties of Individual Electrospun Nanofibers*. International Journal of Pharmaceutics, 2013, 455(1), 338–347.

399. Hay, E. D., *Cell Biology of Extracellular Matrix*. 1991, Springer. 468.

400. Khadka, D. B., D. T. Haynie, *Protein-and Peptide-based Electrospun Nanofibers in Medical Biomaterials*. Nanomedicine: Nanotechnology, Biology and Medicine, 2012, 8(8), 1242–1262.

401. Kjellen, L., U. Lindahl, *Proteoglycans: Structures and Interactions*. Annual Review of Biochemistry, 1991, 60(1), 443–475.

402. Rosso, F., et al., *From Cell–ECM Interactions to Tissue Engineering*. Journal of Cellular Physiology, 2004, 199(2), 174–180.

403. Schaefer, L., R. M. Schaefer, *Proteoglycans: From Structural Compounds to Signaling Molecules*. Cell and Tissue Research, 2010, 339(1), 237–246.

404. Roughley, P. J., *Articular Cartilage and Changes in Arthritis: Noncollagenous Proteins and Proteoglycans in the Extracellular Matrix of Cartilage*. Arthritis Res, 2001, 3(6), 342–347.

405. Liang, D., B. S. Hsiao, B. Chu, *Functional Electrospun Nanofibrous Scaffolds for Biomedical Applications*. Advanced Drug Delivery Reviews, 2007, 59(14), 1392–1412.

406. Hasan, M. D., et al., *Electrospun Scaffolds for Tissue Engineering of Vascular Grafts*. Acta Biomaterialia, 2013.

407. Metter, R. B., et al., *Biodegradable Fibrous Scaffolds with Diverse Properties by Electrospinning Candidates from a Combinatorial Macromer Library*. Acta Biomaterialia, 2010, 6(4), 1219–1226.

408. Greiner, A., J. H. Wendorff, *Electrospinning: a fascinating method for the preparation of ultrathin fibers*. Angewandte Chemie International Edition, 2007, 46(30), 5670–5703.

409. Reneker, D., et al., *Electrospinning of nanofibers from polymer solutions and melts*. Advances in applied mechanics, 2007, 41, 43–346.

410. Agarwal, S., J. H. Wendorff, A. Greiner, *Use of electrospinning technique for biomedical applications*. Polymer, 2008, 49(26), 5603–5621.

411. Zeng, J., et al., *Poly (vinyl alcohol) nanofibers by electrospinning as a protein delivery system and the retardation of enzyme release by additional polymer coatings*. Biomacromolecules, 2005, 6(3), 1484–1488.

412. Xie, J., C.-H. Wang, *Electrospun microand nanofibers for sustained delivery of paclitaxel to treat C6 glioma* in vitro. Pharmaceutical research, 2006, 23(8), 1817–1826.

413. Gouma, P., *Nanostructured polymorphic oxides for advanced chemosensors*. Rev. Adv. Mater. Sci, 2003, 5(123), 122.

414. Sawicka, K., A. Prasad, P. Gouma, *Metal oxide nanowires for use in chemical sensing applications*. Sensor Letters, 3, 2005, 1(4), 31–35.

415. Kim, T. G., T. G. Park, *Surface functionalized electrospun biodegradable nanofibers for immobilization of bioactive molecules.* Biotechnology progress, 2006, 22(4), 1108–1113.

416. Aussawasathien, D., J.-H. Dong, L. Dai, *Electrospun polymer nanofiber sensors.* Synthetic Metals, 2005, 154(1), 37–40.

417. Choi, J., et al., *Electrospun PEDOT: PSS/PVP nanofibers as the chemiresistor in chemical vapor sensing.* Synthetic Metals, 2010, 160(13), 1415–1421.

418. Laxminarayana, K., N. Jalili, *Functional nanotube-based textiles: pathway to next generation fabrics with enhanced sensing capabilities.* Textile Research Journal, 2005, 75(9), 670–680.

419. Kessick, R., G. Tepper, *Electrospun polymer composite fiber arrays for the detection and identification of volatile organic compounds.* Sensors and Actuators B: Chemical, 2006, 117(1), 205–210.

420. Qi, Q., et al., *Humidity sensing properties of KCl-doped ZnO nanofibers with super-rapid response and recovery.* Sensors and Actuators B: Chemical, 2009, 137(2), 649–655.

421. Qi, Q., et al., *Influence of crystallographic structure on the humidity sensing properties of KCl-doped TiO$_2$ nanofibers.* Sensors and Actuators B: Chemical, 2009, 139(2), 611–617.

422. He, Y., et al., *Humidity sensing properties of BaTiO$_3$ nanofiber prepared via electrospinning.* Sensors and Actuators B: Chemical, 2010, 146(1), 98–102.

423. Spichiger-Keller, U. E., *Chemical sensors and biosensors for medical and biological applications.* 2008, John Wiley & Sons.

424. Wang, X., et al., *Electrospun nanofibrous membranes for highly sensitive optical sensors.* Nano letters, 2002, 2(11), 1273–1275.

425. Wang, X., et al., *Synthesis and electrospinning of a novel fluorescent polymer PMMA-PM for quenching-based optical sensing.* Journal of Macromolecular Science, Part A, 2002, 39(10), 1241–1249.

426. Deng, C., et al., *Highly fluorescent TPA-PBPV nanofibers with amplified sensory response to TNT.* Chemical physics letters, 2009, 483(4), 219–223.

427. Wang, X., et al., *Electrostatic assembly of conjugated polymer thin layers on electrospun nanofibrous membranes for biosensors.* Nano letters, 2004, 4(2), 331–334.

428. Tao, S., G. Li, J. Yin, *Fluorescent nanofibrous membranes for trace detection of TNT vapor.* J. Mater. Chem., 2007, 17(26), 2730–2736.

429. Yoon, J., S. K. Chae, J.-M. Kim, *Colorimetric sensors for volatile organic compounds (VOCs) based on conjugated polymer-embedded electrospun fibers.* Journal of the American Chemical Society, 2007, 129(11), 3038–3039.

430. Ding, B., M. Yamazaki, S. Shiratori, *Electrospun fibrous polyacrylic acid membrane-based gas sensors.* Sensors and Actuators B: Chemical, 2005, 106(1), 477–483.

431. Luoh, R., H. T. Hahn, *Electrospun nanocomposite fiber mats as gas sensors.* Composites science and technology, 2006, 66(14), 2436–2441.

432. Im, J. S., et al., *Improved gas sensing of electrospun carbon fibers based on pore structure, conductivity and surface modification.* Carbon, 2010, 48(9), 2573–2581.

433. Viswanathamurthi, P., et al., *Vanadium pentoxide nanofibers by electrospinning.* Scripta Materialia, 2003, 49(6), 577–581.

434. Schindler, M. S., H. Y. Chung, *Nanofibrillar structure and applications including cell and tissue culture,* 2010, Google Patents.

435. Kim, J. R., et al., *Electrospun PVDF-based fibrous polymer electrolytes for lithium ion polymer batteries.* Electrochimica Acta, 2004, 50(1), 69–75.

436. Han, S. O., et al., *Electrospinning of ultrafine cellulose fibers and fabrication of poly (butylene succinate) biocomposites reinforced by them.* Journal of applied polymer science, 2008, 107(3), 1954–1959.

437. Li, Z., C. Wang, *Applications of Electrospun Nanofibers*, in *One-Dimensional nanostructures*. 2013, Springer. 75–139.
438. Ozden, E., Y. Z. Menceloglu, M. Papila, *Engineering chemistry of electrospun nanofibers and interfaces in nanocomposites for superior mechanical properties.* ACS applied materials & interfaces, 2010, 2(7), 1788–1793.
439. Gibson, P., H. Schreuder-Gibson, D. Rivin, *Transport properties of porous membranes based on electrospun nanofibers.* Colloids and Surfaces A: Physicochemical and Engineering Aspects, 2001, 187, 469–481.
440. Gibson, P., H. Schreuder-Gibson, D. Rivin, *Electrospun fiber mats: transport properties.* AIChE journal, 1999, 45(1), 190–195.
441. Kirstein, T., et al., *Wearable computing systems–electronic textiles.* 2005, CRC Press: Boca Raton, FL, USA.
442. Van Langenhove, L., et al., *Smart textiles.* Stud. Health Technol. Inform, 2004, 108, 344–352.
443. Marculescu, D., et al., *Electronic textiles: A platform for pervasive computing.* Proceedings of the IEEE, 2003, 91(12), 1995–2018.
444. Huang, C.-T., et al., *Parametric design of yarn-based piezoresistive sensors for smart textiles.* Sensors and Actuators A: Physical, 2008, 148(1), 10–15.
445. Van Langenhove, L., C. Hertleer, *Smart clothing: a new life.* International Journal of Clothing Science and Technology, 2004, 16(1/2), 63–72.
446. Barr, M. C., et al., *Direct monolithic integration of organic photovoltaic circuits on unmodified paper.* Advanced Materials, 2011, 23(31), 3500–3505.
447. Köhler, A. R., L. M. Hilty, C. Bakker, *Prospective impacts of electronic textiles on recycling and disposal.* Journal of Industrial Ecology, 2011, 15(4), 496–511.
448. Bedeloglu, A. C., et al., *A photovoltaic fiber design for smart textiles.* Textile Research Journal, 2010, 80(11), 1065–1074.
449. Hadimani, R., et al., *Continuous production of piezoelectric PVDF fiber for e-textile applications.* Smart Materials and Structures, 2013, 22(7), 075017.
450. Krebs, F. C., *Fabrication and processing of polymer solar cells: a review of printing and coating techniques.* Solar Energy Materials and Solar Cells, 2009, 93(4), 394–412.
451. Small, S. M., A. Roseway, *The Printing Dress: You are what you Tweet.*
452. Nusser, M., V. Senner, *High-tech-textiles in competition sports.* Procedia Engineering, 2010, 2(2), 2845–2850.
453. Nelson, G., *Microencapsulation in textile finishing.* Review of progress in coloration and related topics, 2001, 31(1), 57–64.
454. Starešinič, M., B. Šumiga, B. Boh, *Microencapsulation for Textile Applications and Use of SEM Image Analysis for Visualization of Microcapsules.* Tekstilec, 2011, 54.
455. Starešinič, M., *Izdelava prototipa varnostnega oblačila "Safety Vest."* Tekstilec, 2011, 54.
456. Jourand, P., H. De Clercq, R. Puers, *Robust monitoring of vital signs integrated in textile.* Sensors and Actuators A: Physical, 2010, 161(1), 288–296.
457. Coosemans, J., B. Hermans, R. Puers, *Integrating wireless ECG monitoring in textiles.* Sensors and Actuators A: Physical, 2006, 130, 48–53.
458. Bahadir, S. K., V. Koncar, F. Kalaoglu, *Wearable obstacle detection system fully integrated to textile structures for visually impaired people.* Sensors and Actuators A: Physical, 2012, 179, 297–311.
459. Benito-Lopez, F., et al., *Pump less wearable microfluidic device for real time pH sweat monitoring.* Procedia Chemistry, 2009, 1(1), 1103–1106.
460. Chang, J., et al., *Piezoelectric Nanofibers for Energy Scavenging Applications.* Nano Energy, 2012, 1(3), 356–371.

461. Chen, X., et al., *1.6 V Nanogenerator for Mechanical Energy Harvesting Using PZT Nanofibers.* Nano Letters, 2010, 10(6), 2133–2137.
462. Kim, H. S., J. H. Kim, J. Kim, *A Review of Piezoelectric Energy Harvesting Based on Vibration.* International Journal of Precision Engineering and Manufacturing, 2011, 12(6), 1129–1141.
463. Thavasi, V., G. Singh, S. Ramakrishna, *Electrospun nanofibers in energy and environmental applications.* Energy & Environmental Science, 2008, 1(2), 205–221.
464. Song, M. Y., et al., *Electrospun TiO2 electrodes for dye-sensitized solar cells.* Nanotechnology, 2004, 15(12), 1861.
465. Priya, A. S., et al., *High-performance quasi-solid-state dye-sensitized solar cell based on an electrospun PVDF–HFP membrane electrolyte.* Langmuir, 2008, 24(17), 9816–9819.
466. Onozuka, K., et al., *Electrospinning processed nanofibrous TiO2 membranes for photovoltaic applications.* Nanotechnology, 2006, 17(4), 1026.
467. Kim, I.-D., et al., *Dye-sensitized solar cells using network structure of electrospun ZnO nanofiber mats.* Applied physics letters, 2007, 91(16), 163109.
468. Zhang, W., et al., *Facile construction of nanofibrous ZnO photoelectrode for dye-sensitized solar cell applications.* Applied physics letters, 2009, 95(4), 043304.
469. Sudhagar, P., et al., *The performance of coupled (CdS: CdSe) quantum dot-sensitized TiO_2 nanofibrous solar cells.* Electrochemistry Communications, 2009, 11(11), 2220–2224.
470. Archana, P. S., et al., *Structural and Electrical Properties of Nb-Doped Anatase TiO2 Nanowires by Electrospinning.* Journal of the American Ceramic Society, 2010, 93(12), 4096–4102.
471. Wu, S., et al., *Effects of poly (vinyl alcohol)(PVA) content on preparation of novel thiol-functionalized mesoporous PVA/SiO_2 composite nanofiber membranes and their application for adsorption of heavy metal ions from aqueous solution.* Polymer, 2010, 51(26), 6203–6211.
472. Joshi, P., et al., *Electrospun carbon nanofibers as low-cost counter electrode for dye-sensitized solar cells.* ACS applied materials & interfaces, 2010, 2(12), 3572–3577.
473. Sundmacher, K., *Fuel cell engineering: toward the design of efficient electrochemical power plants.* Industrial & Engineering Chemistry Research, 2010, 49(21), 10159–10182.
474. Kim, Y. S., et al., *Electrospun bimetallic nanowires of PtRh and PtRu with compositional variation for methanol electrooxidation.* Electrochemistry Communications, 2008, 10(7), 1016–1019.
475. Li, M., G. Han, B. Yang, *Fabrication of the catalytic electrodes for methanol oxidation on electrospinning-derived carbon fibrous mats.* Electrochemistry Communications, 2008, 10(6), 880–883.
476. Xuyen, N. T., et al., *Hydrolysis-induced immobilization of Pt (acac) 2 on polyimide-based carbon nanofiber mat and formation of Pt nanoparticles.* Journal of Materials Chemistry, 2009, 19(9), 1283–1288.
477. ThiáXuyen, N., et al., *Three-dimensional architecture of carbon nanotube-anchored polymer nanofiber composite.* Journal of Materials Chemistry, 2009, 19(42), 7822–7825.
478. Lin, H.-L., et al., *Preparation of Nafion/poly (vinyl alcohol) electro-spun fiber composite membranes for direct methanol fuel cells.* Journal of Membrane Science, 2010, 365(1), 114–122.
479. Wang, L., et al., *Electrospinning synthesis of C/Fe_3O_4 composite nanofibers and their application for high performance lithium-ion batteries.* Journal of Power Sources, 2008, 183(2), 717–723.
480. Huang, C. T., et al., *Single-InN-Nanowire Nanogenerator with Upto 1 V Output Voltage.* Advanced Materials, 2010, 22(36), 4008–4013.
481. Chang, C., et al., *Direct-write piezoelectric polymeric nanogenerator with high energy conversion efficiency.* Nano letters, 2010, 10(2), 726–731.

482. Hansen, B. J., et al., *Hybrid nanogenerator for concurrently harvesting biomechanical and biochemical energy.* ACS nano, 2010, 4(7), 3647–3652.
483. Whittingham, M. S., *Electrical energy storage and intercalation chemistry.* Science, 1976, 192(4244), 1126–1127.
484. Yazami, R., P. Touzain, *A reversible graphite-lithium negative electrode for electrochemical generators.* Journal of Power Sources, 1983, 9(3), 365–371.
485. Thackeray, M., et al., *Lithium insertion into manganese spinels.* Materials Research Bulletin, 1983, 18(4), 461–472.
486. Manthiram, A., J. Goodenough, *Lithium insertion into $Fe_2 (SO_4)_3$ frameworks.* Journal of Power Sources, 1989, 26(3), 403–408.
487. Padhi, A. K., K. Nanjundaswamy, J. B.d. Goodenough, *Phospho-olivines as positive-electrode materials for rechargeable lithium batteries.* Journal of the Electrochemical Society, 1997, 144(4), 1188–1194.
488. Yang, C., et al., *Polyvinylidene fluoride membrane by novel electrospinning system for separator of Li-ion batteries.* Journal of Power Sources, 2009, 189(1), 716–720.
489. Gao, K., et al., *Crystal structures of electrospun PVDF membranes and its separator application for rechargeable lithium metal cells.* Materials Science and Engineering: B, 2006, 131(1), 100–105.
490. Cho, T.-H., et al., *Battery performances and thermal stability of polyacrylonitrile nano-fiber-based nonwoven separators for Li-ion battery.* Journal of Power Sources, 2008, 181(1), 155–160.
491. Gu, Y., D. Chen, X. Jiao, *Synthesis and electrochemical properties of nanostructured LiCoO2 fibers as cathode materials for lithium-ion batteries.* The Journal of Physical Chemistry B, 2005, 109(38), 17901–17906.
492. Endo, M., et al., *Recent development of carbon materials for Li ion batteries.* Carbon, 2000, 38(2), 183–197.
493. Kim, C., et al., *Fabrication of Electrospinning-Derived Carbon Nanofiber Webs for the Anode Material of Lithium-Ion Secondary Batteries.* Advanced Functional Materials, 2006, 16(18), 2393–2397.
494. Winter, M., et al., *Insertion electrode materials for rechargeable lithium batteries.* Advanced Materials, 1998, 10(10), 725–763.
495. Subramanian, V., H. Zhu, B. Wei, *High rate reversibility anode materials of lithium batteries from vapor-grown carbon nanofibers.* The Journal of Physical Chemistry B, 2006, 110(14), 7178–7183.
496. Zou, G., et al., *Carbon nanofibers: synthesis, characterization, and electrochemical properties.* Carbon, 2006, 44(5), 828–832.
497. Li, C., et al., *Porous carbon nanofibers derived from conducting polymer: synthesis and application in lithium-ion batteries with high-rate capability.* The Journal of Physical Chemistry C, 2009, 113(30), 13438–13442.
498. Ji, L., X. Zhang, *Fabrication of porous carbon nanofibers and their application as anode materials for rechargeable lithium-ion batteries.* Nanotechnology, 2009, 20(15), 155705.
499. Ji, L., X. Zhang, *Manganese oxide nanoparticle-loaded porous carbon nanofibers as anode materials for high-performance lithium-ion batteries.* Electrochemistry Communications, 2009, 11(4), 795–798.
500. Ji, L., A. J. Medford, X. Zhang, *Porous carbon nanofibers loaded with manganese oxide particles: Formation mechanism and electrochemical performance as energy-storage materials.* Journal of Materials Chemistry, 2009, 19(31), 5593–5601.
501. Ji, L., X. Zhang, *Fabrication of porous carbon/Si composite nanofibers as high-capacity battery electrodes.* Electrochemistry Communications, 2009, 11(6), 1146–1149.

502. Ji, L., et al., *Formation and electrochemical performance of copper/carbon composite nanofibers.* Electrochimica Acta, 2010, 55(5), 1605–1611.
503. Ji, L., et al., *In-Situ Encapsulation of Nickel Particles in Electrospun Carbon Nanofibers and the Resultant Electrochemical Performance.* Chemistry-A European Journal, 2009, 15(41), 10718–10722.
504. Yu, Y., et al., *Tin nanoparticles encapsulated in porous multichannel carbon microtubes: preparation by single-nozzle electrospinning and application as anode material for high-performance Li-based batteries.* Journal of the American Chemical Society, 2009, 131(44), 15984–15985.
505. Yu, Y., et al., *Encapsulation of Sn@ carbon Nanoparticles in Bamboo-like Hollow Carbon Nanofibers as an Anode Material in Lithium-Based Batteries.* Angewandte Chemie International Edition, 2009, 48(35), 6485–6489.
506. Choi, S.-S., et al., *Electrospun PVDF nanofiber web as polymer electrolyte or separator.* Electrochimica Acta, 2004, 50(2), 339–343.
507. Choi, S., et al., *Electrochemical and spectroscopic properties of electrospun PAN-based fibrous polymer electrolytes.* Journal of the Electrochemical Society, 2005, 152(5), p. A989-A995.
508. Kim, J.-K., et al., *Preparation and electrochemical characterization of electrospun, microporous membrane-based composite polymer electrolytes for lithium batteries.* Journal of Power Sources, 2008, 178(2), 815–820.
509. Ding, Y., et al., *The ionic conductivity and mechanical property of electrospun P (VdF-HFP)/ PMMA membranes for lithium ion batteries.* Journal of Membrane Science, 2009, 329(1), 56–59.
510. Ahn, Y., et al., *Development of high efficiency nanofilters made of nanofibers.* Current Applied Physics, 2006, 6(6), 1030–1035.
511. Kim, C., K. Yang, *Electrochemical properties of carbon nanofiber web as an electrode for supercapacitor prepared by electrospinning.* Applied physics letters, 2003, 83(6), 1216–1218.
512. Kim, G., W. Kim, *Highly porous 3D nanofiber scaffold using an electrospinning technique.* Journal of Biomedical Materials Research Part B: Applied Biomaterials, 2007, 81(1), 104–110.
513. Lin, D., et al., *Enhanced photocatalysis of electrospun Ag– ZnO heterostructured nanofibers.* Chemistry of Materials, 2009, 21(15), 3479–3484.
514. Niu, H., et al., *Preparation, structure and supercapacitance of bonded carbon nanofiber electrode materials.* Carbon, 2011, 49(7), 2380–2388.
515. Choi, S.-H., et al., *Facile synthesis of highly conductive platinum nanofiber mats as conducting core for high rate redox supercapacitor.* Electrochemical and Solid-State Letters, 2010, 13(6), p. A65-A68.
516. Wee, G., et al., *Synthesis and electrochemical properties of electrospun V2O5 nanofibers as supercapacitor electrodes.* Journal of Materials Chemistry, 2010, 20(32), 6720–6725.
517. Fang, J., et al., *Evolution of fiber morphology during electrospinning.* Journal of applied polymer science, 2010, 118(5), 2553–2561.
518. Kim, D.-K., et al., *Electrospun polyacrylonitrile-based carbon nanofibers and their hydrogen storages.* Macromolecular Research, 2005, 13(6), 521–528.
519. Hong, S. E., et al., *Graphite nanofibers prepared from catalytic graphitization of electrospun poly (vinylidene fluoride) nanofibers and their hydrogen storage capacity.* Catalysis today, 2007, 120(3), 413–419.
520. Srinivasan, S., et al., *Reversible hydrogen storage in electrospun polyaniline fibers.* international journal of hydrogen energy, 2010, 35(1), 225–230.

521. Rose, M., et al., *High surface area carbide-derived carbon fibers produced by electrospinning of polycarbosilane precursors.* Carbon, 2010, 48(2), 403–407.

522. Im, J. S., S.-J. Park, Y.-S. Lee, *Superior prospect of chemically activated electrospun carbon fibers for hydrogen storage.* Materials Research Bulletin, 2009, 44(9), 1871–1878.

523. Im, J. S., et al., *Hydrogen storage evaluation based on investigations of the catalytic properties of metal/metal oxides in electrospun carbon fibers.* international journal of hydrogen energy, 2009, 34(8), 3382–3388.

524. Im, J., et al., *The effect of embedded vanadium catalyst on activated electrospun CFs for hydrogen storage.* Microporous and Mesoporous Materials, 2008, 115(3), 514–521.

525. Bergshoef, M. M., G. J. Vancso, *Transparent nanocomposites with ultrathin, electrospun nylon-4, 6 fiber reinforcement.* Advanced Materials, 1999, 11(16), 1362–1365.

526. Li, G., et al., *Inhomogeneous toughening of carbon fiber/epoxy composite using electrospun polysulfone nanofibrous membranes by in situ phase separation.* Composites science and technology, 2008, 68(3), 987–994.

527. Zhang, J., T. Lin, X. Wang, *Electrospun nanofiber toughened carbon/epoxy composites: effects of polyetherketone cardo (PEK-C) nanofiber diameter and interlayer thickness.* Composites science and technology, 2010, 70(11), 1660–1666.

528. Zhang, J., X. Wang, T. Lin, *Synergistic effects of polyetherketone cardo (PEKC)/carbon nanofiber composite on eposy resins.* Composites science and technology, 2011, 71, 1–060.

529. Kim, J.s. and D. H. Reneker, *Mechanical properties of composites using ultrafine electrospun fibers.* Polymer composites, 1999, 20(1), 124–131.

530. Fong, H., *Electrospun nylon 6 nanofiber reinforced BIS-GMA/TEGDMA dental restorative composite resins.* Polymer, 2004, 45(7), 2427–2432.

531. Pinto, N., et al., *Electrospun polyaniline/polyethylene oxide nanofiber field-effect transistor.* Applied physics letters, 2003, 83(20), 4244–4246.

532. Yang, S., C. F. Wang, S. Chen, *A Release-Induced Response for the Rapid Recognition of Latent Fingerprints and Formation of Inkjet-Printed Patterns.* Angewandte Chemie, 2011, 123(16), 3790–3793.

533. Morimoto, T., et al., *Electric double-layer capacitor using organic electrolyte.* Journal of Power Sources, 1996, 60(2), 239–247.

534. Han, L., A. L. Andrady, D. S. Ensor, *Polymer nanofiber-based electronic nose,* 2011, Google Patents.

535. Wang, Y., J. J. Santiago-Avilés, *Large negative magnetoresistance and two-dimensional weak localization in carbon nanofiber fabricated using electrospinning.* Journal of applied physics, 2003, 94(3), 1721–1727.

536. Zhou, Y., et al., *Fabrication and electrical characterization of polyaniline-based nanofibers with diameter below 30 nm.* Applied physics letters, 2003, 83(18), 3800–3802.

537. Wang, Y., I. Ramos, J. Santiago-Avilés, *Electrical characterization of a single electrospun porous SnO2 nanoribbon in ambient air.* Nanotechnology, 2007, 18(43), 435704.

538. Zhou, Z., et al., *Development of carbon nanofibers from aligned electrospun polyacrylonitrile nanofiber bundles and characterization of their microstructural, electrical, and mechanical properties.* Polymer, 2009, 50(13), 2999–3006.

539. Pinto, N. J., et al., *Electrospun polyaniline/polyethylene oxide nanofiber field-effect transistor.* Applied physics letters, 2003, 83(20), 4244–4246.

540. Natali, D., M. Caironi, *Charge Injection in Solution-Processed Organic Field-Effect Transistors: Physics, Models and Characterization Methods.* Advanced Materials, 2012, 24(11), 1357–1387.

541. Sokolov, A. N., et al., *Chemical and engineering approaches to enable organic field-effect transistors for electronic skin applications.* Accounts of chemical research, 2011, 45(3), 361–371.

542. Huang, Z.-M., et al., *A review on polymer nanofibers by electrospinning and their applications in nanocomposites.* Composites science and technology, 2003, 63(15), 2223–2253.

543. Park, G., et al., *An outlier analysis framework for impedance-based structural health monitoring.* Journal of Sound and Vibration, 2005, 286(1), 229–250.

544. Jung, S.-B. and S.-W. Kim, *Improvement of scanning accuracy of PZT piezoelectric actuators by feed-forward model-reference control.* Precision Engineering, 1994, 16(1), 49–55.

545. Shen, D., et al., *The design, fabrication and evaluation of a MEMS PZT cantilever with an integrated Si proof mass for vibration energy harvesting.* Journal of Micromechanics and Microengineering, 2008, 18(5), 055017.

546. Chen, X., et al., *Potential measurement from a single lead zirconate titanate nanofiber using a nanomanipulator.* Applied physics letters, 2009, 94(25), 253113.

547. Agnew, B., *NIH plans bioengineering initiative.* Science, 1998, 280(5369), 1516–1518.

548. Pan, N., P. Gibson, *Thermal and moisture transport in fibrous materials. 2006.* Cambridge: Woodhead Pub.

549. Martin, J. R., G. E. Lamb, *Measurement of thermal conductivity of nonwovens using a dynamic method.* Textile Research Journal, 1987, 57(12), 721–727.

550. Farnworth, B., *Mechanisms of heat flow through clothing insulation.* Textile Research Journal, 1983, 53(12), 717–725.

551. Mecheels, J. *Concomitant heat and moisture transmission properties of clothing.* in *3rd Shirley Institute Seminar, Textiles in Comfort.* 1971.

552. Crank, J., *The mathematics of diffusion.* 1979, Oxford university press.

553. Wehner, J. A., B. Miller, L. Rebenfeld, *Dynamics of water vapor transmission through fabric barriers.* Textile Research Journal, 1988, 58(10), 581–592.

554. Li, Y., B. Holcombe, *A two-stage sorption model of the coupled diffusion of moisture and heat in wool fabrics.* Textile Research Journal, 1992, 62(4), 211–217.

555. Kissa, E., *Wetting and wicking.* Textile Research Journal, 1996, 66(10), 660–668.

556. Ito, H., Y. Muraoka, *Water transport along textile fibers as measured by an electrical capacitance technique.* Textile Research Journal, 1993, 63(7), 414–420.

557. Wissler, E. H., *Steady-state temperature distribution in man.* Journal of applied physiology, 1961, 16(4), 734–740.

558. Downes, J., B. Mackay, *Sorption kinetics of water vapor in wool fibers.* Journal of Polymer Science, 1958, 28(116), 45–67.

559. Li, Y., Z. Luo, *An improved mathematical simulation of the coupled diffusion of moisture and heat in wool fabric.* Textile Research Journal, 1999, 69(10), 760–768.

560. Ogniewicz, Y., C. Tien, *Analysis of condensation in porous insulation.* International Journal of Heat and Mass Transfer, 1981, 24(3), 421–429.

561. Motakef, S., M. A. El-Masri, *Simultaneous heat and mass transfer with phase change in a porous slab.* International Journal of Heat and Mass Transfer, 1986, 29(10), 1503–1512.

562. De Vries, D., *Simultaneous transfer of heat and moisture in porous media.* Transactions, American Geophysical Union, 1958, 39, 909–916.

563. Wang, Z., et al., *Radiation and conduction heat transfer coupled with liquid water transfer, moisture sorption, and condensation in porous polymer materials.* Journal of applied polymer science, 2003, 89(10), 2780–2790.

564. Schlangen, H., *Experimental and numerical analysis of fracture processes in concrete.* 1993.

565. Sarhadov, I., M. Pavluš, *Models of Heat and Moisture Transfer in Porous Materials.*

566. Gagge, A., *An effective temperature scale based on a simple model of human physiological regulatory response.* Ashrae Trans., 1971, 77, 247–262.

567. Stolwijk, J. A., *A mathematical model of physiological temperature regulation in man.* Vol. 1855, 1971, National Aeronautics and Space Administration.

568. Li, Y., B. Holcombe, *Mathematical simulation of heat and moisture transfer in a human-clothing-environment system.* Textile Research Journal, 1998, 68(6), 389–397.
569. Ghaddar, N., K. Ghali, B. Jones, *Integrated human-clothing system model for estimating the effect of walking on clothing insulation.* International journal of thermal sciences, 2003, 42(6), 605–619.
570. Murakami, S., S. Kato, J. Zeng, *Combined simulation of airflow, radiation and moisture transport for heat release from a human body.* Building and environment, 2000, 35(6), 489–500.
571. Antunano, M., S. Nunneley, *Heat stress in protective clothing: validation of a computer model and the heat-humidity index (HHI).* Aviation, space, and environmental medicine, 1992, 63(12), 1087–1092.
572. Schwenzfeier, L., et al. *Optimization of the thermal protective clothing using a knowledge bank concept and a learning expert system.* in *The sixth biennial conference of the European Society for Engineering and Medicine.* 2001.
573. Barry, J. J., R. W. Hill, *Computational modeling of protective clothing.* Int Nonwovens J, 2003, 12, 25–34.
574. Bhattacharjee, D., V. Kothari, *Prediction Of Thermal Resistance of Woven Fabrics. Part II: Heat transfer in natural and forced convective environments.* Journal of the Textile Institute, 2008, 99(5), 433–449.
575. Parsons, K., *Computer models as tools for evaluating clothing risks and controls.* Annals of Occupational Hygiene, 1995, 39(6), 827–839.
576. Prasad, K., W. H. Twilley, J. R. Lawson, *Thermal Performance of Fire Fighters' Protective Clothing: Numerical Study of Transient Heat and Water Vapor Transfer.* 2002, US Department of Commerce, Technology Administration, National Institute of Standards and Technology.
577. Choi, K.-J. and H.-S. Ko, *Research problems in clothing simulation.* Computer-Aided Design, 2005, 37(6), 585–592.
578. Breen, D. E., House, D. H., Wozny, M. J. *Predicting the drape of woven cloth using interacting particles.* in *Proceedings of the 21st Annual Conference on Computer graphics and interactive techniques.* 1994. ACM.
579. Wang, C. C., Yuen, M. M. *CAD methods in garment design.* Computer-Aided Design, 2005, 37(6), 583–584.
580. Yi, L., et al., *P-smart—a virtual system for clothing thermal functional design.* Computer-Aided Design, 2006, 38(7), 726–739.

INDEX